Complex Systems

This book explores the exciting field of complexity. It features in-depth coverage of important theoretical ideas, including fractals, chaos, non-linear dynamics, artificial life and self-organisation. It also provides overviews of complexity in several applied areas, including parallel computation, control systems, neural systems and ecosystems. Some of the properties that best characterise complex systems, including algorithmic richness, non-linearity and abundant interactions between components, are examined. In this way the book draws out themes, especially the ideas of connectivity and natural computation, that reveal deep, underlying similarities between phenomena that have formerly been treated as completely distinct. The idea of natural computation is particularly rich in fresh approaches applicable to both biology and computing. Analogies such as the DNA code as life's underlying program, or organisms as automata, are very compelling. Conversely, biologically inspired ideas such as cellular automata, genetic algorithms and neural networks are at the forefront of advanced computing.

TERRY BOSSOMAIER and DAVID GREEN hold Chairs in Information Technology at Charles Sturt University, Australia. They are co-authors of *Patterns in the Sand*, a popular account of the science of complexity, and co-editors of *Complex Systems – from Biology to Computation*.

Complex Systems

Edited by
Terry R. J. Bossomaier and David G. Green
Charles Sturt University, Australia

PUBLISHED BY THE PRESS SYNDICATE OF THE UNIVERSITY OF CAMBRIDGE
The Pitt Building, Trumpington Street, Cambridge, United Kingdom

CAMBRIDGE UNIVERSITY PRESS
The Edinburgh Building, Cambridge CB2 2RU, UK   http://www.cup.cam.ac.uk
40 West 20th Street, New York, NY 10011–4211, USA   http://www.cup.org
10 Stamford Road, Oakleigh, Melbourne 3166, Australia
Ruiz de Alarcón 13, 28014 Madrid, Spain

© Cambridge University Press 2000

This book is in copyright. Subject to statutory exception
and to the provisions of relevant collective licensing agreements,
no reproduction of any part may take place without
the written permission of Cambridge University Press

First published 2000

Printed in the United Kingdom at the University Press, Cambridge

Typeface Computer Modern   *System* LaTeX

A catalogue record for this book is available from the British Library

Library of Congress Cataloguing in Publication data

Complex systems / edited by Terry Bossomaier and David Green.
    p. cm.
  Includes index.
  ISBN 0 521 46245 2 (hc.)
    1. Chaotic behavior in systems.   2. Computational complexity.
  3. Fractals.   4. Artificial intelligence.   I. Bossomaier, Terry.
  II. Green, David (David Geoffrey), 1949– .
  Q172.5.C45C64   1999
  003'.857–dc21   98–39480   CIP

ISBN 0 521 46245 2 hardback

Contents

1	Introduction *T. R. J. Bossomaier & D. G. Green*	page 1
2	Self-organisation in complex systems *D. G. Green*	11
3	Network evolution and the emergence of structure *D.A. Seeley*	51
4	Artificial life: growing complex systems *Z. Aleksić*	91
5	Deterministic and random fractals *John E. Hutchinson*	127
6	Non-linear dynamics *D. E. Stewart & R. L. Dewar*	167
7	Non-linear control systems *M. R. James*	249
8	Parallel computers and complex systems *G.C. Fox & P.D. Coddington*	289
9	Are ecosystems complex systems? *R. H. Bradbury, D. G. Green & N. Snoad*	339
10	Complexity and neural networks *Terry Bossomaier*	367
Index		407

1
Introduction

TERRY R. J. BOSSOMAIER & DAVID G. GREEN

Charles Sturt University,
New South Wales,
AUSTRALIA

1.1 Overview

Imagine a violent storm sweeping across a beach. The wind picks up debris, swirls it around madly, dumping it somewhere else. Devastation remains, structure and order are gone. But one of the great waves of interest to break over science in the last decade has been the delicate balance between order and disorder: the patterns which appear out of nowhere through self-organisation; the fine structure of chaotic dynamics; the long range order which appears at phase transitions.

Fractal geometry in the 1970s moved from abstract mathematics to real world applications: simple rules could generate complex structures of great symmetry over an infinite range of scales. Chaos caught many peoples' imagination in the 1980s: simple rules could produce arbitrarily complex patterns which appear quite random. As the millennium draws to a close, more and more of these phenomena seem like different aspects of an underlying paradigm: complexity. Not just a new scientific theory or body of data, it challenges fundamental attitudes to the way we do science.

Although it might be intuitively apparent that a system is complex, defining complexity has proved difficult to pin down with numerous definitions on record. As yet there is no agreed theory of complexity. Much of the mathematics is intractable and computer simulation plays a major part. Thus in this book we do not try to present a uniform theory of complexity, nor do we try to describe or characterise many complex systems. We try to present the major theoretical tools and mathematics which help us along the way to these much more lofty goals.

It is convenient to separate sources of complexity into two mechanisms, although in any complex system aspects of both will usually be found:

Iteration: fractals and chaos result from repetition of simple operations. These generating rules produce complex phenomena. There are many interesting questions to ask about how to describe the processes, how to measure the resulting complexity, whether we can work backwards from the reult to the rules and so on.

Interaction: the other major source of complexity is the interaction of many autonomous, *adaptive* agents. Again, there are many questions to ask about the agents, the nature of the interaction and the circumstances in which complex surface phenomena result.

Introduction

The book has roughly three phases. To begin with we look at computer models of a whole range of self-organising phenomena and artificial life. Computer models have shown how immensely complex behaviour can result from the simplest of rules. As the internet connects the developed world and intelligent agents are becoming a necessity to find anything, we could be in for all sorts of interesting phenomena in cyberspace. Finally, we study ecology, the quintessential complex adaptive system.

We then tackle four fundamental methodologies, increasing the mathematical sophistication as we go:

Graphs and connectivity: graph theory goes back at least as far as Euler. But conventional theory is not really what we need. The essential tools relate to directional graphs, particularly random ones. The results here tend to be obtained by computer simulation rather than analytical mathematics. The behaviour of complex systems often has elements which are determined *solely* by the connectivity among the components and not on the strength or dynamics of the interaction. Doug Seeley presents a wide ranging and speculative overview of the applicability of graph-theoretic methods to understanding complexity (see chapter 3).

Fractals: With fractals we commence the first of the theoretical studies of the iterative route to complexity. A little more mathematics is needed to understand the whole of the chapter, but the ideas are explained qualitatively throughout. Fractals reappear in other chapters including the structure of chaotic attractors and the spatial and temporal properties of nerve cells.

Dynamical systems Dynamical systems theory treats the evolution of deterministic systems. When the 'time' is a discrete variable, it describes evolution under iteration. It includes both regular and chaotic motion, bifurcations, strange attractors — all the phenomena conjured up by the phrase 'chaos theory'. This chapter's tour of dynamical systems theory emphasises numerical analysis, an essential underpinning for successful computation.

Control Having seen the diversity of complex systems, one might then ask how does one control them. Biology has solved this problem and we have seen something of a paradigm shift in robotics in recent years towards copying the strategies used by animals. In some cases such a strategy may not exist in any explicit form, but be merely learnt, on the job as it were. Matthew

James gives us a state of the art description of control theory with particular emphasis on non-linear systems and reflects on the growing significance of neural network methods for solving control problems.

In the final phase we turn to a metaphor for physics and the natural world, the idea of *physics and biology as computation*. The world's most powerful man-made computers are now parallel machines with numerous interactive processes. Geoffrey Fox as director to the C^3P project at Caltech was one of the pioneers of university programmes in complexity. He addresses the relationship between parallel computing and complexity.

Lastly we turn to the most complex system we know, the human brain. Here we have all the ideas of complexity fused into one: many adaptive agents, diverse and subtle connectivity, but above all the embodiment of computation by dynamic systems. As we write, a ground swell of change is building in neuroscience, as attention shifts from the single neuron to the activity of populations. Concurrently frenetic buidling and testing of artificial neural networks is taking us closer and closer to understanding nature's computers.

It is unlikely that a complete book on complexity could be written this century. We hope to give you some tools to go ahead and contribute to the development of this fascinating field.

Of course what we would all like to see is a general theory of complex systems or complexity. Despite several promising candidates the selection process is still under way. Maybe there is no universal theory, but there are certainly common paradigms and methods which have proved to be useful across a wide area.

1.1.1 Simple building blocks

Although many natural phenomena may result from the interaction of complex entities, the details of the components may be unimportant. In the discussion of neural networks, the individual neuron turns out to be a highly sophisticated biological system. But the collective properties of neurons may be captured by spin-glass models, in which the neuron is simplified to a binary quantity, as discussed by Terry Bossomaier in chapter 10. The archetypical complexity from simplicity is of course, the cellular automaton, invented by von Neumann but classified by Wolfram. David Green and Zoran Aleksic in the chapters on self-organisation (chapter 2) and artificial life (chapter 4) deal with cellular automata in detail. There is a fundamental lesson here for studying any complex system: look for the simplest possible building block.

1.1.2 Simulation

The essence of complexity is that the outcome should not be obvious from the simple building blocks. Although in some cases global properties may be deducible directly, progress in complexity has occurred primarily through computer simulation. Most chapters make some reference to simulation and Geoffrey Fox in chapter 8 addresses the relationship between parallel computation and complexity.

1.1.3 Criticality

The *edge of chaos* has become a popular phrase, applying as it does to many disparate phenomena. Near a phase change systems exhibit long rage order and are delicately balanced for massive change in one or other direction. This notion of criticality underlies many complex phenomena, and again appears in the chapters by Green and Aleksic. It has also been invoked as important in the dynamics of neural networks, and corresponds to well-established results in graph theory as will be found in the relevant chapters.

1.1.4 Information

Information pops up in lots of places: it appears in the study of chaotic trajectories and fractals, in the control of non-linear systems and in the computational capabilities of neural networks. It is a powerful and all-pervasive concept. Shannon's measure of information is pervasive from chaos to understanding neural organisation, but a new measure is moving to centre-stage. This is the idea of algorithmic complexity.

Green briefly explores the idea of physics as computation, and computation is the driving force for algorithmic complexity. The time and space required to perform a calculation are in complexity terms perhaps less relevant than the length of the program required.

To see this, imagine the problems of an animal attempting to build a model of its environment to help it find food and the other necessities of its life. Many insects have genetically determined, relatively simple behaviour patterns. They are essentially quite predictable. At the other extreme of animal evolution, prediction of the behaviour of human beings is at best unreliable. We can think of this in terms of the algorithmic information involved in computing the behaviour at each successive time step. As agents become more complex, more and more computation is needed. Imagine now that there is some limit to the possible behaviour of an agent, a finite-size brain if you like. Then if the environment is very complicated, ultimately the agent will just end up making random choices.

Thus the capacity of an adaptive system to model its environment may be crucial to its survival. Bossomaier takes up this issue for neural computation at the end of the book.

1.2 Outline

In general each chapter begins with an overview of the topic, but goes into more detail where in some chapters tertiary experience of mathematics would be advantageous. The book falls loosely into three parts: paradigms of complexity; tools for modelling and controlling complexity; examples.

1.2.1 Paradigms of complexity

We begin the book with a study of the major paradigms and tools for describing complex systems, beginning with interaction and continuing with iteration.

1.2.1.1 Sources of order and organisation

We begin with one of the most fundamental issues of complexity: the nature of order and organisation. These properties pervade the whole of the natural world. They are also apparent in many formal systems. Evolution, adaptive systems, cellular automata, all fall within this broad area.

1.2.1.2 Graph theory

Touched upon in the first chapter is the role of connectivity between agents in defining the complexity of a system. A full analysis in terms of graph theory is now presented.

Introduction

1.2.1.3 Artificial life

Artificial life (Alife) began as a recreation with Conway's game of LIFE, but is now a respectable intellectual discipline with annual international conferences. Alife is strongly linked to self-organisation and is a testbed for the study of complexity. Does it also suggest a paradigm shift in biological theory? If we can create Alife models which through ultra-simple rules successfully mimic some aspect of biological behaviour, do we have something akin to the simple models beloved of physicists? Or do we merely have an alternative way of generating a particular pattern of activity? To take a popular example, flocking behaviour of *boids* seems to mimic flocking behaviour of birds rather well. Yet a boid is a simple entity with just three rules of interactions with its neighbouring birds in the flock. Can we deduce from this anything of the algorithm used by birds flying in formation? These are the philosophical issues of Alife which transcend the pretty pictures which can be generated on even the smallest computer. Thus the Alife chapter devotes subsubstantial attention to the nature of the Alife paradigm in research.

1.2.1.4 Fractals

Fractals – patterns that repeat on different scales – provide a suitable bridge between the earlier chapters and those that follow. On the one hand fractals embody organic issues, such as connectivity and emergent patterns, which are discussed in the preceeding chapters. On the other hand they also possess computational properties, such as iteration and sensitivity, which come to the fore in the chapters that follow. The 'strange attractors' of many chaotic processes have fractal properties.

1.2.1.5 Non-linear systems

Non-linear behaviour is one of the cornerstones of complexity. Such systems are typically far from equilibrium and exhibit a range of important properties such as sensitivity to initial condition and other hallmarks of chaos. Interactions between objects are often non-linear, and contribute to the unpredictable behaviour of large systems.

1.2.2 Tools for modelling and controlling complexity

We now move to two important areas for complexity research. Both of these furnish tools: computer simulation is essential to much complexity research, but parallel computers introduce complexity issues in their own right. Control theory, usually thought of as the prerogative of systems engineers, runs through many other topics of complexity research.

1.2.2.1 Parallel computing

It is now generally accepted that parallel computers are not only the most cost effective but also the most powerful machines and those with the greatest potential for growth. From the point of view of the application writer, parallelism is a nuisance and she would rather it was hidden. But parallel computation highlights some very significant ideas in compex systems.

1.2.3 Examples

We conclude the book with two carefully chosen examples. Ecology is the quintessential complex systems and most chapters of the book bear on ecology in one way or another. Neural networks are arguably the most complex systems known to man. They embody all the ideas of the book and are of ever increasing importance in dealing with the complexity of the natural and man-made world.

1.2.3.1 Are ecosystems complex systems?

With environmental hazards and destruction constantly in the news, we hardly need reminding of the complexity of ecosystems. But the discipline of ecology has suffered from repeated soul-searching for its philosophical foundations and in this chapter the impact of complexity ideas is assessed.

Observation and understanding are first steps in any scientific field, but in ecology, control is also of practical importance. The Great Barrier Reef is one of the seven wonders of the world, and despite the tourist hype, nobody is likely to be disappointed as they snorkel amongst brightly coloured flora and fauna. Yet reefs are delicate, and constant observation and perhaps intervention is essential to preserve these breathtaking enviroments. But intervention within Australia has not been entirely successful: one of the worst enviromental pests is the cane toad, now immortalised in a movie, deliberately introduced to eat a beetle but quickly finding more attractive prey.

1.2.3.2 Neural networks

The human brain has more neurons than there are people on the planet or ants in the largest ant colony, and is unquestionably the most complex system of which we have detailed knowledge. Thus we end the book with complexity in the nervous system. Many ideas of earlier chapters reappear here, and it seems fitting that we should complete the book with the greatest challenge faced by scientists today: the working of our brain.

2
Self-organisation in complex systems

DAVID G. GREEN

Charles Sturt University,
Albury, NSW 2640, AUSTRALIA

2.1 Introduction

How does a spider, with no knowledge of geometry, create the symmetry of its web? How do millions of nearly identical cells organise themselves into living creatures? Questions such as these remain some of the deepest mysteries, and greatest challenges, of modern science. The mere term 'organism' expresses the fundamental roles that interactions, self-organisation and emergent behaviour play in all biological systems.

Organisation arises from several sources. External constraints, such as biochemical gradients limiting cell activity, or money supply controlling economic activity, have been intensively studied in many disciplines. In this chapter we address the less understood question of *self-organisation*. That is, how do intrinsic properties, interactions between elements and other factors act to create order within complex systems?

The question of self-organisation is one of the central issues in the study of complexity. Early pioneers such as von Bertalanffy (1968) recognized the importance of internal interactions and processes as sources of organization and order. The resulting 'general systems theory' drew on analogies to point to the existence of common processes in superficially different systems. The need to plan complex military operations during World War II, and especially the coordination of many inter-dependent elements, provided the origins for systems analysis, operations research and related bodies of theory. In the years following the War these ideas contributed to the development of important fields, such as engineering and business management. However it was many years before the importance of systems thinking filtered into mainstream science, especially biology, which even now remains largely reductionist.

Perhaps the greatest puzzle in self-organisation is the question of emergent properties. That is, how do features that characterise a system as a whole emerge from the interplay of the parts that go to make up the system? Plants and animals, for instance, are clearly more than simply heaps of individual cells. This issue led to the idea of 'holism' (Koestler, 1967) in which objects are regarded not only as discrete objects, but also as building blocks within larger structures.

The issue of self-organisation remains as great a challenge today as it ever has been. Though there remains much yet to learn about self-organisation, we can at least identify some of the processes involved. Here we shall briefly examine four broad classes, which include both natural and artificial processes:

- inherent computation;

- non-linear and non-equilibrium processes;
- emergence; and
- evolution.

Though they are by no means exhaustive (and there is a good deal of overlap), these classes account for most of current research and progress on self-organisation in complex systems.

2.2 Inherent computation

Many systems are 'programmed' to change or behave according to fixed rules. In general we can usually represent such systems as automata (Hopcroft & Ullman, 1969), or as interacting networks of automata (e.g. see cellular automata in chapter 4). This is especially the case for systems that are discrete in structure and iterative in behaviour. For instance organisms are composed of discrete cells and cell replication is a discrete, iterative process. To illustrate the general idea we shall briefly consider two biological examples: growth and animal behaviour.

2.2.1 L-systems

Living things often grow in such a way as to produce fractal (i.e. self-similar) patterns (chapter 5). Examples include passageways in the lungs and the branches of a tree. The self-similarity on different scales arises because growth often involves iteration of simple, discrete processes (e.g. branching). These repetitive processes can often be summarised as sets of simple rules.

A popular way of representing growth processes is to use L-systems. L-systems are computational models of growth (Herman & Rozenberg, 1975). The name 'L-system' is short for 'Lindenmayer System', after Aristid Lindenmayer, who was a pioneer of the approach (Lindenmayer, 1971). Formally an *L-system* is a set of syntactic rules and symbols (i.e. a 'formal grammar') that represents discrete steps and units in a growth process. A simple L-system contains four elements:

- *VARIABLES* are symbols denoting elements that can be replaced;
- *CONSTANTS* are symbols denoting elements that remain fixed.
    For example, the expression

```
<subject> <verb> <predicate>
```

consists of grammatical variables. We can replace each variable by constants (English words or phrases) to produce sentences in English, such as 'The cat sat on the mat' or 'The dog ate the bone'. Constants and variables together constitute the *alphabet* of the system.

- *START* words ('axioms') are expressions defining the system's initial state. For instance, in the above examples from English, we might start from the single variable `<sentence>`.
- *RULES* ('syntax') define how the variables are to be replaced by constants or other variables.

For instance the syntax

$$
\begin{aligned}
\text{< sentence >} &\rightarrow \text{< subject >< verb >< predicate >} & (2.1)\\
\text{< subject >} &\rightarrow \text{the cat} \\
\text{< subject >} &\rightarrow \text{the dog} \\
\text{< verb >} &\rightarrow \text{sat} \\
\text{< verb >} &\rightarrow \text{ate} \\
\text{< predicate >} &\rightarrow \text{the mat} \\
\text{< predicate >} &\rightarrow \text{the bone}
\end{aligned}
$$

suffices to generate the sentences given in the above example.

The language $L(G)$ generated by a grammar G is the set of all words that can be obtained from the axioms by applying rules in a finite number of steps. A *word* that contains no variables is termed a *sentence*. Note that *alphabets*, *words* and *sentences*, as defined for formal grammars, are only vaguely related to their usual meanings for natural language. We can classify languages by properties of their syntax.

A syntax is said to be *regular* if every rule is of the form

$$
\begin{aligned}
A &\rightarrow aB \\
A &\rightarrow a
\end{aligned} \qquad (2.2)
$$

where A, B are variables and a is constant. A syntax is said to be *context-free* if every rule is of the form

$$A \rightarrow X \qquad (2.3)$$

where A is a variable and X, Y are any permissible expressions. On the other hand, *context-sensitive* grammars include rules of the form

$$AX \to Y \tag{2.4}$$
$$XA \to Y$$

where X and Y are any permissible expressions.

2.2.2 Example – Fibonacci numbers

Consider the simple grammar, defined as follows

$$
\begin{aligned}
\text{variables} &: \text{A,B} \\
\text{constants} &: \text{none} \\
\text{start} &: \text{A} \\
\text{rules} &: \text{A} \to \text{B} \\
&: \text{B} \to \text{AB}
\end{aligned}
\tag{2.5}
$$

This L-system produces the following sequence of strings

```
Stage 0 : A
Stage 1 : B
Stage 2 : AB
Stage 3 : BAB
Stage 4 : ABBAB
Stage 5 : BABABBAB
Stage 6 : ABBABBABABBAB
Stage 7 : BABABBABABBABBABABBAB
```

Counting the length of each string yields the famous Fibonacci sequence:

$$1\ 1\ 2\ 3\ 5\ 8\ 13\ 21\ 34\ \ldots$$

This simple sequence also demonstrates the way in which iterative properties of L-systems lead to self-similarity and hence fractal growth patterns. If we denote stage n above as S_n, then observe that for $n \geq 2$ we have

$$S_n = S_{n-2}S_{n-1}. \tag{2.6}$$

Provided the underlying grammar is context-free, then self-similarity

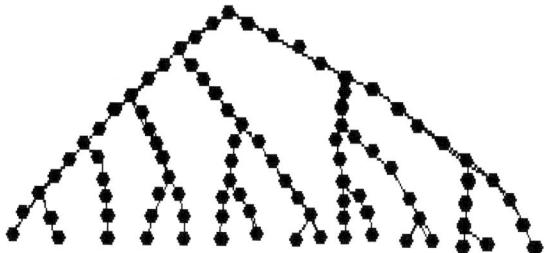

Fig. 2.1. Growth of the alga *Chaetomorpha linum*.

of this kind is assured whenever some stage of a process can be represented in terms of earlier stages. Generalising (by induction) leads to the following general result.

Theorem 2.1 *Let L(G) be a context-free language that is generated from the syntax G. If there exists an integer m, such that S_m can be expressed in terms of S_{m-i}, where $i < m$, then this same relationship holds for S_n, for all $n > m$.*

2.2.3 Example – algal growth

To use L-systems to model biological growth patterns we must define the *semantics* of the underlying grammar. That is, we assign meanings to the symbols and rules. For example, to describe the pattern of cell lineages found in the alga *Chaetomorpha linum* (fig. 2.1), we let the symbols denote cells in different states (Herman & Rozenberg, 1975). This growth process can be generated from an axiom A and growth rules

$$
\begin{aligned}
A &\to DB \\
B &\to C \\
C &\to D \\
D &\to E \\
E &\to A
\end{aligned}
\qquad (2.7)
$$

The pattern generated by this model matches the arrangement of cells in the original alga (fig. 2.2).

Self-organisation in complex systems

```
Stage  0 :                    A
Stage  1 :            D               B
Stage  2 :            E               C
Stage  3 :            A               D
Stage  4 :         D     B            E
Stage  5 :         E     C            A
Stage  6 :         A     D     D      B
Stage  7 :      D  B     E     E      C
Stage  8 :      E  C     A     A      D
Stage  9 :      A  D  D  B  D  B      E
Stage 10 :   D  B  E  E  C  E  C      A
Stage 11 :   E  C  A  A  D  A  D  D   B
```

Fig. 2.2. L-system model for growth of the alga from fig. 2.1. See text for details.

```
Leaf1 {              ; Name of the lsystem, "{" indicates start
  angle 8            ; Set angle increment to (360/8)=45 degrees
  axiom x            ; Starting character string
  a=n                ; Change every "a" into an "n"
  n=o                ; Likewise change "n" to "o" etc. ...
  o=p
  p=x
  b=e
  e=h
  h=j
  j=y
  x=F[+A(4)]Fy       ; Change every "x" into  "F[+A(4)]Fy"
  y=F[-B(4)]Fx       ; Change every "y" into  "F[-B(4)]Fx"
  F=@1.18F@i1.18
  }                  ; the final } indicates the end
```

Fig. 2.3. L-system model (using FRACTINT) for a compound leaf with alternating branches. A semi-colon indicates that the rest of a line is a comment. Program by Adrian Mariano.

2.2.4 Example – a compound leaf (or branch)

The public domain computer program FRACTINT provides many examples of L-systems. It also allows users to design and draw their own. The example (fig. 2.3) shows a model (as used by FRACTINT) that draws a leaf (fig. 2.4).

2.2.5 Some practical aspects of L-system models

The above simple examples illustrate some common features, and limitations, of L-system models:

Fig. 2.4. Picture of a compound leaf with alternating branches generated by FRACTINT using the model defined in fig. 2.3.

- L-systems employ two distinct kinds of *semantics* (i.e. the meaning of symbols). In one approach a model gives a complete description of the structure at any growth stage (as in the Fibonacci sequence). The alternative approach models only the new elements that appear at each stage (as in the algal model).
- In designing the syntax (i.e. rules) of L-systems, there was at first a bias towards *context-free* models, and especially *regular* grammars, rather than *context-sensitive* grammars. This bias reflected the origins of syntactic models in logic and formal language theory. Another important factor was that much is known about the properties of context-free languages. For instance, it can be shown that parsing methods always exist and there are simplifying rules, such as Chomsky's Normal Form (Hopcroft & Ullman, 1969).

Theorem 2.2 *Any context-free language can be generated by a grammar in which all production rules are of the form*

$$A \rightarrow BC \qquad (2.8)$$
$$A \rightarrow a$$

where A, B, C are variables and a is a constant.

- Some of the above models (e.g. the alga and compound leaf) introduce intermediate states simply to ensure that the timing of particular events (e.g. branching) is correct. In most cases no physical meaning (e.g. internal changes in cell chemistry) has been demonstrated for these intermediate states. If the aim is to capture the true nature of the processes involved, then context-sensitive models are often more appropriate, as the following example illustrates.

2.2.6 Example – cell differentiation during growth

A classic question in developmental biology is the so-called 'French Flag Problem' (Herman & Rozenberg, 1975). That is, how do arrays of dividing cells synchronise their behaviour to produce sharp boundaries between (say) bands of skin colour, or body segments? Models that treat cells as context-free automata (Herman & Rozenberg, 1975) can generate realistic banding patterns. However, as mentioned above these models assume arbitrary changes in internal cell states. Furthermore, experimental studies show that inter-cellular interactions *do* play a role in cell differentiation (Carroll, 1990, Karr & Kornberg, 1989, Small & Levine, 1991). That is, the process is usually context-sensitive. Moreover it is often associated with biochemical gradients, e.g. changes in concentration across a surface or within a limb.

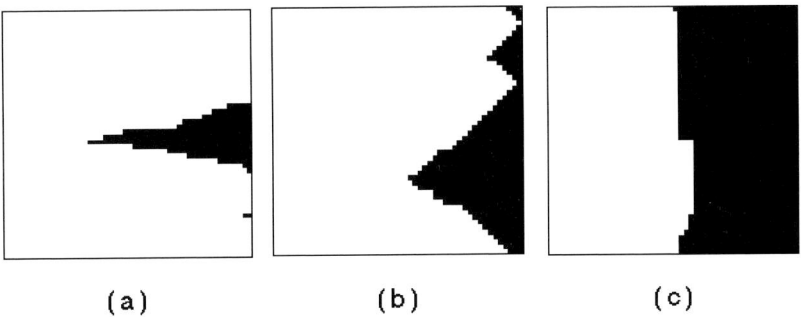

Fig. 2.5. Simulated development of cell banding. In each case successive vertical lines of cells grow from left to right, paralleling a biochemical gradient. (a) Probability is low of a cell appearing in a new state or of affecting its neighbours; (b) cells with neighbours in a new state always produce offspring in that new state; (c) as in (b), but state changes can also propagate sideways *after* cells form.

Simulation (fig. 2.5) using cellular automata models demonstrates the mechanism suggested by experiment. Suppose that cells form in layers perpendicular to a biochemical gradient, and that the probability of a cell spontaneously changing state increases along the gradient. Without inter-cellular interactions, no edge forms (fig. 2.5a). Even if adjacent cells do interact only a ragged edge forms (fig. 2.5b). Only if a change in state can propagate along an entire line of newly formed cells does the edge become sharp (fig. 2.5c).

In recent years research by, amongst others, Rozenberg (1986), Lindenmayer (1989) and Prusinkiewicz (1991) on L-systems has addressed most of the issues raised above. Prusinkiewicz (1996) describes various classes of L-systems now used, including: parametric, context sensitive, stochastic, and differential. For instance adding parameters to L-systems makes it possible to model fine details of plant architecture and hence to examine similarities and differences between growth forms in different plant species (fig. 2.6).

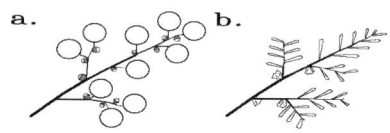

a. *Eucalyptus alpina*:

$<$ bud : leaf, angle $>$	$\rightarrow$	$<$ meristem $><$ bud : stem, angle $>$
$<$ bud : stem, angle $>$	$\rightarrow$	$<$ stem $><$ bud : leaf, angle $+$ increment $>$
$<$ meristem $>$	$\rightarrow$	$<$ leaf $><$ fruit $><$ bud : juvenile, 0 $>$
$<$ bud : juvenile, angle $>$	$\rightarrow$	$<$ bud : stem, angle $>$ /delay $=$ 1 year

b. *Hakea salicifolia*:

$<$ bud : leaf, angle $>$	$\rightarrow$	$<$ meristem $><$ bud : stem, angle $>$
$<$ bud : stem, angle $>$	$\rightarrow$	$<$ stem $><$ bud : leaf, angle $+$ increment $>$
$<$ meristem $>$	$\rightarrow$	$<$ leaf $><$ bud : juvenile, 0 $>$
$<$ bud : juvenile, angle $>$	$\rightarrow$	$<$ fruit $><$ bud : stem, angle $>$ /delay $=$ 1 year

Fig. 2.6. Growth patterns for two tree species. Simplified models are given beneath the diagrams. Names in angle brackets denote growth structures and arrows denote growth processes. The states of particular structures are indicated by colons. The models differ syntactically in the timing of fruit production and semantically in the shapes of leaves and fruit.

2.2.7 Animal behaviour

The idea of representing organisms as automata is as old as computing. Perhaps the strongest tradition stems from artificial intelligence, especially robotics. There have been some attempts to interpret the organisation of animal behaviour in grammatical terms (Westman, 1977), but the idea of animals as living automata has mostly been confined to simulations of animals in an environment. This idea finds its simplest expression in so-called 'turtle geometry'.

2.2.8 Example – turtle graphics

To use L-systems for generating graphical images requires that the symbols in the model refer to elements of a drawing on a computer screen. For example, the program FRACTINT uses 'Turtle Graphics' to produce screen images (fig. 2.4). It interprets each constant in an L-system model as a turtle command.

Turtle geometry, introduced by Seymour Papert (Papert, 1973), deals with patterns produced by the path of an imaginary turtle moving around on a plane. We can describe the turtle's path by a sequence of symbols representing the moves that the turtle makes as it moves. These sequences form words in a formal language, which we can define by a grammar such as the following:

```
Constants   = {nF, nB, aR, aL, Stop },
Variables   = {<Path>, <Design>, <Arm>, ...},
Start       = <Path>
```

where (omitting details of numerical syntax)

```
nF    denotes    "n steps Forward"
nB    denotes    "n steps Back"
aR    denotes    "Turn a degrees Right"
aL    denotes    "Turn a degrees Left"
```

and the basic production rules are:

$$<Path> \rightarrow nF <Path> \qquad (2.9)$$
$$<Path> \rightarrow aR <Path>$$
$$<Path> \rightarrow nB <Path>$$
$$<Path> \rightarrow aL <Path>$$
$$<Path> \rightarrow STOP$$

In this grammar, the variable <Path> denotes part (yet to specified) of the turtle's trail. The transitions represent moves made by the turtle. At any time, the completed portion of the turtle's path is specified by a sequence of individual movements, such as

```
4F  90R  F  90R  F  90R  <Path>
```

Turtle geometry is frequently used in computer graphics. Models that form complex patterns are obtained by augmenting the above grammar with new variables to denote particular pattern elements, and with new

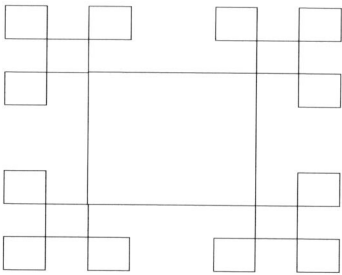

Fig. 2.7. A simple design produced by turtle geometry.

rules governing the structure of those patterns elements. For this reason the list of variables in the above definition is left open-ended. For example, the following rules use the variables <Design>, <Arm>, etc. to describe the formation of a simple design (fig. 2.7).

$$
\begin{aligned}
<\text{Path}> &\rightarrow <\text{Design}> \text{ stop} \\
<\text{Design}> &\rightarrow 4 <\text{Arm}> \\
<\text{Arm}> &\rightarrow 4F\ 3 <\text{Corner}>\ 1F \\
<\text{Corner}> &\rightarrow 2F\ 3 <\text{Turn}> \\
<\text{Turn}> &\rightarrow 90R\ F
\end{aligned}
\tag{2.10}
$$

From turtle geometry it is only a small step to syntactic models that describe the organisation of animal behaviour. The chief requirement is to make behaviour context-sensitive: that is the animal must interact with its environment. Models of this kind have been quite successful and have received considerable stimulus in recent years through the advent of artificial life as a discipline in its own right (Langton, 1990, Langton, 1992).

Some of the most notable work has been that of Hogeweg and Hesper (1981, 1983, 1991, 1993). An important theme in their research has been to demonstrate that organisation and complex behaviour often arise out of interactions between simple systems. Their bumblebee model (1983, 1991), for instance, shows that the observed social organisation of bumblebees arises as a natural consequence of the interaction between simple properties of bumblebee behaviour and of their environment. For example, one rule they invoke is the TODO principle (Hogeweg & Hesper, 1991). That is, bumblebees have no intended plan of action, they simply do whatever there is *to do* at any given time

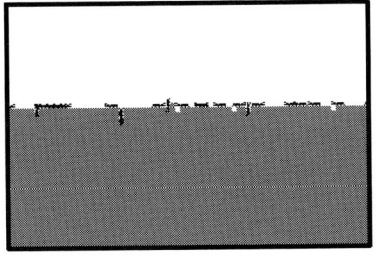 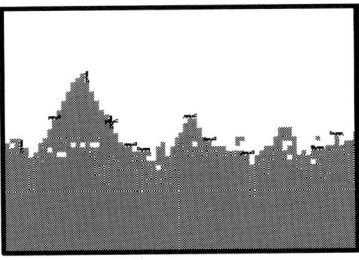

Fig. 2.8. A simulated ant colony (Xantfarm). The ants wander around picking up and dropping 'sand grains'. The figure shows the model at an early (left) and advanced (right) stage. Notice how non-directed activity of the ants produces mounds and other large-scale features.

and place. Hogeweg and Hesper have found that the TODO principle applies in the behaviour of many organisms. For instance ants sort eggs and corpses (Hogeweg, 1993) by following simple TODO rules, such as

```
IF you find a single  egg  THEN
                           pick it up (unless carrying one).
IF you find a heap of eggs THEN
                           drop any egg you are carrying.
```

Simply by following these rules, the ants quickly sort a random distribution of eggs into piles. Following a similar set of rules leads them to build an entire ant colony by sorting sand grains (cf. fig. 2.8), without knowing the overall design of the colony.

The evident power of simple heuristics, such as the TODO principle, teaches us the important lesson that global behaviour patterns, and social organisation, can emerge out of local interactions. Organisms do not necessarily need to have an over-riding plan nor do they require awareness of the large-scale. Grand patterns and processes can emerge as the nett effect of small-scale, local behaviour.

The above ideas lead to many computational heuristics, such as search and sorting algorithms (e.g. bubble sort). However they find their most compelling applications in artificial intelligence, especially robotics. Rodney Brooks (1986, 1991, 1992), for instance, has shown that neither central processing, nor intelligence, is needed for the operation of a successful robot. In his robot insects, for instance, each leg is individually controlled and responds separately according to a set of simple rules (e.g. IF leg is pointing back, THEN lift leg and move it forward). Interacting feedback loops coordinate the legs to control walking and other fundamental tasks.

2.3 Non-linear and non-equilibrium processes

An important source of self-organisation in complex systems arises from the dynamics of non-linear and non-equilibrium processes. Non-linear dynamics are discussed in depth elsewhere (chapter 6), so we discuss them only briefly here.

Thermodynamics deals with the macroscopic physical properties of large ensembles of particles. In a gas, for instance, we cannot deal with the detailed trajectories of individual molecules, but we can make predictions about the properties of the entire system. If we introduce cold water into a tub of hot water, for instance, the hot and cold water do not remain at their original temperatures. The whole tub quickly adjusts to an equilibrium temperature that lies somewhere between the two extremes.

The second law of thermodynamics generalises the above observation. It replaces 'heat' with the more general notion of 'entropy' – i.e. disorder. The second law states that entropy always increases within closed systems. This is really a statement about average behaviour. High entropy means uniformity, differences ('order') are damped by interactions within the system. For water in a tub the water molecules exchange energy via what are essentially elastic collisions. Thus water everywhere in the tub soon acquires the same temperature. What the water does *not* do is to separate spontaneously into areas of hot and cold water. That would be to decrease the entropy in the system.

Difficulties arise when the second law is applied to the real world. If entropy must increase, then how is it possible (say) for all the variety of the living world to persist? The usual answer to the above question is that living systems are *open* systems, not closed, so the law does not apply locally. However this answer is somewhat unsatisfying. In effect *all* systems are open systems, since everything interacts with its surroundings to some degree.

To illustrate the problem consider a body of gas out in space. According to the second law it should dissipate its heat and become a uniformly expanding shell of gas. However besides elastic collisions, the molecules also interact via gravitational attraction. If the body of gas is dense enough it does not dissipate out into space, but instead coalesces into stars, planets, and all the variety of objects that we see around us. Overall the system still radiates heat outwards into the rest of the universe, but internally entropy has definitely not increased.

Some of the most notable ideas have been those of the Nobel Prize winner Ilya Prigogine. In an influential work (Prigogine, 1980) he discussed the 'physics of becoming'. This idea concerns ways in which non-equilibrium processes lead to self-organisation in physical systems. Prigogine focuses especially on 'dissipative structures'. These are thermodynamic systems that produce energy internally and exchange it with an external environment. This definition includes organisms, many chemical systems (which Prigogine focused on particularly), and the coalescing gas clouds mentioned above. Prigogine showed that these *open*, dissipative systems behave differently from the *closed* systems of classical thermodynamics. He showed how it is that entropy can actually decrease (i.e. order increases) in dissipative systems.

First, Prigogine points out that dissipative systems are inherently non-linear and may be far from equilibrium. He goes on to show that systems far from equilibrium are liable to behave in unstable or chaotic ways. Under these circumstances, irregularities are likely to be reinforced, instead of being dampened. So minor irregularities grow into large-scale patterns and lead to an increase in order within the system concerned. A good example of this effect, which Prigogine himself cites, is the chemical reaction discovered by the Russians B. P. Belousov and A. M. Zhabotinsky (fig. 2.9).

The point is that the second law describes only average behaviour. It assumes that local irregularities get smoothed out. However in non-equilibrium systems, that is not the case. Irregularities can be reinforced and grow into large scale patterns.

Haken (1978, 1981, 1983) describes cooperative effects (which he calls '*synergetics*') in a wide range of non-linear complex systems. He attributes coherence in lasers, and similar systems, to a process that he calls 'enslavement'. In lasers, for example, pumping weakly causes them to act like lamps. The atoms emit photons independently of each other, giving rise to incoherent light. Above a certain pumping threshold, however, the internal atomic dipole moments become ordered, giving rise to extremely coherent 'laser' light. At the start of such a transmission, many different wavelengths of light are present. A slight preference for a particular wavelength progressively aligns ('enslaves') excited electrons so that they emit light at a single wavelength. At a critical point this process cascades, resulting in high intensity, coherent output from the laser.

Fig. 2.9. Simulation of patterns produced by the B-Z reaction. Minor irregularities at the centre of each set of concentric circles sets off a series of cycles of alternating chemical reactions that create expanding rings.

Haken also introduced the idea of an *order* parameter for cooperative systems. He notes that, as in the example of the laser, these systems normally display phase changes in properties and behaviour at some critical point. As we shall see later, Haken's order parameter indicates the extent of interactions (connectivity) within a system. The resulting phase changes constitute a universal property that occurs with changing connectivity in any kind of complex system.

2.4 Emergence

In the preceding sections we have seen two ways in which order emerges in complex systems. In the first case properties of the system, such as plant growth or animal behaviour, amount to computational processes. The iteration of those processes causes large scale patterns to emerge. In the second case, non-equilibrium processes lead a system to diverge from near uniformity. Local irregularities can grow or coalesce, patterns may collapse and the system virtually crystallises into a new state.

Although superficially different, both of the above processes share an underlying pattern. That is, the system diverges from its initial state and after a transient period settles into some attractor state. This attractor may be a simple equilibrium or cycle; or may be a strange attractor if the process is chaotic. So settling into the basin of an attractor seems to be a general way for properties and patterns to emerge.

In this section we shall now see that this assertion is true. What is more we shall see that the answer helps to answer one of the popular puzzles about emergence, namely when is the whole greater than the sum of its parts?

The really crucial question in multi-object systems is whether local interactions *do* grow into large-scale patterns. As we shall see this question is intimately bound up with the issue of how well connected a system is. We now look at this question, after which we shall then see how the answers relate to emergent features in three specific cases: embryo development, landscape ecology and the origin of life.

2.4.1 Universality of digraph structure

In many complex systems, organisation *emerges* out of interactions between the elements. Now interactions can be regarded as a form of non-linearity. However the interactions dealt with by control systems (chapter 7) usually involve relatively small ensembles of elements. To learn how the whole becomes greater than the sum of its parts, we need to understand when and how interactions envelop an entire system. Here we shall look at some basic features of interactions within large ensembles. We shall then see how they lead to emergent properties in applied contexts, using biological development and landscape ecology as examples.

Although complex systems may vary enormously in their form and function, most share a common underlying pattern: they consist of ensembles of interacting elements. Interactions between elements define the *connectivity*, which is a universal attribute of complex systems. Despite superficial differences, most complex systems display patterns of connectivity that are essentially similar (Green, 1993).

We can show the above to be true by proving that the connectivity patterns underlying the ways we represent complex systems are all equivalent. Although we use many different ways to model complex systems, almost all of these models rest on a small set of underlying representations. These representations include:

- *directed graphs* – (e.g. decision trees, cluster analysis);
- *matrix models* – (e.g. linear systems, Markov processes);
- *dynamical systems* – (i.e. differential equations); and

- *cellular automata* – see Wolfram (1984, 1986).

Between them, these representations underlie models of a vast range of biological systems and processes. In each of the above representations the patterns of connectivity reduce to objects with relationships between them. That is they exhibit an underlying directed graph (digraph) structure.

Theorem 2.3 *The patterns of dependencies in matrix models, dynamical systems, cellular automata, semigroups and partially ordered sets are all isomorphic to directed graphs.*

The proof proceeds by constructing digraphs for each of the systems in question. For instance, in a matrix model (e.g. a Markov chain) we treat the rows and columns as nodes of the graph and insert an edge from a particular row to a given column wherever the associated matrix entry is non-zero. Likewise for dynamical systems we treat the variables as nodes and insert edges wherever the partial derivative is non-zero. For cellular automata we treat the cells as nodes and insert edges wherever one cell belongs to the neighbourhood of the other. The remainder of the formal proof (Green, 1993) follows in like fashion.

The universal nature of connectivity patterns in directed graphs is further demonstrated by the following result (Green, 1993), which shows that the behaviour of a system can also be treated as a directed graph.

Theorem 2.4 *In any array of automata, the state space forms a directed graph.*

The proof is similar to the above. We define states of the automata to be the nodes. For arrays we take ensembles of states – configurations – to be the nodes. We define an edge to exist wherever the system makes a transition from one state (or ensemble) to another.

The significance of the above two theorems is that properties of digraphs are inherent in both the structure and behaviour of virtually *all* complex systems. The most important of these properties is a phase change in connectivity that occurs in random digraphs as the number of edges increases (Erdos & Renyi, 1960, Seeley & Ronald, 1992). This phase change therefore underlies a vast range of criticality phenomena associated with complex systems. Formerly many forms of criticality were treated as different phenomena. It now appears that most of them have a common underlying cause. Thus at a very fundamental level,

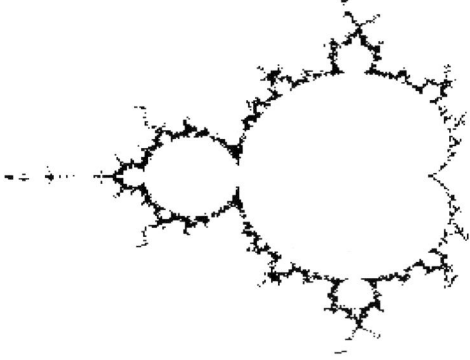

Fig. 2.10. The Mandelbrot set (represented here in black) is a critical region that marks a phase change in the behaviour of an automaton. In the internal white region, all points spiral in to a neighbourhood of the origin within a finite number of steps. For the outer white area, every point spirals outwards (away from the origin) indefinitely.

nuclear fission is related to the spread of an epidemic, which in turn is related to plant migration and so forth.

Having identified the fundamental importance of connectivity in digraphs for complex systems, we can now see how wide-ranging are its consequences. For instance it is immediately apparent that the familiar patterns seen in Mandelbrot and Julia sets reflect phase changes in the state space of the algorithm concerned. To understand this more formally we need only observe that in any tesselation $T(z, \delta)$ of the complex plane z (with tile size δ), there exists a region $|z| < R_1$ such that iteration of the algorithm maps the entire region to the tile containing the origin ($|z| < \delta$) in a finite number of steps, so the set of connected 'state tiles' is always finite. At the other extreme there is a region $|z| > R_2$ for which the set of connected state tiles is always infinite. That is, iteration of the algorithm maps each point further and further away from the origin. Between these two extremes lies a critical region in which some tiles (represented, say, by their centroid) spiral inwards under the algorithm, whereas others spiral ever outwards (fig. 2.10).

Despite the similarities we must of course recognise that differences in the nature of connectivity *do* exist between different kinds of complex systems. In dynamical systems, for instance, not only is the dependence of one variable on another important, but also its magnitude and sign. In cellular automata only cells within a local neighbourhood are ever connected. Also, the neighbourhood function is usually symmetric. That is, if the graph contains the edge (a,b), then it also contains the edge (b,a). Finally, notice that in the state space of an automaton, any number of edges may lead to a given node (state), but only one edge leads from it. This observation leads to the following simple, but important, result.

Theorem 2.5 *Any deterministic automaton, or cellular automaton (on a finite grid), with a finite number of states ultimately falls into either a fixed state, or a finite limit cycle.*

2.4.2 Genetic nets

One of the greatest mysteries in biology is how the genetic code influences the development and growth of organisms. How do genes turn growth on and off at the correct time? Kauffman (1991, 1992) has addressed these problems by viewing genes as simple automata that may be turned on or off. He proposed a simple model that suggests how such processes might occur. It shows the possibility that the complexity of growth may result from relatively simple mechanisms.

Kauffman (1991, 1992) represents control of development as a boolean switching network. Genes are represented as simple automata that may exist in either of two states: ON or OFF. Furthermore, the genes transmit their current state to other genes. Let us suppose here that each gene communicates with k other genes (possibly including itself). Finally, suppose that the state of the gene at any time $T+1$ is a simple function of the inputs that it receives from the k other genes at time T. Since each gene exists in either of 2 states, then k inputs can take 2^k possible forms. Each of these 2^k inputs can yield either 0 or 1 as a result, so there are 2^{2^k} possible functions in all. For instance, suppose that there are two inputs x, y to gene z. Then there are 16 possible functions (table 2.1).

Boolean networks consist of nodes in which the state (0 or 1) of each node is a boolean function of the states of other nodes. If each node accepts inputs from two other nodes ($K = 2$), then each node can be represented as one of the functions in table 2.1. For instance, suppose that the 'wiring' in the net consists of the following functions.

$$\begin{aligned} x' &= f_9(x,x) \\ y' &= f_9(y,z) \\ z' &= f_{12}(x,y) \end{aligned} \quad (2.11)$$

Table 2.1. *All 16 possible boolean functions of two input variables x,y*

Inputs									Functions								
x	y	0	1	2	3	4	5	6	7	8	9	10	11	12	13	14	15
0	0	0	0	0	0	0	0	0	0	1	1	1	1	1	1	1	1
0	1	0	0	0	0	1	1	1	1	0	0	0	0	1	1	1	1
1	0	0	0	1	1	0	0	1	1	0	0	1	1	0	0	1	1
1	1	0	1	0	1	0	1	0	1	0	1	0	1	0	1	0	1

Note. The functions are coded (0–15) by treating the outputs as binary numbers (e.g. function 10 = 1010).

Table 2.2. *State changes through time of a simple boolean network*

T	0	1	2	3	4	5	6	7	8	9	10	11	12
X	1	0	1	0	1	0	1	0	1	0	1	0	1
Y	1	1	0	0	1	1	0	0	1	1	0	0	1
Z	0	0	1	0	1	1	0	0	1	1	0	0	1

where the functions are as defined in table 2.1. Now suppose that the system starts off in some state, say $x = 1$, $y = 1$, $z = 0$. Then we can apply the above functions iteratively to obtain subsequent states of the system.

Notice that x depends only on its own past state and cycles with period 2 (table 2.2). The overall state (x, y, z) of the system at time $T = 5$ is the same as its state at time $T = 1$. Thereafter the system cycles with period 4 time steps. This behaviour is typical of what happens in any randomly constructed net. Because there is only a finite number of possible states, any such system must eventually settle into a stable limit cycle (theorem 2.5). The above example has only one limit cycle. That is, whatever initial configuration we choose for x, y and z the system always settles into the same limit cycle. However this is not always the case. Large nets may have several different cycles that can occur, depending on the initial configuration.

By simulating many thousands of nets of this kind, Kauffman (1991, 1992) built up a picture of the sorts of cyclic behaviour to expect. The resulting distribution of cycle lengths is highly skewed, with short cycles predominant. Moreover the median cycling period m is proportional to $k^{0.3}$ when all functions are allowed, and to $k^{0.6}$ when tautology and contradiction are omitted. This is surprising, considering the vast numbers of possible states that such systems can exhibit. Furthermore the number of different limit cycles that can occur for a given net is proportional to $\sqrt{k}$. Thus a net of 1 million elements would typically have a limit cycle of about 5000 time steps.

Finally, if we disturb a net (by randomly changing the values of some components) then the chance of the system changing from its current cycle to a different one is less than 0.05. That is, once established, the cycles tend to resist change.

What is the relevance of these results to biological development? Suppose that each state of the system represents a particular suite of genes that are turned on (i.e. operating). Then those genes are causing the cells in which they occur to produce various substances and those substances (chiefly proteins) cause various processes to occur. The cyclic behaviour of the system means that after a finite time, the system of genes will begin to repeat its behaviour over again. For instance, if a particular state of the system causes a cell to divide, then after completing its cycle, the system will come back again to the stage where the cell has to divide. In this way the cyclic activity of the genes would cause reproduction to occur at regular intervals.

2.4.3 Landscape ecology

Studies of landscape ecology, both theoretical and field oriented, have usually focused on the effects of environmental variations (e.g. soils) that impinge on ecosystems. However, important properties of ecosystems also emerge from biotic interactions within a landscape (see chapter 9). The chief biotic interactions between sites in a landscape are 'space-filling' processes, such as seed dispersal or animal migration, and 'space-clearing' processes, such as fires, storms and other causes of mortality as shown by Green (1989, 1990, 1993, 1994). Ecological modellers, such as Roff (1974), Hogeweg (1988) and Turner et al. (1989a, 1989b), have become increasingly aware of the need to incorporate spatial interactions in their models.

The global effects of ecological processes in a landscape are often different from their local effects. For instance, communities can be locally unstable, yet globally stable. Both field studies and theoretical work (e.g. Hogeweg (1988)) show that competitive exclusion does not always occur within a landscape; competing populations can often coexist. Short distance dispersal promotes the formation of clumped species distributions (fig. 2.11). The effects of this clumping include promoting the persistence of species that would otherwise be eliminated by superior competitors

Self-organisation in complex systems

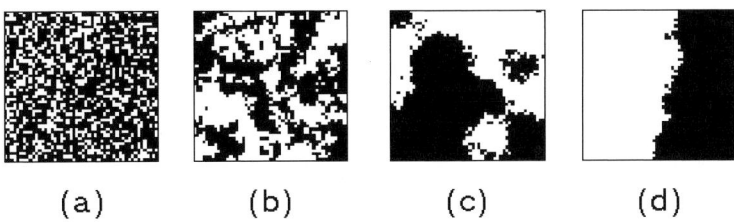

Fig. 2.11. Some spatial patterns that can emerge in cellular automata simulations of forests ecosystems The initial distribution was random in each case. (a) 'Global' dispersal — seeds spread everywhere; (b) 'Local' dispersal — seeds reach neighbouring cells only; (c) Local dispersal with fire; (d) As in (c), but on an environmental gradient.

and resisting invasion from outsiders because of a super-abundance of local sources of seeds or other propagules as in Green (1989, 1990).

Other emergent effects of landscapes include critical phase changes in connectivity between sites (fig. 2.12). Such effects are direct consequences of the 'double jump' (chapter 3) associated with the underlying digraph structure, described earlier. For example, patchy fuels must exceed a critical density within a landscape if a fire is to spread. Similarly, the density of suitable growing sites in a landscape must exceed a density threshold if an invading or introduced species is to spread. Seed dispersal acts as a conservative process in forest communities (Green, 1989). Because they possess an overwhelming majority of seed sources, established species are able to out-compete invaders. By clearing large regions, major fires enable invaders to compete with established species on equal terms. Conversely seed dispersal also enables rare species to survive in the face of superior competitors by forming clumped distributions. This process implies a possible mechanism for the maintenance of high diversity in tropical rainforests (Green, 1989).

2.4.4 Hypercycles and the origin of life

An important question concerning the origins of life is how essential features of living systems arose during prebiotic, molecular evolution. Any such mechanism must not only achieve self-replication, but also be capable of accumulating information by accommodating longer and longer molecules. This need poses a fundamental difficulty (Eigen, 1992). Eigen and Schuster (1979) showed that self-replication is limited by an 'error threshold'. As molecules become longer the error rate during

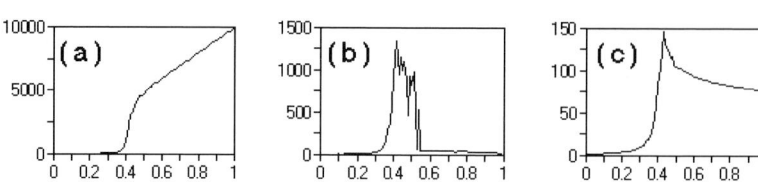

Fig. 2.12. Critical changes in connectivity of a CA grid as the proportion of 'active' cells increases. (a) Average size of the largest connected subregion (LCS). (b) Standard deviation in the size of the LCS. (c) Traversal time for the LCS. Each point shown is the result of 100 iterations of a simulation in which a given proportion of cells (in a square grid of 10,000 cells) are marked as 'active'. Note that the location of the phase change (here ~ 0.4) varies according to the way we define connectivity within the model grid.

replication increases. Beyond a certain critical molecule size (which depends on the error rate), errors cascade, so accurate replication becomes impossible. Thus it would appear that self-replicating strings must be short.

Eigen and Schuster (1979) suggested that the above problem is solved via *hypercycles*. These are cyclic sequences of molecules that each catalyse self-replication for other molecules in the sequence (e.g. A catalyses B; B catalyses C; C catalyses A). Although each element (A, B, C) in the process must still be short, the hypercycle itself can become indefinitely complex.

Several authors (Bresch et al., 1980, Maynard Smith, 1979) quickly pointed out that 'parasites' pose a problem for the hypercycle theory. Suppose that as well as catalysing A, species C above also catalyses a species D, which does not catalyse any other chemical species. If D receives more support from C than does A, then D will increase at the expense of A, thus breaking the cycle. For some time it seemed that these parasitic chemical species posed a fatal flaw in the theory of hypercycles. However Boerlijst and Hogeweg (1991a, 1991b) showed that incomplete mixing of the spatial substrate overcomes the parasite problem. In their simulations (which use a cellular automaton representation), the hypercycles led to waves that spiralled outwards from foci, which themselves gradually shift. Instead of dominating the spirals, parasitic species are forced to the margins of the spirals and removed.

The above results have several important implications. First, they show that spatial self-structuring plays an important role in evolution. Secondly, they illustrate that *global* stability can emerge in spite of *local* instability. Thirdly, selection occurs at the level of the spirals and can contradict selection at the level of individual molecules (Boerlijst & Hogeweg, 1991a).

2.5 Evolution – natural and artificial

The term 'evolution' refers to the process of overall change in complex systems. In biology it refers explicitly to mechanisms by which new species emerge. We begin here by summarising the processes involved in the evolution of species. We then show that these belong to a much wider class of processes that act in many kinds of complex systems, including evolutionary computation.

2.5.1 Genetics

The mechanics by which biological species change and new species appear are governed by processes involving the *genes*. Genes determine the character and identity of a species. During (sexual) reproduction, offspring acquire genetic information from both parents in equal proportions. This genetic information, which determines the biologic 'design' of the individual, is contained in a set of chromosome pairs (one from each parent). The chromosomes contain DNA sequences that govern the formation of proteins and control many aspects of growth. A *gene* is a sequence of nucleotides that have a particular function. The site where a gene occurs on the chromosome is called its *locus*. Different forms of a gene (e.g. blue versus brown eyes) are called *alleles*. Since both parents contribute genetic material, two alleles are found at each locus. Together they determine an individual's *genotype*. Species with two alleles at each locus are termed *diploid*; *haploid* species have only one set of chromosomes and therefore only a single allele at each locus.

Genetic control over observable *traits* (e.g. height, eye colour) is often complicated. Some genes control many different traits; on the other hand, genes at many different loci can affect a single trait. Most attention in genetic theory has focused on traits that are controlled by a small number of genes. The traits that are expressed (e.g. eye colour) determine the individual's *phenotype*. If the two alleles that occur at each locus are the same, then the individual is called *homozygous* with respect to that trait; if the alleles are different, then the individual is *heterozygous*. In heterozygous individuals, the resulting phenotype often expresses traits corresponding to only one of the two alleles. An allele that is expressed in this way is called *dominant*, whereas the allele coding for a suppressed trait is called *recessive*. If two alleles are both expressed phenotypically when they occur together, then they are called *co-dominant*.

2.5.2 Example – blood groups

The ABO blood groupings are governed by two dominant alleles (A and B) and one recessive allele (O). This means that the blood groups A and B consist of both homozygous individuals (AA, BB) and heterozygous individuals (AO, BO), whereas group O contains only homozygous individuals (OO). The alleles A and B are codominant when they occur together and are expressed phenotypically as a fourth blood group, known as type AB.

A central problem in genetics is to determine how the frequencies of different genotypes change within a population. A result of fundamental importance to this question is the *Hardy–Weinberg law*:

Theorem 2.6 *Let $A_1, ..., A_n$ denote all of the alleles that may occur at a particular locus, and let $p_1, ..., p_n$ denote their initial frequencies within a given population. Furthermore, assume that these frequencies are the same for both sexes, that individuals mate randomly (so far as the genotypes are concerned), that the population is large enough for the gene frequencies to be treated as continuous random variables, and that mutation, migration and other such processes may be ignored. Then*

- *the frequency of each genotype reaches an equilibrium level in a single generation;*
- *the equilibrium frequency of each homozygous genotype $A_i A_i$ is p_i^2;*
- *the equilibrium frequency of each heterozygous genotype $A_i A_j$ is $2 p_i p_j$.*

The Hardy–Weinberg law shows that with random mating, and adequate mixing, gene frequencies settle rapidly to equilibrium levels. Moreover, it implies that reproduction by itself does *not* change gene frequencies within a large population.

2.5.3 Species evolution

Since the time of Darwin and Wallace, it has been generally accepted that natural selection is the predominant mechanism by which species evolve. In the intervening time the theory has been fleshed out and missing details (e.g. genetic variation) have been filled in. Briefly, this 'neo-Darwinian' theory can be summarised as follows.

- More individuals are born in a population than can be supported by the environment, so they must compete with one another for food and other resources.
- Within a population, the individuals vary in their 'fitness', that is their ability to survive and reproduce.
- As a result of competition, fitter individuals produce more offspring, so the frequency of their genes increases within a population.
- Natural selection removes unfit individuals from a population, so over time the population becomes better and better adapted to its environment.
- New species arise when a population becomes isolated for long enough that individuals are no longer capable of breeding with members of the parent species (it becomes 'reproductively isolated').

The results described earlier about criticality and connectivity can help to explain some aspects of species evolution. Let us consider here how connectivity within a landscape affects speciation. Sites in a landscape are 'connected' if the local populations interbreed with each other (i.e. share genetic information). Dispersal is essential to maintain genetic homogeneity within populations. Should the connectivity provided by dispersal between sites fall below a critical level, then a regional population effectively breaks down into isolated subpopulations.

Isolation has long been regarded as the chief method by which speciation occurs. The effect of cataclysms is to lay waste large regions, so isolating the surviving groups from one another and promoting speciation.

We can simulate this process (fig. 2.13) using a cellular automaton model to represent the 'landscape' as a square grid of cells (here 400 'sites'). Randomly selected cells are occupied ('active'); other cells represent unused territory. The active cells contain a real number G to represent the phenotype of a hypothetical gene. Each cycle of the model represents a turnover of generations: random perturbations of each cell's 'gene' mimic mutation; averaging with a randomly selected neighbour

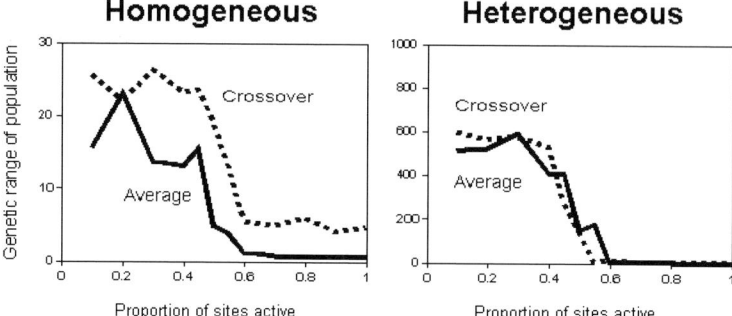

Fig. 2.13. Simulated evolution within a landscape. The range of gene values (G), after 10 000 'generations' of a population that initially is: (left) homogeneous ($G = 0$ everywhere); and (right) heterogeneous ($-100 < G < 100$) in response to the proportion P of active sites. Solid lines indicate sexual reproduction by averaging; dashed lines indicate crossover. See the text for further explanation.

(if one is active) mimics sexual reproduction. An alternative 'sexual' mechanism is 'crossover' in which a neighbour's gene may replace the local one. Note that here 'mutation' is simply a uniform random number r, so the 'mutation rate' is the radius of the distribution of r (here set at 1). Tests (not shown) confirm that its only effect on the results is to set the scale for the genetic range.

Simulation (fig. 2.13) shows that in a fully connected landscape, reproduction acts as a spatial filter that restricts genetic drift in uniform populations (fig. 2.13a) and forces heterogeneous populations to converge (fig. 2.13b). However, if connectivity falls below the critical level, then genetic drift proceeds unimpeded in initially uniform populations (fig. 2.13a) and heterogeneous populations do not converge, but continue to drift apart (fig. 2.13b). Note that the critical region in these simulations (40–60% coverage) is specific to the neighbourhood function used and will vary according to the pattern of dispersal.

These results show that the genetic make-up of a population is highly sensitive to changes in landscape connectivity. They suggest that major landscape barriers are not necessary for speciation to occur, it suffices that overall landscape connectivity drop below the critical threshold long enough, or often enough. Conversely, they also hold a warning for conservation: populations whose distributions become restricted to small, homogeneous areas risk losing their natural genetic diversity.

2.5.4 Genetic algorithms

Genetic algorithms (Goldberg, 1989, Holland, 1975) solve complex problems of search and optimisation by emulating evolution. The 'standard' genetic algorithm (GA) works as follows.

- Parameters, assumptions and other mutable features of the problem are each expressed as 'genes'. The combination of genes for each model defines its 'genotype'. For discrete parameters, possible values are represented as alleles. Numerical parameters are allowed to vary freely. However, unlike biological genes, alleles that are paired together by GA algorithms usually represent different features entirely. So dominant and recessive properties do not exist in the standard algorithm.
- The algorithm operates not on a single model, but on an entire population of models, each represented by its set of genes.
- Each 'generation' is an iteration of the algorithm. It consists of 'reproduction', in which new models are generated from existing ones, and 'selection', in which poor models are removed. The important procedures in the reproduction phase are point mutation, by which individual parameters are changed, and crossover, in which features of separate models are mixed in their 'offspring' (fig. 2.14).

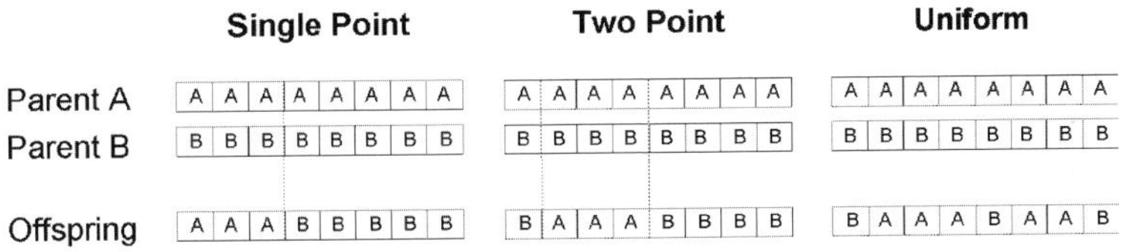

Fig. 2.14. Crossover in genetic algorithms, showing three separate mechanisms that can be invoked. Each box denotes a separate 'gene', which may be (say) a parameter in a model. In 'crossover', offspring are formed by selecting genes from one or other of the parents.

Genetic algorithms (GA) have several practical uses. For instance GAs deal with a whole population of examples, so they are ideal for exploring the variety in a given system. However the most common role is optimisation. One advantage is that they can often be used where many other methods cannot. For example, the assumptions of linear programming, such as continuity and convexity, are not necessary. Nor do GAs require a single object function to optimise: they can be used with all manner of complex constraints and criteria.

Because genetic algorithms operate on whole populations of models, they are inherently parallel. This property makes them ideal for dealing with problems of high combinatorial complexity, such as the travelling salesman problem. However, it also means that they can easily become computationally intensive, so parallel implementations are often desirable.

When Holland introduced genetic algorithms a crucial question was whether fitness in the population would increase. A positive answer to this question was provided by the 'schema theorem' (Holland, 1975), also known as the 'fundamental theorem of genetic algorithms'. A *schema* is a string of symbols that denotes a pattern of similarities within a set of genotypes. As an example suppose that the encoding is binary. Then genotypes can be represented as strings of 0 and 1: for example 11010 or 00010. In the examples just given the two strings match in the last three positions. So we can represent the common pattern by a similarity template **010. Such a template is called a schema (plural 'schemata'). Here the symbol '*' (a linguistic variable) means 'either 0 or 1'. In general an alphabet for such schemata would be $A = \{0, 1, *\}$. Then a schema is any string a^n, where each a is one of the above symbols and n is the length of the genotype.

In the above example notice that as well as **010 the above strings 11010 and 00010 also share the schemata **01*, **0*0, ***10, **0**, ***1*, ****0, and *****. More generally the genotype 11010 can be represented by 32 possible schemata, such as 11*1*, 1*0**, ****0, and 1***0. Similarly the schema 1**10 would denote any of the four sequences: 11110, 11010, 10110, 10010.

Two measures of a schema are its order and its length (Goldberg, 1989). The *length* $\lambda(s)$ of a schema s is the distance from the first to the last constants in the string. The *order* $o(S)$ of a schema S is the number of constants in the string. Table 2.3 provides some examples to make these definitions clearer.

The schema theorem shows that in a large population the frequency of any schema can be expected to increase at essentially a geometric rate from one generation to the next. Using the notation of Goldberg (Goldberg, 1989), the theorem can be stated as follows.

Theorem 2.7 *Let S denote a schema and let $m(S,t)$ denote its frequency at generation t. Then the frequency $m(S, t+1)$ of S at generation $t+1$ satisfies the relation*

Table 2.3. *Examples of simple schemata*

Schema	Length	Order
11010	4	5
11*1*	3	3
1*0**	2	2
***0	0	1
1***0	4	2

$$m(S, t+1) \geq m(S,t)\frac{f(S)}{\bar{f}}[1 - p_c\frac{\lambda(S)}{L-1} - o(S)p_m] \qquad (2.12)$$

where:

$f(S)$ *is the average fitness of all strings representing schema S;*
$\bar{f}$ *is the average fitness of the entire population;*
p_c *is the probability of crossover;*
p_m *is the probability of mutation;*
$\lambda(S)$ *is the length of the schema;*
$o(S)$ *is the order of the schema;*
L *is the length of the string.*

Many variants of the basic genetic algorithm have been introduced. For example the 'generational' approach replaces an entire generation all at once. In contrast a *steady-state* GA replaces individuals one at a time. Many of the introduced variations are attempts to improve the performance of the algorithm. One of the most common problems – *premature convergence* – occurs if a particular genotype drives out all others before the evolving population reaches a global optimum. This problem is likely to occur if (say) the population is small and if the fitter individuals are allowed to make too many copies of themselves in a single generation.

Another important issue in applications of genetic algorithms is *encoding* (also called the *representation problem*). That is, how are elements of a problem encoded as a set of 'genes'? For simple problems the answer is obvious – just treat each parameter as a separate 'gene'. However many problems cannot be reduced to a simple set of parameters. For example, the *travelling salesman problem* (TSP) requires that a salesman visit every one of a set of cities in an order that minimises the total distance

	Code	Resulting tour
Parents		
	1 2 2	A B C D
	2 3 1	B A C D
Potential offspring		
	1 2 1	A C D B
	2 2 2	B D C A
	2 3 2	B A D C

Fig. 2.15. Encoding the travelling salesman problem. See text for details.

that he must travel. Here the obvious genome is a list of the cities in the order that they are to be visited. However, this representation leads to the problem of how to perform crossover on lists: cutting and joining two lists may produce a list with some cities omitted and others visited twice.

A general solution to the above issue is to treat the lists as phenotypes, not genotypes. That is, perform crossover not on the lists themselves, but on codes for building the lists. For instance suppose that we have four cities – A, B, C, D. We can encode tours for the TSP by writing a list of three numbers in which the first number indicates the starting point and the remaining numbers tell us, at any stage of selection, how many places to skip in the list of remaining cities. Only three alleles are needed because the final city in the list is fixed as soon as the first three are chosen. So the string (1,2,2) would yield the list (A,C,B,D) and the string (2,3,1) would yield the list (B,A,C,D). The important point is that any product of crossover between these lists is also a valid list, as figure 2.15 shows.

2.5.5 Evolutionary and adaptive programming

As its name implies, *evolutionary programming* (EP) produces computer programs by emulating evolutionary processes. This area of research stems from the need to solve complex problems where it is a suitable combination of processes, not simply values, that needs to be found.

The starting point is normally a population of programs, which then mutate and reproduce through a series of generations. Selection (culling unsuccessful programs) is determined by a set of fitness functions and/or criteria. Perhaps the most crucial issue is to represent programs in ways that allow sensible, and fruitful, variations to occur. Add random lines to (say) a working piece of FORTRAN code and the result is almost certain to fail. An important achievement of Thomas Ray's Tierra model (chapter 4) was its demonstration that code can be mutated to produce viable results (Ray, 1991).

Another requirement is to provide precise criteria of 'fitness'. Most algorithms apply selection by culling the least fit programs from each generation. In this respect, EP is often treated as a form of search and optimisation. Important concerns therefore are to devise systems that allow programs to explore the 'fitness landscape' both widely and efficiently, without becoming trapped by local maxima.

One of the confusing aspects of this area of research, which is still in its infancy at the time of writing, is the proliferation of terms to connote variations on the theme. Work in this area also blurs into closely related areas, notably artifical life, and artificial intelligence. Fogel (1995) gives a detailed overview of the history of the field.

At least two distinct approaches to EP have emerged: *direct methods* and *genetic algorithms*.

Direct methods apply mutation and selection directly to members of a population of programs. Two common representations are neural networks (see chapter 10) and finite state automata (FSA). In FSA, for example, each program consists of a set of 'states' and 'transitions' between them. For example, Fogel's pioneering experiments (Fogel *et al.*, 1966) adapted FSA's to predict future values of time series. Mutations consist of changing the set of states or altering details of the transition pathways. Normally each program derives from a single parent, so there is no crossover of features.

Koza's Genetic programming (1992a, 1992b) takes this idea one step further. It evolves populations of programs that are intended to achieve some result. The programs are represented in a systematic way. For instance, the GARP algorithm ('Genetic Algorithm for Rule Production') (Stockwell, 1992) generates non-procedural, predictive models that consist of sets of discrete rules to partition a sample space (e.g. for predicting species distributions). Crossover here consists of selecting random subsets from the parents' pooled set of rules.

As with the TSP problem described earlier, a crucial first step in genetic programming is to encode a procedure in a way that allows crossover to yield valid offspring. Tree representations are useful for this purpose. The root of the tree denotes the starting point for processing and each node encodes a discrete operation. The branching structure indicates the order of processing. In mutation we can change values of

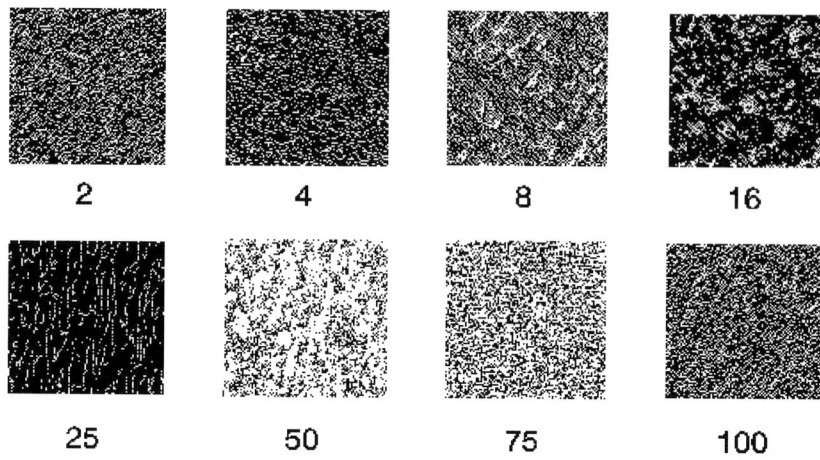

Fig. 2.16. Changes in system configuration that accompany the transition from 'cool' to 'hot' cellular automata. Each CA starts from a random configuration and includes more rules (selected at random) than the one before it. The numbers below each image indicate the approximate percentage P of configurations (of states for a cell and its neighbours) for which the model changes the state of the cell concerned. 'Cool' models are those for which P is small; 'hot' models have high values for P.

parameters at individual nodes, or even add or subtract whole branches. Crossover can likewise involve flipping nodes or branches.

Two major questions of increasing interest in EP have been:

- What systems are capable of evolving?
- What conditions are necessary and sufficient for systems to be capable of universal computation?

As will be seen in other chapters (e.g. chapter 4), much of the research on the above questions has focused on cellular automata (CA). Wolfram (1984, 1986) identified various classes of behaviour in CAs. Langton (1990, 1992) identified transitions between these classes in terms of an order parameter λ, which represents increasing connectivity within the state spaces. Several authors, such as Kauffman (1991, 1992) and Langton (1990, 1992) have identified the transition from ordered to chaotic behaviour (the 'edge of chaos') as especially significant. They suggest that the critical region in behaviour has several important properties:

- systems capable of universal computation lie within the critical region;
- many systems with 'interesting' behaviour fall within the critical region (e.g. fig. 2.16 and the game LIFE (chapter 4);
- there is a balance between 'frozen' and 'fluid' components (Kauffman, 1992), that is systems form both stable and varying configurations.

The above special properties have led to intense interest in systems near the edge of chaos. This includes speculation (Langton, 1992) that evolution includes selection for organisms whose genetic makeup lies within this critical region. Kauffman (see above) has stressed the importance of the genetic network in evolution. He argues that networks lying in the fluid region are unstable, and that those in the ordered region cannot adapt.

2.5.6 Evolution in complex systems

To conclude this discussion of evolution, let us draw together some of the ideas presented above into what may be the beginnings of a general theory about evolution and self-organisation in complex systems.

Several authors, e.g. (Freeman, 1975), have suggested that chaos is an important source of novelty and variety in living systems. The results given earlier about the universality of digraphs (theorems 2.3 and 2.4) suggest that critical phase changes in the behaviour and structure of complex systems may provide mechanisms by which nature exploits chaos.

Two different evolutionary mechanisms have been suggested that involve the above phase changes. First, as mentioned earlier, Kauffman (1992) and Langton (1990, 1992) have suggested that the 'edge of chaos' – the phase change between ordered and chaotic behaviour for automata – plays a central role in evolution. They suggest that because systems near the edge have the richest, most complex behaviour, such systems can adapt better and therefore have a selective advantage over other systems. They therefore suggest that living systems have evolved so as to lie close to the edge of chaos. Most of these arguments concern the behaviour of automata, and of living things (regarded as automata).

Green (1993, 1994) derives a second theory from observations of phase changes in the structure of complex systems, rather than their behaviour. As we saw earlier, the transition here is from disconnected to fully connected systems. An important property of this transition is the variability associated with the phase change itself (Green, 1993). Put simply, we might regard the transition as a 'chaotic edge', rather than an 'edge of chaos'. It is possible (Green, 1994) that this chaotic edge provides a source of novelty in many natural phenomena (especially in biology). Rather than evolve to sit on the phase change, external stimuli cause the system to flip across it and back. In the connected phase 'selection' operates to inhibit variation; in the disconnected phase variation has free reign. The earlier discussions of evolution and ecology in a landscape show how this process may work in real systems. In each case the systems normally exist in the connected phase, for which variation is inhibited. However external stimuli (comets, fires etc.) act to flip the system to a disconnected state. As connectivity builds up again the system 'crystallises' into a new configuration.

Although the above ideas describe particular mechanisms, there is still much to learn about the overall processes by which complex systems adapt and evolve. Several authors have addressed this issue. For instance, Holland (1995) has suggested that seven basic elements are involved in the emergence of order in complex adaptive systems. These elements include the following four properties.

- *Aggregation* – Common properties shared by different objects make it possible to groups them in categories (e.g. 'cars', 'trees').
- *Non-linearity* – Non-linear interactions can make the collective behaviour of an aggregate different from what we would expect from simply averaging the behaviour of the components.
- *Flows* – Networks of linked objects lead to flows (e.g. material or information) within which iteration (e.g. cycles, the multiplier effect) produces distinctive features.
- *Diversity* – The pattern of interactions within a system tends to produce a characteristic diversity of elements (e.g. ecological niches).

Holland also identifies three mechanisms.

- *Tagging* – Adding or forming tags (e.g. flags, headers) help to distinguish and identify aggregates.
- *Internal models* – Adaptive systems tend to embed models of their environment within their structure or behaviour (e.g. plants grow towards the light, ants follow pheromone trails).
- *Building blocks* – Internal models tend to be built out of simple elements, e.g. a 'room' consists of ceiling, floor and walls, possibly with windows and doors.

2.6 Conclusion

In a short account such as this it is inevitable that many important ideas are touched on only briefly, or even omitted altogether. Some, but not all, of these ideas are taken up in later chapters. However, the understanding of self-organisation remains one of the grand challenges for complexity research in the new millenium.

One of the exciting aspects of such work is that we are now beginning to identify fundamental processes by which order emerges in many different systems. The existence of such processes reveals deep similarities between a vast range of systems that have formerly been viewed and studied as completely distinct, e.g. why is an epidemic like an atomic bomb? (Bossomaier & Green, 1995). It is beginning to become apparent that there are general principles to explain how complex organisation comes about. Whether we can ever realise the vision of the early researchers and develop a general theory of complex systems remains to be seen.

Bibliography

Boerlijst, M., & Hogeweg, P. (1991a). Self-structuring and selection: spiral waves as a substrate for prebiotic evolution. *Pages 255–276 of:* Langton, C. G., Taylor, C., Farmer, J. D., & Rasmussen, S. (eds), *Artificial Life II.* Addison-Wesley, New York.

Boerlijst, M., & Hogeweg, P. (1991b). Spiral wave structure in prebiotic evolution: hypercycles stable against parasites. *Physica D*, **48**, 17–28.

Bossomaier, T. R. J., & Green, D. G. (1995). *Patterns in the Sand – Computers, Complexity and Life.* Allen and Unwin, Sydney.

Bresch, C., Niesert, U., & Harnasch, D. (1980). Hypercycles, parasites and packages. *J. Theoret. Ecology*, **85**, 399.

Brooks, R. A. (1986). *Achieving Artificial Intelligence through Building Robots.* AI memo 899. MIT.

Brooks, R. A. (1991). Intelligence without representation. *Artificial Intelligence*, **47**, 139–159.

Brooks, R. A. (1992). *Intelligence without Reason.* Ai memo. MIT.

Carroll, S. B. (1990). Zebra patterns in fly embryos: activation of stripes or repression of interstripes. *Cell*, **60**, 9–16.

Eigen, M. (1992). *Steps towards Life: a Perspective on Evolution.* Oxford University Press, Oxford.

Eigen, M., & Schuster, P. (1979). *The Hypercycle: a Principle of Natural Self-Organization.* Springer, Berlin.

Erdos, P., & Renyi, A. (1960). On the evolution of random graphs. *Mat. Kutato. Int. Kozl.*, **5**, 17–61.

Fogel, D. B. (1995). *Evolutionary Computation : Toward a New Philosophy of Machine Intelligence.* IEEE Press, New York.

Fogel, L. J., Owens, A. J., & Walsh, M. J. (1966). *Artificial Intelligence through Simulated Evolution.* John Wiley and Sons, New York.

Freeman, W.J. (1975). *Mass Action in the Nervous System*. Acad. Press, New York.
Goldberg, D. E. (1989). *Genetic Algorithms in Search, Optimization, and Machine Learning*. Addison-Wesley, Reading, Mass.
Green, D. G. (1989). Simulated effects of fire, dispersal and spatial pattern on competition within vegetation mosaics. *Vegetatio*, **82**, 139–153.
Green, D. G. (1990). Landscapes, cataclysms and population explosions. *Mathematical and Computer Modelling*, **13**, 75–82.
Green, D. G. (1993). Emergent behaviour in biological systems. *Pages 24–35 of:* Green, D.G., & Bossomaier, T.R. J. (eds), *Complex Systems – from Biology to Computation*. IOS Press, Amsterdam.
Green, D. G. (1994). Connectivity and the evolution of biological systems. *Journal of Biological Systems*, **2**, 91–103.
Haken, H. (1978). *Synergetics*. Springer-Verlag, Berlin.
Haken, H. (1981). *The Science of Structure: Synergetics*. Van Nostrand Reinhold, New York.
Haken, H. (1983). *Advanced Synergetics*. Springer-Verlag, Berlin.
Herman, G. T., & Rozenberg, G. (1975). *Developmental Systems and Formal Languages*. North-Holland, Amsterdam.
Hogeweg, P. (1988). Cellular automata as a paradigm for ecological modeling. *Applied Math. Comput*, **27**, 81–100.
Hogeweg, P. (1993). As large as life and twice as natural: bioinformatics and the artificial life paradigm. *Pages 2–11 of:* Green, D. G., & Bossomaier, T. R. J. (eds), *Complex Systems – from Biology to Computation*. IOS Press, Amsterdam.
Hogeweg, P., & Hesper, B. (1981). Two predators and a prey in a patchy environment: an application of MICMAC modeling. *J. Theoret. Biol*, **93**, 411–432.
Hogeweg, P., & Hesper, B. (1983). The ontogeny of the interaction structure in bumblebee colonies: a mirror model. *Behav. Ecol. Sociobiol*, **12**, 271–283.
Hogeweg, P., & Hesper, B. (1991). Evolution as pattern processing: Todo as a substrate for evolution. *Pages 492–497 of:* Meyer, J. A., & Wilson, S. W. (eds), *From Animals to Animats*. MIT Press.
Holland, J. (1975). *Adaptation in Natural and Artificial Systems*. University of Michigan Press, Ann Arbor.
Holland, J. H. (1995). *Hidden Order: How Adaptation Builds Complexity*. University of Michigan Press, Ann Arbor.
Hopcroft, E., & Ullman, J. D. (1969). *Formal Languages and their Relation to Automata*. Addison-Wesley, Reading, Mass.
Karr, T. L., & Kornberg, T. B. (1989). Fushi tarazu protein expression in the cellular blastoderm of drosophila, detected using a novel imaging technique. *Development*, **105**, 95–103.
Kauffman, S. A. (1991). Antichaos and adaptation. *Scientific American*, **265**, 64–70.
Kauffman, S. A. (1992). *Origins of Order: Self-Organization and Selection in Evolution*. Oxford University Press, Oxford.
Koestler, A. (1967). *The Ghost in the Machine*. Hutchinson, London.
Koza, K. E. (1992a). *Genetic Programming II*. MIT Press, Cambridge MA.

Koza, K. E. (1992b). *Genetic Programming: on the Programming of Computers by Natural Selection.* MIT Press, Cambridge MA.

Langton, C. G. (1990). Computation at the edge of chaos: phase transitions and emergent computation. *Physica D,* **42(1–3)**, 12–37.

Langton, C. G. (1992). Life on the edge of chaos. *Pages 41–91 of:* Langton, C. G., Taylor, C., Farmer, J .D., & Rasmussen, S. (eds), *Artificial Life II.* Addison-Wesley, New York.

Lindenmayer, A. (1971). Developmental systems without cellular interactions in develoment, their languages and grammars. *Theoret. Biol,* **30**, 455–484.

Maynard Smith, J. (1979). Hypercycles and the origin of life. *Nature,* **280**, 445–446.

Papert, S. (1973). *Uses of Technology to Enhance Education.* Logo Memo No. 8. MIT Artificial Intelligence Lab.

Prigogine, I. (1980). *From Being to Becoming.* W. H. Freeman and Co, San Francisco.

Prusinkiewicz, P. (1996). L-systems: from theory to plant modeling methodology. Michalewicz, M. (ed), *Computational Challenges in the Life Sciences.* CSIRO Division of Information Technology, Melbourne.

Prusinkiewicz, P., & Lindenmayer, A. (1989). Developmental models of multicellular organisms. *Pages 221–249 of:* G., Langton C. (ed), *Artificial Life.* Addison-Wesley, New York.

Prusinkiewicz, P., & Lindenmayer, A. (1991). *The Algorithmic Beauty of Plants.* Springer-Verlag, Berlin.

Ray, T. (1991). An approach to the synthesis of life. *Pages 41–91 of:* Langton, C. G., Taylor, C., Farmer, J. D., & Rasmussen, S. (eds), *Artificial Life II.* Addison-Wesley, New York.

Roff, D. A. (1974). Spatial heterogeneity and the persistence of populations. *Oecologia,* **15**, 245–258.

Rozenberg, G., & Salomaa, A. (1986). *The Book of L.* Berlin: Springer-Verlag. *Collection of papers on L-systems.*

Seeley, D., & Ronald, S. (1992). The emergence of connectivity and fractal time in the evolution of random digraphs. *Pages 12–23 of:* Green, D. G., & Bossomaier, T. R. J. (eds), *Complex Systems – from Biology to Computation.* IOS Press, Amsterdam.

Small, S., & Levine, M. (1991). The initiation of pair-rule stripes in the drosophila, blastoderm. *Current Opinion in Genetics and Development,* **1**, 255–260.

Stockwell, D. R. (1992). *Machine Learning and the Problem of Prediction and Explanation in Ecological Modelling.* Ph.D. thesis, Australian National University.

Turner, M. G., Costanza, R., & Sklar, F. H. (1989a). Methods to evaluate the performance of spatial simulation models. *Ecological Modelling,* **48**, 1–18.

Turner, M. G., Gardner, R. H., Dale, V. H., & O'Neill, R. V. (1989b). Predicting the spread of disturbance across heterogeneous landscapes. *Oikos,* **55**, 121–129.

von Bertalanffy, L. (1968). *General Systems Theory.* George Brazillier, New York.

Westman, R. S. (1977). Environmental languages and the functional basis of behaviour. *Pages 145–201 of:* Hazlett, B. A. (ed), *Quantitative Methods in the Study of Animal Behaviour.* Academic Press, New York.

Wolfram, S. (1984). Cellular automata as models of complexity. *Nature*, **311**, 419–424.

Wolfram, S. (1986). *Theory and Applications of Cellular Automata.* World Scientific, Singapore.

3
Network evolution and the emergence of structure

D.A. SEELEY

University of South AUSTRALIA,
Adelaide,
South Australia,
AUSTRALIA

3.1 Introduction: networks of relationships are everywhere

We live in an ocean of networks. Both in our everyday lives with families or friends, and at work in a technological world, we are participating in various networks. And these networks are somehow reflections of our own bodies and minds.

When we rise in the morning we participate in the network of the family, a network which is a reverberation of our recent generations. When we shower and turn on the toaster we are utilising the provisions of electrical and water utilities which are themselves large networks. Turning on the radio on our trip to work, we are participants in a broadcast network connected by the electromagnetic spectrum, and the traffic network dynamically emerging as other citizens in our community engage in their business. Once at work, we may interact with the Internet with its diverse and always changing quilt of electronic communities. At times we may interact with different workgroups, and with our network of clients in the broader business or academic community. At lunch with a friend from our personal network, and after work with our tennis club, we continue to participate in various networks, variously overlapping in a greater whole.

Our bodies are a finely interwoven tapestry of many networks; the circulatory system providing nourishment to the vast network of body cells, the nervous system, the immune system, the lymphatic system and others subtly cooperating together in sustaining our lives. While the vast network of concepts, ideas, impressions and attitudes in our minds forms a vast interconnected network, the cortex of our brains is the network par excellence, an enormous loom of interconnections continually changing. Our place on the planet is defined by the networks which are the local watershed, the local ecosystem, parts of greater networks of relationships in the water of the world. Here we participate in the vast webs of food chains, both natural and artificial, which are in turn part of the genetic organisation of life into species in the 'great chain of being'.

For many of us involved in the communication of scientific and technical ideas, dependency graphs or 'nodes and arrows' diagrams have come to pervade our thinking and our imaginations. The prevalence of such diagrams in the interfaces of modern software suggests that these diagrams will quickly spread to the popular imagination as well.

The natural networks have always been there. It is our exposure to technological networks which is providing new tools of perception, such as dynamic systems, fractals and the edge of chaos. These tools of perception appear as we project more of ourselves into the great mirror of electronic and computer technology, where we begin to recognise different aspects of ourselves, not made evident from that earlier great projection system, human language.

Why are these networks there? Without interactions with others there would not be a world; we would each remain in isolation. Chains of interactions therefore naturally arise, giving the world a 'cause and effect' quality. These chains would also remain isolated if it were not for the fact that many of them return again and again to effect us, forming circuits both small and vast. Within the relative constancy of such circuits, systems emerge, identifiable as units because of the definition which the connectivity of their own links and circuits of interactions provide, in contrast to the connections 'outside'.

Because of the ubiquity of networks of relationships, it is important to examine how such networks come into being and sustain themselves. This chapter is about some initial forays into the conceptualisation, observation and analysis of the life of networks. It began with the curiosity and seminal work of one of the most productive mathematicians of the modern era, Paul Erdos. Some of us believe that such activity will eventually reward us with rich insights which will permeate all of our lives.

When Erdos and Renyi (1960) published their seminal paper on random graphs some twenty years ago, they opened up a rich area of investigation now called graphical evolution (Palmer, 1985). The phenomena they uncovered reveal some surprising and sudden phase changes in connectivity. Although a substantial literature has subsequently emerged for the evolution of random graphs, similar material for directed graphs has been sparse, and has largely addressed regularly connected digraphs. In this chapter, similar behaviour is reported for directed graphs using analytic and simulation results. Moreover, a further exploration is taken using the natural orderings of 'connection from a viewpoint', orderings which have fractal properties as well. Applications of this approach to the various networks which comprise our world are indicated and some are explored. The significance of these explorations in the emergence of 'emergence' complete the chapter.

3.2 Directed graphs and connectivity

3.2.1 Terminology

In English, there are a number of different words to describe a collection of entities which have various relationships (dependencies) upon each other. Diagrams, networks and dependency graphs are often used to describe this situation. Mathematicians in this century (those identified with graph theory and combinatorics) have tended to draw certain distinctions amongst such collections, using common words in quite specific ways, which at times seem quite arbitrary. Hence, we will distinguish between 'graphs', 'directed graphs' and 'networks', while also retaining the use of the word 'network' to cover all three (fig. 3.1). There are analogous terms associated with each of these three concepts, and a consensus on these terms, even by mathematicians, is only slowly emerging.

In a directed graph, the relevant entities will be called the nodes, while the directed relationships will be called simply arrows, indicating an asymmetric influence or dependency with the sense of the arrowhead for every relationship. Notice that the dependency or influence is not necessarily returned. Sometimes arrows are called arcs or links (fig. 3.1). Directed graphs are also called digraphs. We shall also retain connection as a general term implying some kind of relationship, be it an edge, arrow or link.

In an (undirected) graph or simply, graph, the links have often been called edges and the entities called vertices. These edges indicate that the relationship between entities goes both ways; i.e. it is symmetric. In a simple graph, this symmetry holds for all relationships. We will continue to use 'nodes' to indicate the relevant entity, since vertices seems obscure as a generic term.

Networks, which have been studied for many years in operations research, are usually distinguished from graphs by virtue of the special weightings placed upon either links or nodes or both. Weighted links indicate the strength of the relationship, and signs, when they exist, indicate a direction in a manner similar to directed graphs. Weighted nodes indicate that an entity is measured by some attribute relevant to the study under investigation. We shall use the word link to indicate connections in a network.

We shall consider the existence of a directed connection between two nodes in a digraph, to be a direct connection from the first node to the second. Also, within an undirected graph, a connection between nodes indicates that there is a direct connection in both directions. In an undirected graph, the number of edges which a node is involved in is called the degree of the node. For digraphs, the orientation of arrows implies that some arrows lead into a given node, while some lead outwards. The numbers of arrows involved are termed the in-degree and out-degree respectively.

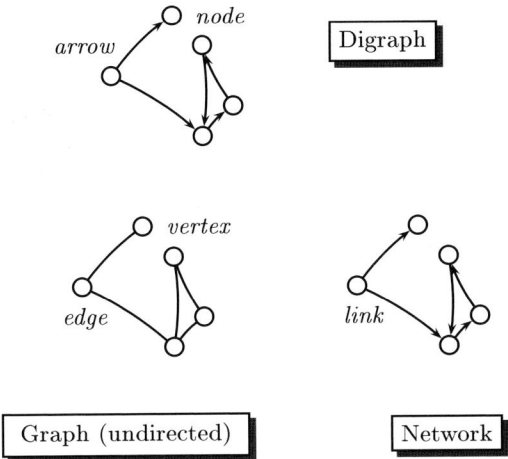

Fig. 3.1. Graphs, digraphs and networks.

A path is a sequence of direct connections wherein the destination of one link is the source of the next link. A path, which returns to the source node of the first link, without visiting any other node more than once, completes a closed loop which is called a cycle for undirected graphs and a circuit for directed graphs. A Hamiltonian cycle or Hamiltonian circuit is a closed path which visits every node in the entire graph exactly once.

Connectivity is a term which indicates just how interdependent the nodes in a network are. Using arrows gives a network an asymmetry which complicates connectivity beyond its simple application in undirected graphs. This is because the paths which connect nodes must have their direction 'sense' continued throughout the path (table 3.1). When there is a simple path in an undirected graph from each node to every other node, then the graph is connected. When there is a (directed) path from each node to every other node in a digraph or network, then the path is said to be strongly connected (occasionally a digraph is said to be connected when the sense of its arrows is ignored). When there is a link joining each node directly to each other node, then the network is totally connected.

However, in what follows we will use the term 'connectivity' as the proportion of possible pairs of nodes for which a path exists. Hence, this value moves between 0 and 1. For undirected graphs, the possible pairs are $1/2n(n-1)$, while for digraphs there are $n(n-1)$ such pairs.

In biological applications, sometimes the term 'patch' has been used to describe pockets of connection within a larger collection of entities. The technical term for a patch is a connected component for symmetric connection and a strongly directed component for digraphs. In the interests of brevity, the term 'patch' will also refer to a strongly connected component, along with 'simple patch' for the undirected sense.

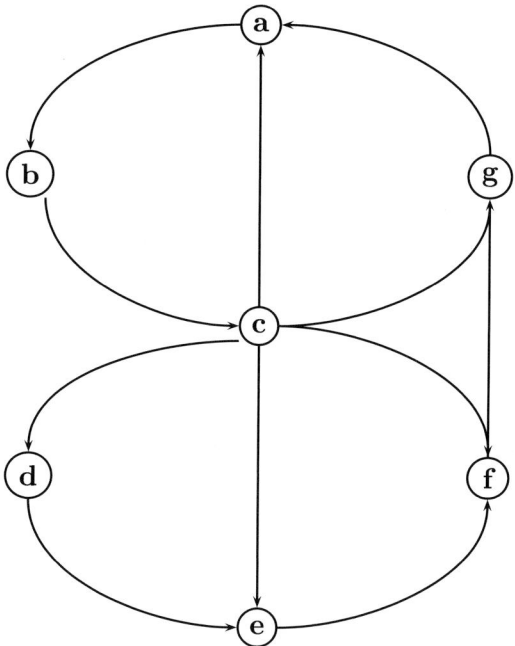

Fig. 3.2. Components of a digraph.

In fig. 3.2, the path $<a, b, c>$ consists of the two arrows linking node a to node b, and linking node b to node c. Hence, $<a, b, c, f, g>$ is also a path, while $<f, g, a, c>$ is not a path because the orientation on the arrow $<a, c>$ is wrong. An example of a patch (or strongly connected component) is the set of nodes $[a, b, c, g]$, where there exists a path between each node. Note as well that this patch contains two circuits, $<a, b, c, a>$ and $<a, b, c, g, a>$. An example of a directed tree from node b is the following $<b, c : c, d, e; c, d; e, f>$ (where the

Table 3.1. *Summary of terminology*

	Graphs	Digraphs	Networks
Entity	node	node	node
Direct Connection	edge	arrow	link
Indirect Connection	simple path	path	path
Closed Loop	cycle	circuit	circuit
Branchings	tree	directed tree to/from a node	trees or directed
Connectivity		connected	strongly connected

semi-colon separates the different branches of the tree, and the colon indicates a branching node).

A frequently occurring structure in networks is that of the 'tree' (see table 3.1). For undirected graphs a tree consists of n nodes with $(n-1)$ edges such that there is a path between all n nodes (this has the consequence that there are no cycles either). Similarly for digraphs with n nodes, a directed tree to/from a node has $(n-1)$ arrows such that no two arrows oppose each other, and such that each other node has an outdegree/indegree of exactly 1. That is, the sense of the arrows is either travelling to a special node, or all travelling away from a special node, and without the existence of any cycles, let alone circuits. A spanning tree is one which includes all of the nodes within the network being referenced.

It is often thought that since undirected graphs are 'simpler' than digraphs, that they are somehow more fundamental as a model. However, a symmetric connection is not more fundamental than an asymmetric one, since it really signifies the presence of a direct connection in both directions. Influence or dependency really indicates some flow of causality. Hence, symmetric connections are more complex. Therefore, digraphs are a more fundamental model than undirected graph models.

3.2.2 Directed graphs as a fundamental model

In an important paper, reminiscent of general systems theory, Green (1993) pointed out that 'the patterns of dependencies in matrix models, dynamical systems and cellular automata are all isomorphic to directed graphs'. Matrix models include both linear equations and the matrix descriptions of Markov processes. Dynamical systems consist of systems of linear equations and are used for models of all time-based dependencies where the state space is continuous. This result can also be extended to other deterministic automata where links correspond to state transitions. This isomorphism has been exploited for some time in the study of Markov chains and queuing theory. The range of these models spans almost all mathematical modelling whether it is of physical or biological systems. Social and economic systems are covered by these as well, except for many econometric models used for forecasting which often exclude interaction effects between variables.

Green further points out that the state space of finite, deterministic automata can be described by directed graphs. Nodes represent system state, while arrows represent possible transitions, and paths represent state trajectories through time. Furthermore, he shows that such systems 'must fall into either a fixed state, or a finite, limit cycle.'

It is worth noting that analogous results have been known for some time for the highly useful discrete Markov chain model wherein the system will either end up in an 'absorbing state' or in a strongly connected component of states. Within this patch, states will have a long run or steady-state probability of occurring. Although not a limit cycle, the system will stay within the patch returning to each state roughly with a recurrence time which is the inverse of each steady-state probability. It may well be that there are analogous results for stochastic automata.

It is possible that there are intrinsic relationships between the dependency graph of a model and its corresponding behaviour in state space, i.e. that there are complementary transformations between the two representations. Certainly non-linear terms make such a mapping problematic. But it may be that there is a relationship based upon the directed tree and circuit structures contained within the dependency graph. It certainly seemed that this should be the case when the author (Seeley, 1966) developed a result linking the spanning trees of the dependency graph with the steady-state probabilities of Markov chains, many years ago. In the proof, the dynamics of such systems was an interplay between the existing circuits and trees in the dependency graph. Eventually for steady-state, all of the circuits and all of the trees which did not span the system lost their influence.

In the 1970s, many people, including the author (Seeley, 1973), inspired by Conway's game of Life (Gardiner, 1970) and the efforts of

Codd (1968) developed systems to explore various cellular automata (see also chapter 4) which exhibited complex behaviour reminiscent of biological systems, including self-replication and universal computation. Much of this effort transpired at M.I.T. inspired by the efforts of the unpublished work of Ed Fredkin.

Wolfram has undertaken a very comprehensive study and analysis of the behaviour of deterministic cellular automata (Wolfram, 1984a). He has examined the state transition networks for the large number of configurations which a given automaton will enter, and characterised their behaviour as a combination of rooted (directed) trees and cycles (circuits). The occurrence of cycles and their length within these transition networks has helped him to categorise different automata. Moreover, he has explored the analogues between the behaviours of these classes and those of dynamical systems (Wolfram, 1985). This strongly suggests that a unifying directed graph representation underlies the behaviour of both dynamic systems and deterministic automata. This possibility is explored in more detail in §3.3.5.

Speculations aside, there is strong evidence to suggest that directed graphs can be considered an essential or fundamental model. That is, within any model which purports to describe the dynamics of some aspect of the world there resides a core, directed graph model. Static models merely describe the stopping states of dynamic situations which could be modelled more comprehensively by a dynamic model. Therefore, any observations which apply to the nature of directed graph models should apply as well to a corresponding dynamic model. Given the universality of dynamic models, directed graph models could be considered universal as well, a not-surprising result given the opening statement that we live in an ocean of networks. It is in this light that we now look at observations which can be made upon examining the dynamic growth of networks themselves.

3.3 The evolution of random graphs and digraphs

3.3.1 Graphical evolution

The evolution of random graphs refers to the changes in structural properties which a graph (undirected) of n nodes undergoes as successive edges are added. In Edgar Palmer's book *Graphical Evolution* (1985), he dramatically describes the evolution of random graphs as flowing in the following manner:

'Threshold functions (are described) which facilitate the careful study of the structure of a graph as it grows and specifically reveal the mysterious circumstances surrounding the abrupt appearance of the Unique Giant Component which systematically absorbs its neighbours, devouring the larger first and ruthlessly continuing until the last Isolated Nodes have been swallowed up, whereupon the Giant is suddenly brought under control by a Spanning Cycle.'

There is a substantial literature on the evolution of random (undirected) graphs. Recent books by Bollobas (1985) and Palmer (1985) outline the overall results.

Graphical evolution can be viewed as having five phases:

(i) Initial branchings (trees)
At first there are only isolated edges in the (potentially vast) space of nodes. After more edges have been added small trees start to appear. More structures occur, with multiple branches and trees with several edges appearing, until...

(ii) The initial growth of cycles and patches
Cycles start occurring when a tree has an edge added to it which connects two of the already existing nodes in the tree. Cycles implicitly bound connected components or patches, although some patches may not have spanning cycles within them. During this phase, the largest components become of the order of $\log n$ nodes.

(iii) The double jump, and the emergence of a giant component
When the number of edges has grown to almost exactly $1/2n$, a giant component of order $n^{2/3}$ suddenly appears for almost all graphs. Hence, the connectivity of the overall graph suddenly and very rapidly increases from significantly below $1/2$ to significantly above $1/2$. It is this sudden and huge increase in connectivity which is referred to as the 'double jump'.

(iv) The trickle of isolated nodes
After the formation of the giant component a small proportion of nodes are either completely isolated or belong to small trees of order at most $\log n$ nodes. Such nodes and trees are gradually absorbed by the giant until only the giant remains and the graph is connected.

(v) Graph is connected and Hamiltonian cycles abound
At almost exactly $1/2n \log n$ edges, the evolving graph becomes connected. Just before connectivity there exist only a few isolated nodes. At this point or very shortly afterwards, a Hamiltonian

cycle will occur. Often many occur at this point, and rapidly multiply in numbers thereafter.

Another interesting result is that for most of the graphical evolution, the distribution of node degrees is distributed in a Poisson fashion.

3.3.2 Simulation studies of digraph evolution

Myself and some colleagues (Michael Baker, Simon Ronald, Charles Berner) have undertaken the study of random directed graphs both by computer simulation and by analysis. Previously, only a few results for regularly connected graphs (with chemical applications) and for spanning trees (for reliability) have been reported in the literature, until we reported some initial results recently (Seeley & Ronald, 1993). However, analytical results have proven very difficult for us to achieve in any kind of tractable form (Ronald, 1992).

Our simulation algorithm adds arrows one at a time, in a Monte Carlo fashion, to the evolving digraph, enabling time-based (as a metaphor for arrows added) measurements to be performed. An alternative scheme, which is also used for analytic studies of undirected graphical evolution, is by assigning to each node a certain probability of connection to other nodes. This probability is associated with the average number of connections each node has on the average. It can be interpreted as a kind of 'temperature' for the graph. Notice, that at this stage probabilistic removal of nodes has not been included in the study.

Since a reduced form of a digraph that ignores arrow orientations reveals a simple, undirected graph, the phenomena observed in random (undirected) graphs will be in the background to digraph evolution. The main complication for digraphs is that arrow orientations must 'make sense' in forming directed paths, components and strong connection. Hence, it is reasonable to expect that most values of results for simple graphs, in terms of the number of connections, will be somewhat more than double for digraphs. In this vein, it is useful to consider the multiplicative effects of emerging connectivity while the giant is forming. Figure 3.3 shows two strongly connected components (without their arrowheads) in an evolving digraph. If each component has n_1 and n_2 nodes respectively, then their individual contributions to overall connectivity adds to $n_1(n_1-1)+n_2(n_2-1)$. Once the additional connections are made (dotted lines) their combined contribution to the overall connectivity is $(n_1+n_2)(n_1+n_2-1)$, an increase in connectivity of $2n_1n_2$. It is this multiplicative term which is the major contributor to the explosive effects at the double jump.

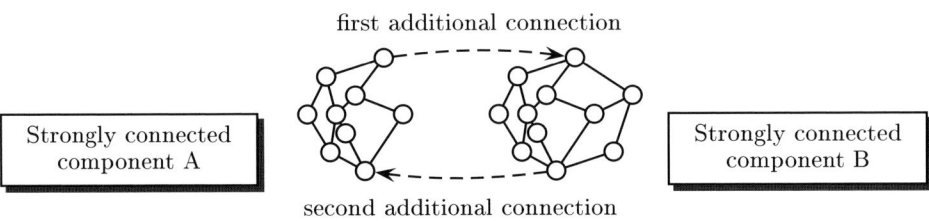

Fig. 3.3. Multiplicative mechanism at the double jump.

It is not surprising that our simulation studies reveal that sudden phase changes also occur for random directed graphs. Our investigations have focused on the emergence of small directed trees or 'forks' and initial circuits, the occurrence of the double jump point (connectivity avalanche), the (strong) connection point, and on the emergence of a (directed) Hamiltonian circuit. Some examination of the emergence of circuits and circuit chords has also been done. Some of the empirical work is summarised in table 3.2. Keep in mind that tractable formulae for these results have proven difficult to achieve.

In the following summary of our more substantive simulation measurements, the connectivity point values were determined by a sample of 1000 runs for each number of nodes in most cases; however, some of the double jump point values were based upon much smaller (*) samples. The values are the mean number of connections which must be included in order for the point to occur. However, the number of simulation runs for the double jump point was 100. The number of connections required to reach this point appears empirically to be close to $p/2N$.

There are a number of other measurements which we have sampled but are yet to provide systematic results for. These include the observation of circuit formation for small numbers of N; these required lengthier runs because they are computationally much more difficult to detect than branchings. When the major focus was on the emergence of a Hamiltonian circuit, the computations were not applied until after the connection point was detected. However, some informal observations were made of numbers of circuits of various lengths, in the build-up to the double jump point for very small N. The super-saturation process of

Table 3.2. *Empirical results*

No. of nodes	Double jump point	Connectivity point	No. of runs
10	16	29	1000
20	34	77	1000
30	52	129	1000
40	70	185	1000
50	86	245	1000
100	172	570	1000
150	259	948	100
200	351	1294	1000
250	436	1664	40
300	517*	2072	1000
350	612*	2287	30
400	709*	2885	1000
500	–	3723	1000

large circuits before the connection point was also observed informally. Another class of observations were the first occurrences of a variety of features, such as the first double and triple branchings, the first few occurrences of circuits of various sizes, and the dynamic evolution of out-degrees of nodes.

3.3.3 The connectivity avalanche and implosion

The most dramatic result of Erdos and Renyi is the double jump, wherein for undirected graphs, 'the order of the largest component jumps dramatically as the number of edges approaches one half the number of nodes'. Having watched numerous animations of this double jump, and studied the enormous changes in connectivity around this point, we believe that the tremendous changes at the double jump should be called the connectivity avalanche.

Observing the animations, the different phases of digraph evolution are visually apparent (fig. 3.4). At first isolated connections appear, followed by small clusters or patches of nodes, then these patches become strongly connected within themselves. Next the patches coalesce into larger, strongly connected patches. Often the patches arise in association with a circuit. This process accelerates until just before the double jump there are a few large patches which include the majority of nodes. At the double jump, the coalescence of these patches produces the 'giant' strongly connected component. It is at this point that the overall connectivity of the graph skyrockets enormously. This is because of the large increase in nodes between which there are now directed paths; this has a multiplicative effect of the strong connection.

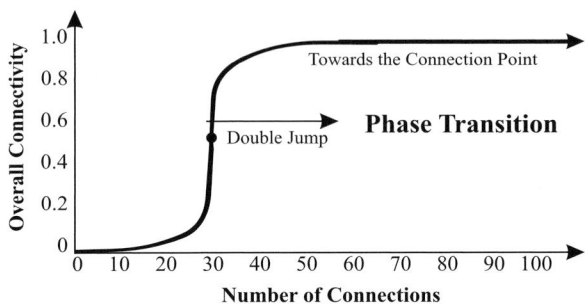

Fig. 3.4. The connectivity avalanche.

Once the giant strongly connected component has fully emerged, there remain only isolated nodes, occasionally connected partially to each other. Unlike the situation with undirected graphs however, some of these isolated nodes either have connections to the giant but lack return connections to themselves, or the opposite wherein nodes in the giant connect to them, but they have no connections to the giant. Occasionally there remains a lonely, singleton without any connections. There are many observations of explosive phenomena and extinction phenomena which could be associated with the phenomenon of the double jump. The notion of a 'chain reaction' in nuclear fission, and in the spread of epidemics are obvious candidates. In living systems (including electronically mediated ones), the emergence of communities which function as individual entities may well be linked to a double jump in connectivity. Other more speculative areas to explore for the presence of a connectivity avalanche are for example, in punctuated evolution terms the Cambrian explosion of multi-cellular organisms on the Earth (Kauffman, 1989), or in the massive explosion of space-time (Nagels, 1985) in the inflationary episode of the Big Bang.

It is important to notice that digraph devolution can also be investigated by taking a strongly connected graph and serially removing (breaking) connections. The opposite of the connectivity avalanche is a connectivity implosion wherein there is a sudden and huge collapse in the digraph's connectivity with only a few 'broken' connections. Potential candidates for the connectivity collapse are as diverse as mass extinctions, the collapse of organisational infrastructure, the decline of complex societies, and the collapse of the immune network at the onset of a variety of modern illnesses (Gislason, 1991). It is conceivable that growth in the number of fundamental relationships within such systems can be modelled by digraph or undirected graphical evolution.

3.3.4 Super-saturation with circuits at the connection point

Even though the giant component proceeds to 'gobble up' the various isolated nodes and trees remaining after the connectivity avalanche, the evolution of the digraph must include many more connections before these outliers are included and the whole digraph becomes strongly connected. We call this event the (strong) connection point. This point is significantly beyond the point of the avalanche (see the above table of simulation results).

There is a very strong interplay between strong connection and spanning circuits (circuits which include all of the nodes of some component). Hence, the point of strong connection emerges when the evolving digraph is 'pregnant' with circuits which are almost Hamiltonian. In fact, it would be accurate to say that the evolving digraph is super-saturated with very large circuits at this point. Moreover, we have found that in about 65% of cases, the connection point is accompanied by the emergence of Hamiltonian circuits. In fact not just one circuit usually, but a very large number of Hamiltonian circuits arise with the same breakthrough connection.

3.3.5 Phase changes and the edge of chaos

An overview of graphical evolution reveals a number of distinctive phase changes. Some of these are sharp and distinct like the connectivity avalanche at the double jump, or the emergence of a Hamiltonian circuit, and the first occurrences of nodes with specific degrees, or the first occurrences of circuits of specific sizes. But the phase change par excellence is the connectivity avalanche.

The phase change which occurs at the connectivity avalanche is from a system only sparsely connected to a system which is richly connected. The sharpness of the phase change at the double jump is extreme; the evolution graph is almost vertical for a significant range of connectivity values. There is also extreme sensitivity at this point as well; that is, a few connections included in a sparsely connected digraph, and the system flips to rich connectivity, a few connections removed from a richly connected digraph, and the system flips into sparse connectivity. The number of connections required in a digraph of N nodes to put it near the double jump point appears to be intimately related to the growth of the number of circuits in the evolving digraph, but this has been very difficult to measure, even with empirical simulation.

There appears to be a strong overlap here with Bak and Chen's discussion of 'self-organised criticality' (Bak & Chen, 1991). Adding a grain of sand to a sandpile is similar to adding a new connection to an evolving digraph. The addition of each grain makes both a structural connection to other grains in the pile, but also disturbs the subtle balance between structure and gravity in ways that suggest differing sets of indirect connectivity. It may be possible to show that this criticality point is directly related to a double jump point phenomenon of random network connections. Bak and Chen make the intriguing observation that the impact of adding grains at the criticality point, causes sand to fall off the table in amounts which exhibit a chaotic structure following a power law which exhibits 'flicker noise'. Bak and Chen then suggest a nexus between power laws and self-organised criticality; phenomena with finger prints in time (flicker noise) and space (fractal patterns). It would be intriguing to determine whether the changes in connectivity produced by evolution around the double jump point in digraph evolution, exhibit the same fingerprints. This could be examined both monotonically as new connections are added, or with a variant network model wherein connections are either added or removed with a certain probability.

Simulating evolution both for its own sake and as a problem-solving tool emerged in the 1960s (Seeley, 1971). Kauffman (1984, 1991) describes experiments he performed in 1965 using binary networks of Boolean functions which connected nodes in the network of 100 nodes. This was an abstract model of the way in which genes could combine in networks. Instead of observing very long sequences of distinct states (the possibilities are 2100), a kind of discrete chaos, his system rapidly (typically in 10 steps) found itself in a small cycle of perhaps 3 or 4 states. He termed this phenomenon 'anti-chaos' and suggested that it was at the heart of self-organisation. In fact, very similar behaviour can be observed with pseudo-random number generators which are characterised by 2 or 3 parameters selected by number theoretic considerations (typically involving certain kinds of prime numbers). If these parameters are chosen arbitrarily, as we ask our students in discrete simulation to do in a small number space, the number sequences observed quickly get into small cycles, unless the hint about prime numbers is given. Note as well that the node degree in Kauffman's experiments seems to range around 2 in order to produce interesting behaviour. Four is definitely too high and that just above 1 is definitely too low, in order to produce the emergence of state-cycles in the dynamics of binary network behaviour. This strongly suggests that a relationship exists between the average node degree in Kauffman's binary nets, and the average degree necessary in random digraphs for the double jump to occur (about $p/2$).

Kauffman felt that the above results, which showed 'the ease with which self-organisation emerges in networks' (quoted by Levy (1992)) applied to the autocatalytic reactions of RNA, enzymes and proteins which are at the core of life, by supporting both self-replication and metabolism. His approach with self-organising circuits paralleled the work by Eigen and Schuster (1979) on the 'hypercycle', a connected network of functionally coupled, self-replicating entities. This is what is also described as 'autopoiesis' by Maturana and Varela (1980). This phenomenon is related to what we describe later (§3.4.3) as Hamiltonian coherence in the emergence of structure from local viewpoints where a global order emerges with a network connectivity that is dense enough.

In studying the transition between stability and chaos in the behaviour of cellular automata (CAs), as first categorised by Wolfram (1984b), Langton (1992) developed a measure which separates these behaviours in a very sharp and distinct manner. Using as his lambda parameter, the probability that a given cell will remain alive in the next step (the proportion of rules which do not lead to a quiescent state), he found a narrow band for lambda between ordered behaviour (periodic and static) and chaotic behaviour. Just at this sharp phase transition there exist cellular automata classified by Wolfram as Class IV (such as Conway's game of 'life') which can produce self-maintaining and self-reproducing patterns which are capable of universal computation. This location in the complexity of CA rules is known as 'the edge of chaos'. On one side of this sharp edge, behaviour evolves to one of a number of separate repeating circuits (Class II) or to dead-ends (Class I), while on the other side behaviour yields very long apparently chaotic patterns which do not repeat (Class III) . It is believed by Langton and others working in the fields of complex systems and artificial life, that life-like behaviour exists near this edge. Table 3.3 compares digraph evolution with Wolfram's classes of finite-state automata and some simple distinctions of dynamical systems.

It is conceivable that there is a direct relationship between the connec-

Table 3.3. *Digraph evolution and classes of systems behaviour*

Stage in evolution	Dynamics	Cellular automata class
small trees	fixed points,	**Class I**; moves to an absorbing state.
small circuits and connected components	limit cycles	**Class II**; moves to one of separate stable and periodic structures
vicinity of the double jump point	–	**Class IV**; localised, irregular patterns which can grow and shrink, and some which propagate
strong connection and beyond	chaotic	**Class III**; complex, apparently non-repeating patterns, sometimes with hints of periodicity

tivity of the corresponding state transition graph which can be formed from the rules of a CA (see Wolfram(1994, p.159) for example), and hence to the edge of chaos. But how do we get from a dependency graph to the description of sequential behaviour? If we were examining the states of a single system such as occur in Markov chains, the time-dependent (or space for that matter) behaviour of the system plays itself out in the paths of state transitions derived from the dependency graph of the chain. Absorbing states, and cyclical behaviour are possible in a fashion similar to the Class I and Class II behaviours of CAs. When there are strongly connected components in the dependency graph, these components 'absorb' system behaviour as well in a fashion reminiscent of strange attractors. When the entire system is strongly connected, then definite steady-state probabilities arise, a kind of stochastic 'order' from the chaos of possible state trajectories.

Even this 'steady-state order' can be described by digraph means. It turns out (Seeley, 1966) that the steady-state probabilities in Markov chains are directly proportional to the distinct spanning trees which exist in the dependency graph, and to the multiplicative effects of the (transition probability) weightings on each spanning tree's arrows. If the transition probabilities were all equal, then the steady-states would depend directly on the number of spanning trees to each state. The effects of being in an initial state vanish, perhaps in a manner analogous to the effects of initial configurations in CAs. It is plausible that dynamic relationships which can be derived from a dependency graph, also hold for cellular automata in a fashion similar to Markov chains.

Hence, in a CA the behaviour of a set of rules and a cell neighbourhood produce a long-run characteristic behaviour which can also be derived from the connectivity of an associated dependency graph, i.e. the state transition diagram. The lengths of transient strings of states, and the periods of observed cycles have been studied in detail by Wolfram (1994, p. 43). He has further observed that state transition diagrams have the general form of 'cycles fed by trees' (1994, p. 461). In these transition diagrams, the nodes are unique configurations of states in the cellular array, while the arrows (transitions) are the application of the CA's rules to produce a new state.

It is plausible however that a different type of state transition network, call it the rule transition network, could characterise a given set of rules and neighbourhood. Instead of considering the automaton's states to be one of the possible configurations, a state and hence a node in this rule network would correspond to the immediate array of cell states that are possible in the neighbourhood of a generic cell. The arrows would correspond to the application of the single rule which applied to a specific state. Integrating the effects of the neighbourhood connections is not immediately obvious, however. The resulting digraph would be a static characterisation of the cellular automaton. It is our hypothesis that the connectivity properties of such a rule transition network would describe the qualitative features of the dynamic behaviour of the configurations of the cellular array. That the production rules of deterministic finite-state automata could be characterised by digraphs which show the reachability of various states from others (i.e. their connectivity) is clearly related to Langton's lambda parameter. In fact, the connectivity issue in rule networks could make possible some fine-tuning in the prediction of automata behaviour by refining the lambda measure.

Hence, it is possible that there is a sequential unfolding of the connectivity in the CA rules (the rule transition network) which describes the CA behaviour (state transition network). Moreover, when examining the space of transition rules for CAs, the cellular automata could then be mapped onto a digraph evolution. It is our second hypothesis that Class IV behaviour in cellular automata emerge from rule transition networks which are poised around the double jump point of digraph evolution.

3.4 The emergence of structure

3.4.1 Observation and modelling the local viewpoint

When examining the relationship between connectivity in a model's digraph and the dynamic behaviour of the system it models, it has proven useful to take the local viewpoint of the individual nodes of the digraph. That is, we examine the connections which a given node can make, not just directly, but indirectly via paths of connections to other nodes, possibly returning again to the original node. Hence, local connectivity includes the notion of indirect connectivity.

In oriented (directed) paths of length greater than one, sequences of nodes connect the origin node to nodes other than those which the origin is directly connected to. As the graph evolves through additional links, rapid and significant changes occur to the paths leading from an origin node. We shall say that the origin node 'observes' these changes. An observation then consists of the network of possible paths leading from a particular node at some point in the digraph's evolution. A corresponding diagram highlights the nodal viewpoint in an obvious manner. In fig. 3.5, a's observations, before the arrow between e and r is included, is that of a directed tree leading from itself. However, node t observes a circuit involving only nodes r, s and t itself.

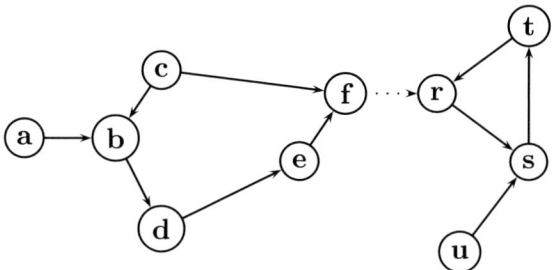

Fig. 3.5. Digraph observation.

Notice that in this approach the overall behaviour of the evolving digraph consists not just in the connectivity of the overall digraph itself, but also as the fusion of the observations of each individual node. It is possible that in this manner, not only can the dynamics of the system behaviour be pursued, but also the interplay between individual observation with the experience of the whole can be kept in the foreground. Such an approach mirrors our work in the design of object-oriented software where nodes are object classes, and directed links represent relationships between classes. Requests from the world are handled by the software as a path of interactions with objects, such paths being regarded as causal descriptions of how the software works. The wholeness of the system consists not just of the collection of nodes (classes) and their links (relationships), but also the activation of these observations (paths of interaction).

Global views tend to make theory abstract and to objectify individuals, ignoring their individual significance. The authority of global and abstract modelling practice can then become barriers to the individual participants. When these are people, they will rebel at such treatment. Just as good design is a blend of global and local perspectives, we take the position that wholeness cannot be examined without blending both local and global perspectives. This viewpoint has important ramifications for the application of such models, as those drawn from complex systems, in the life and social sciences (see discussion in §3.3.1).

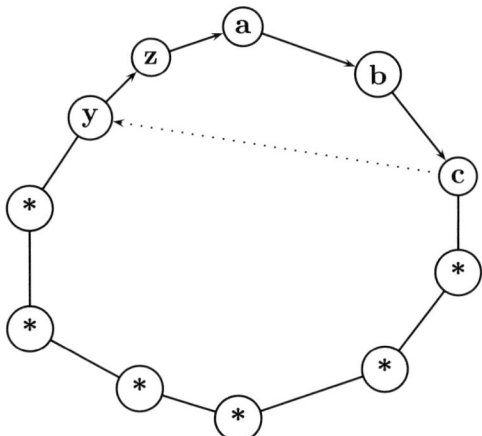

Fig. 3.6. Observation in a circuit.

3.4.2 Paths as orderings

If we take order to indicate the sequence of nodes observed by a given node, then we must also take account of the branchings in the observed trees and the looping in observed circuits. Branchings (which will occur in probabilistic situations) represent alternative orderings and can be regarded as parallel order. When parallel branches converge on a node, these parallel sequences are then fused together into what follows out from such a node. When a circuit is observed, there is the potential to observe an infinite looping of the nodes in the circuit. Circuits and branchings can occur at the same node, inducing an interplay of circuits of possibly different lengths; hence the circuits are observed in parallel.

When paths lead out at various points from a circuit, such descriptions as those above breakdown unless they are tamed by a Hamiltonian circuit which puts all paths into circuits, some of which are embedded into others. One could well imagine a calculus which would attempt to describe such orderings, but we will resist the temptation to do so here. The complexity of this situation can rapidly expand, as our earlier discussion of the double jump has indicated. However, when we examine what happens to these sequences when circuits are present, we find that the circuits impart a commonality to sequences which can be regarded as a kind of 'Hamiltonian coherence'.

3.4.3 Circuits and Hamiltonian coherence

An interesting situation arises with observation as circuits emerge in digraph evolution. First of all, as is the case with strongly connected digraphs, an indirect connection is observed by all of the nodes in the circuit. Even the nodes involved in direct relationship observe their own involvement indirectly. For example, consider the following circuit:

For the nodes a, z, y, etc., the direct connection between nodes b and c will be observed (indirectly) as each entity in the circuit sequence reduces the connectivity of the entity that it is directly connected to. Eventually, the reductions are passed back to C itself, which also observes the original situation again as what we will call an iteration of the pattern. In a way we are left with a dilemma here with the passing of indirect connection back to the initiator of the connection. Does this stop the 'flow of indirect connection', or does this flow circulate around and around, with C observing the connection indirectly again and again? We will take the position that this flow is 'never-ending'.

It is plausible that the indefinite iterations of a circuit provide a stronger observational reference for nodes than that of unclosed paths. The commonality of the nodes sharing the circuit, the commonality of different nodes observing the circuit, and the repetition implicit in the circuit combine to give circuits a reference priority. If we also take larger circuits as a higher priority than smaller ones, then we have a means of making the observation of other nodes coherent (ordered), especially when a Hamiltonian circuit is present. It is as if the never- ending flow of observation around circuits provides a kind of 'standing wave' of connectivity for the observation from a node. Such a standing wave, which includes the self-observation of each node in the circuit, can be regarded as a common background for all of the nodes in the circuit to observe additional connections. In this manner, observation from a node can be seen to be an interplay between circuits and the trees which connect them.

Against such a background, connections which are not explicitly a part of the circuit, come into relief. Superimposed upon a circuit background, these other connections will appear to 'replay' the random digraph evolution over again as other connections are considered. Within such replays or recursions, other independent circuits will form, and so on, creating eddies within eddies of other backgrounds as it were, in the fabric of the observations of individual nodes, a fabric which in general will have both unique and common (with other nodes) components. Combined with the 'standing wave' property of circuits described above, these subcircuits within subcircuits, ... within circuits evokes the observation of connections in a fractal manner. It is as if the Hamiltonian circuit becomes a kind of 'reference beam' for the 'object beam' of other connections to interfere with 'holographically'.

This background of a Hamiltonian circuit seems to us to be akin to the 'autopoietic nexus' postulated by Maturana and Varela as the basis of biological identity (Maturana & Varela, 1980). Given the simulation results of Kauffman on autocatalysis and Hillis on punctuated evolution, mentioned previously in §3.3.5), there are grounds to believe that there is a structural similarity to the 'loop analysis' of Levins (1977), which is used to examine the stability of ecosystems. Furthermore, the hypercycles of self-replicating entities in models of the primordial, prebiotic soup proposed by Eigen (1979) suggests something that is structurally very similar as well. Moreover, the work of Kauffman which shows the emergence of self-organisation in autocatalysis, further suggests that the emergence of directed circuits may be the basis of strange attraction. Mapping these models onto a directed graph background should provide further insight into this possibility.

3.4.4 Fractal emergence of structure

Moreover, the incremental dynamic properties of digraph evolution was also examined as they apply to the observation of individual nodes and globally. This has led to the interesting area of enquiry which is to determine the relative expectations of spontaneous changes in structure in the evolving digraph. The Poisson process plays a key role in this approach.

As indicated earlier (Seeley & Ronald, 1993), the Poisson process can be seen to be fractal in nature since the times between Poisson events, which are distributed according to the negative exponential distribution, are self-similar in time (or an appropriate spatial index). The interdependent aspects of this process, i) complete event independence, ii) memorylessness and iii) self-similarity, reflect a unique quality which we will call spontaneity. Hence the term Poisson spontaneity.

A consequence of this is that it can be shown that the number of connections from a particular node is distributed in a Poisson fashion over the evolution of the digraph for relatively large N (Seeley & Ronald, 1993). In this discussion we take 'epoch' to mean an appropriately scaled time interval. Hence, if we consider r to be the number of global connections C made in a given epoch of the evolution divided by N the number of nodes (i.e. $r = C/N$), then the probability of k connections being made from a particular node during that epoch is given by:

$$p(k) = \frac{rke^{-r}}{k!}. \tag{3.1}$$

Based on the foregoing discussion, my colleagues Michael Baker, Charles Berner and I then asked the question, 'Given the evolution of a random digraph past the Hamiltonian point, how many direct connections are expected between occurrences of a particular configuration, observed against the background of the given digraph?'. The rationale for this question is that Hamiltonian circuits appear so frequently at this point that they present a common background for almost all nodes to observe 'new' phenomena. Furthermore, there is the supposition that with a very large N, this situation provides a kind of steady-state which persists until the late stages of the digraph's evolution to total connectivity.

We shall call the answer to such questions the duration (or 'expectation value') of the configuration. It is assumed that these arrows are randomly drawn from those remaining from the $N(N-1)$ possible direct connections between nodes. We are attempting to determine the durations of simple forking structures and small circuits. Considering the previous discussion of the fractal quality of the Poisson process, by analogy it is reasonable to expect that this applies to the occurrence of structures at nodes. Hence, it is also a reasonable hypothesis that structure observed at nodes emerges in a fractal fashion when considered against variously scaled epochs of connections included in the evolution. We hope that results from ourselves and others will be forthcoming soon.

3.5 Applications of random digraphs

3.5.1 Beyond reductionism

In many respects, our discussion in this chapter reflects similar discussions about information theory and general systems theory (GST), decades ago. Interpretations of each of these endeavours used metaphor in order to establish hypotheses in fields quite different than the source of the original observation. In the case of information theory, important components of the original theory were left out of the picture when it and its companion notion of entropy, were translated into the life sciences. GST attempted to find general principles which governed the behaviour of all systems. It was often criticised for not grounding the translation of these principles into the observable components of the metaphor's target. This work with digraph evolution is certainly in the spirit of GST; however, the translations should be easy to make since models and theories from any field, in order to be effective in the world, must explicitly describe relationships, dynamic relationships. Networks must be used in order to describe the nature of the relationships present in these models, and hence it is a plausible expectation that the evolution of networks will be able to contribute to the characterisation of the behaviour of such models. Complex systems cannot be understood without understanding how the local components interact with each other in order to produce a whole which is much more than the sum of its parts.

What will probably remain somewhat difficult is the characterisation of the particular relationships and interactions in a given model. Positive or negative influence, the magnitude of influence and different orders of influence (e.g. polynomial) are not immediately translatable into a digraph model and yet they have a great bearing on a model's behaviour. However, we suspect ways will be found either to do the translation or to characterise the evolution of networks with such characteristics. Moreover, it is likely that where there exists complexity in the relationships and interactions, the system or model could be approached by using different levels of detail, and looking for the emergence of phase changes at the lower level to provide the interactions of the higher level.

In promoting a model, the evolution of digraphs, as a basis for understanding complex systems, especially one which involves what some have termed artificial life, it is possible that it could be criticised as being reductionist. Indeed, both general systems theory, cybernetics and even the concept of system itself have been criticised on this basis. Although the phenomenon of emergence is sometimes seen as a way out of reductionism in that such phenomena as intelligence and consciousness might hence emerge, it does not address concerns about the underlying basis itself. What we believe does address this issue is the analysis that we have earlier described as point-of-view observation. The whole system is not just seen from looking at a simulation from a global viewpoint, but from the ensemble of individual viewpoints as well. Although its computer animation requirements are more difficult, simulation from viewpoints can be done. What may be more difficult is to develop analytical tools with which to describe system behaviour from a viewpoint. Our exploration of viewpoint analysis within digraph evolution discussed in the previous section is an initial exploration of a possible candidate. However, it is by placing the local as equal to the global that we can go beyond reductionism and fully integrate the individual within our whole systems.

With these considerations in mind, we offer the chart in table 3.4 of correspondences between digraph evolution and system behaviour, as a guide to discerning potential applications of the digraph evolution model. It is offered as an initial guide, with the hope that as it is applied to more and more situations it may be refined and/or at least converge to a useful template.

3.5.2 Living systems

There are a number of ways in which the application of the evolution of random digraphs can be determined. First there are systems in which binary relationships between components are fundamental to the dynamics of the system. Here, digraph evolution can be applied directly unless there is the presence of network variants such as weighted selection of relationships or the presence of a background 'temperature' or its equivalent which dynamically alters the probability of inter-component connection. In such cases, digraph evolution can help identify important

Table 3.4. *Digraph evolution and qualitative system behaviour*

Stage in Evolution	Phenomena	Meta phenomena
small trees	fixed points, absorbing states	'dead ends'
small circuits	archipelago, separate stable islands	stability
connected components	stable subsystems, strange attraction	punctuated evolution as components merge
double jump point	phase transitions, sudden explosive growth or implosion	instability and criticality
giant component	dominance of one component	emergence
strong connection	everything connected indirectly to everything else	community, wholeness
Hamiltonian circuits	coherence and order	commonality, self-organisation, autopoiesis

network phenomena, but the study of specialised simulations will have to be pursued.

Secondly, there will be phenomena such as sharp phase changes, near-critical instability, avalanches (explosions) or collapses (extinctions). In such cases, further insight may become available by putting the observations in a digraph evolution context. Here we consider only a suggestive sampling of applications from biology, physiology and sociology.

Biologically, the growth of communities of organisms (Green, 1993) is a rich field for the application of digraph evolution, along with the stability of various ecosystems. As indicated earlier wherever a collection of organism relies upon inter-component interactions, or wherever there are growth processes, these situations are candidates.

Physiologically, the behaviour of the body's immune system (network) is a rich field of current research within which digraph evolution should apply. Moreover, there may be applications in the cardio-vascular system as well. And of course, the central nervous system abounds with networks, as the work of temporal correlations in the visual cortex (Bossomaier *et al.*, 1993) testifies.

Of course, connectionist theories are currently prevalent in theories of the mind. Creativity for example, may have some partial explanation in the connections made via meaningful associations in the mind. At a certain point, the equivalent of the 'giant component', there appears a new insight.

In geographical studies of the formation of human communities, a concept known as 'role texture' reflects connectivity issues for new participants. The formation of large circuits within such a community may be decisive for the participants' sense of belonging. The gradual dismantling of services in an institution, products from a collection of product lines, and crucial management communication links, may suddenly create a disaster as the overall connectivity collapses.

The emergence of a new paradigm into an intellectual area also mirrors the connectivity avalanche. As more explanations utilise the new ideas, the reinforcing network of ideas that comprise an identifiable area gradually transmit this utilisation, until the sudden increase in overall connectivity using these ideas becomes obvious to all as the emergence of a new way of thinking. This would not be unlike the phases in the acceptance of a new product in a marketing perspective.

3.5.3 Technological systems

A wide range of network models currently exist in computer science, operations research, artificial intelligence, and electronic and communication engineering. This has been significantly enhanced both by hardware in the gradual replacement of mainframe computers by networks and parallel machines (itself a paradigm shift), and by software in such network services as e-mail and the Internet and software which supports group work. Hence, computer network operations, transport system operation, queuing in production and service organisations all have a network essence within which digraph evolution or one of its variants may apply. The growth of mobile phone and fax services, as well as the formation of communities over the Internet, are also important candidates for the application of digraph evolution.

Finally, we propose that random digraph phase changes may underlie the operations of such things as the emergence of higher-level functioning and intelligence which the mobile robots or 'mobots' of Rodney Brooks (1991) exhibit. These mobots work through the rich interconnections between relatively simple components and their operating environment. There may be a mapping from these interconnections to a digraph or network evolution model. Also, the functioning of both artificial and natural neural networks, and the mechanisms underlying successful genetic programming could also be explored from a digraph evolution perspective. Possible connectionist models of Minsky's 'society of mind' (Minsky, 1986, Varela *et al.*, 1991) might also be candidates.

3.5.4 Case study: inside Murphy's laws

Moreover, the author has also noted self-similar relationships in formulations for the steady-state behaviour of queuing networks. As well, within such networks certain chaotic and catastrophic behaviours can emerge, related to the network's circuit structure. The author has been actively engaged in making visual modelling and simulation of business organisations and an associated whole systems perspective effective in the business world. In the course of our consulting activities, near-critical instability and avalanching breakdowns have become evident in the 'fire-fighting' style of management within many organisations.

An in-depth study of this situation by my colleagues and I, while studying the capacity of coal export railway lines, has revealed an apparent core to Murphy's law that 'if a system can go wrong, it will'. It is based upon two levels of chaotic behaviour in common systems involving the flows of discrete items. The first is that under conditions of significant variability, rapid congestion can occur at bottlenecks. The second is a higher-level effect which occurs when such conditions begin to arise in a network, start to spread, and then rapidly cause a complete system shutdown. For example, they have been observed in the occurrence of 'store-and-forward' deadlock in early computer networks. Such systems could be modelled by a variant of digraph evolution.

The most fundamental aspect of the impact of variability process times is the non-linear manner in which a system component can become highly congested, leading to instability and chaos as the component's capacity is reached. This non-linearity is very surprising to our common sense because of the dramatic and sudden manner in which the resulting congestion can become extreme. Our observations of everyday performance can lull us into thinking that another increase in demand can easily be handled in a proportionate manner. However in the presence of variation, once this critical point is passed by the input demand, severe congestion becomes a fact of life.

The critical point in fig. 3.7 is when the input rate to a process reaches the fraction of capacity where unstable behaviour ensues. The point is approximately at the place (the cusp) where the congestion curve changes direction upwards. The position of the critical point is determined by the degree of variability in both the arrival and service process.

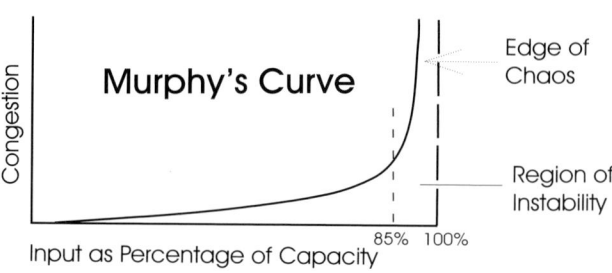

Fig. 3.7. Growth curve.

The less variability, the closer the critical point is to 1.0, the component capacity. Behaviour on the verge of chaos is unexpected when the input rate reaches the critical point. Variability can cause congestion to occur in which the system recovery time is not short enough to anticipate the next burst of demand. In the region of instability, the process can become sufficiently congested that it will cause bottle-necking to occur with respect to adjoining processes in production. Close to full capacity, the process is so unstable that it can be considered chaotic.

Balanced systems are ones where capacity of processing just balances the input of demands. In fig. 3.8 the workstation on the right is running in a so-called balanced manner while subject to significant variation. Hence, it is operating in the critical region. Because unstable behaviour occurs in the critical region, congestion can occur at any time and blowup enough to fill a protecting buffer. When an intervening buffer is filled, this blocks products from leaving the preceding workstation. The latter workstation is then said to be 'blocked'. It is a common assumption by many mangers that this kind of capacity matching of inputs to outputs is an appropriate strategy, derived from the erroneous attitude that every resource could and should be operating at 100% utilisation and efficiency. Such assumptions are very prevalent in decision-making which is based upon overviews and averages, which in turn, provoke many 'fire-fighting' situations which have to be put out by managers, and often cause system disasters.

In fig. 3.9, the spreading of blockages is illustrated. Even if blockages are relatively rare, the circumstances which lead to them, along with the deleterious effect on system performance which they produce, can combine to make the occurrence of more blockages much more probable. Once blockages start to spread, they can be just as uncontrollable as a bush-fire.

Network evolution and the emergence of structure

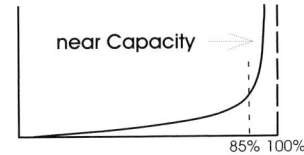

Fig. 3.8. Workstation blockage.

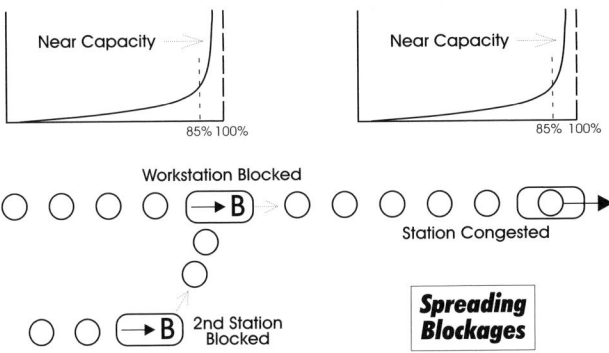

Fig. 3.9. Spreading of blockages.

Blocking can be seen as establishing a connection in a digraph of blockages for a given production system. As blocking increases and the system moves towards the double jump point, the formation of circuits of blocking signals the advent of gridlock. Gridlock (Figure 3.10) then becomes a possibility where the blockages feed back onto themselves, preventing any further production until extra resources are brought in from outside the system to help. Any of these situations may be what constitutes a breakdown for the organisation, it is not just a matter of the loss of productivity, it also depends upon the recovery costs involved, and these can be substantial.

A given organisation can establish its own pragmatic definition of what constitutes a breakdown based both on immediate recovery costs and the loss of future demand from negative market perceptions of it, and will be able to tolerate only a certain frequency of its occurrence. Only when this is clear can system capacity be actually determined. For complex

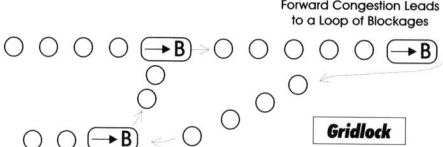

Fig. 3.10. The emergence of gridlock.

systems, this will only be possible with discrete simulation modelling. The management of these breakdowns by proper anticipation of the effects of variation upon the capacity of system components, constitutes the handling of Murphy's law to a significant extent. The observation of this connectivity avalanche phenomenon has led to the following novel insight.

A phenomenon has come to my attention as a result of capacity studies of complex organisations in which significant variations occur. It appears to be not widely appreciated except in an intuitive manner amongst a few operational people who develop a 'big picture' of their organisation. The phenomenon is this: the operating capacity of a complex system subject to significant variation cannot be determined without specifying the reliability of the system's performance as well. In other words, you cannot indicate what the capacity of the system is, without also saying what is the frequency and type of system breakdown that you are willing to tolerate!

3.6 Conclusions and speculations

3.6.1 Network evolution models and variants

The evolution of random digraphs can be a rich treasure trove of phenomena and insights that affect both our understanding of the natural world, and our design of the artificial world as previous discussions in this chapter have identified. Many useful variants of digraph evolution may be explored, including modelling the impact of weights on connections and/or strengths upon individual nodes. The probability of a node's connection could be a function of the number of nodes already connected to it.

Another group of model variants and exploration have been suggested by Green (1993) who describes situations where the level of connectivity for individual nodes is itself a dynamic variable. It has also been suggested (Hogeweg, 1993) that the fitness landscapes of the environment should generate this dynamic either as a dynamic variable or as a state transition system of some kind. Hogeweg has commented that many of our models which form the basis of complex systems assume fixed interaction structures. She contends that they 'are begging the question', and that the concept of life should be understood in terms of the generation of interaction structures rather than as the result of them.

There is a wonderful example of network evolution which many of us are already participating in, the growth of Internet. Communities are growing and decaying rapidly in that environment, and a variant network version of digraph evolution, which modelled both the inclusion and exclusion of connections, could well be used to study this phenomenon. In any case, the digraph evolution example appears to open a rich field for complex systems exploration.

3.6.2 Network evolution is at the core of phase transitions

When you start looking for them, phase transitions are an integral part of the physical world and perhaps the mental world as well. Langton and others describe apparent phase transitions in the evolution of ancient complex societies, emphasising the fact that they are always quick and involve tremendous change (Lewin, 1992, p. 17). Moreover, some evolutionary biologists look at evolution itself as punctuated by sharp, sudden changes (Gould & Eldredge, 1977, Kauffman, 1989). Langton further comments that cell membranes are poised delicately between the solid and liquid states. Gislason (1991) describes an 'avalanche effect' in the build-up leading to the onset of a number of illnesses and immune network breakdowns.

From a theoretical standpoint, Cheeseman *et al.* (1991) used graphical evolution and the emergence of Hamiltonian circuits in order to help understand the phase-like nature of NP-complete problems for algorithmic complexity. The emergence of stochastic deadlock such as referred to in the section on Murphy's law above is also a situation where a rapid change between a functioning system and a grid-locked system occurs. Hillis (1991, 1992) explored a model of enzyme and gene behaviour which demonstrated rapid phase changes in genetic populations, an approach which could prove useful in the analysis of the efficacy of genetic algorithms. Before these sudden changes in fitness which 'punctuated' the evolution of his model, he observed much activity, which did not improve fitness, in the model building up to such jumps. It is possible that underlying such activity is a connectivity avalanche phenomenon, or one of its intermediate stages wherein strong connection is finally made between two large patches. Hillis eventually showed that the occurrence of another power law was at work $1/e^2$, a possible footprint for digraph evolution.

We suggest that whenever there is the occurrence of rapid and sharp phase transitions in the physical or computational worlds, one can find the underlying presence of a variant of digraph evolution. Moreover, the connectivity and network evolution approach should provide the heart of any effective models of the phenomena so described.

3.6.3 The connectivity avalanche as a basis for chaos

Chaos has been observed around what has been called 'near-critical' connectivity. Just below the double jump point, the inclusion of a few more connections will produce a massive increase in connectivity, while conversely, a few connections removed when the digraph is just above the double jump point can drastically reduce overall connectivity. Green (1993, p. 34) conjectures that 'changes in connectivity may be an important source of variety in biology', based upon the observation that 'the variance in patch size is greatest when connectivity reaches a critical level'. Hence fluctuations in the environment will cause sharp changes in 'biotic connectivity' when an ecosystem's structure is near the critical point. It seems clear that the point of critical connectivity is none other than the double jump point.

As discussed in §3.2, the possible role of digraph connectivity in the prediction and classification of self-organising systems, such as those occurring in cellular automata, could be very deep. Moreover, its role in self-organised criticality has also been discussed there. The strong analogies between the various classes of behaviour in cellular automata with that of dynamical systems further suggests that there may be a deep connection between intra-system connectivity and chaotic behaviour.

3.6.4 Circuit coherence and self-organisation

Another intriguing observation emerges from this work on the evolution of random networks, and that is the role that the emergence of circuits plays in self-organisation. Kauffman's work of course is fundamental to this, as is the theoretical work of Maturana and Varela (1980), and Eigen and Schuster (1979). In addition, the gene simulations demonstrating a form of punctuated evolution which Hillis undertook with his Connection Machine (Hillis, 1992) appear to be further corroboration.

When discussing the emergence of autocatalysis in biochemical reactions in fact, Kauffman is quoted by Levy (1992, p. 137) as describing how '... a phase transitions occurred, a sort of connective explosion where suddenly critical mass was achieved', and 'There's a vast explosion of the number of organic molecules ... There's also an even vaster number of legitimate organic reactions by which they can convert to one another'. The correspondence between Kauffman's descriptions and what we have termed the connectivity avalanche in the evolution of random digraphs is clear.

The connections in the corresponding digraph are the legitimate organic conversions. At the double jump point there is a vast increase in the components which can connect to each other. There is also a vast increase in the number of circuits of increasing sizes which builds up to this point. At the double jump circuits large enough to span the entire network become possible. There needs to be more work, either theoretically or computationally, to establish the interplay between the emergence of circuits and the phenomena of the double jump point and the strong connection point. When this is eventually done, these results should provide a strong underlying basis for the various simulation results, given the directed graph or network characterisation of the simulation models.

3.6.5 Emergence: a fundamental paradigm shift

One view of the impact of the 'discipline' of complex systems is that it augurs a shift from the 'top down' to the 'bottom up' in intellectual endeavours, or from a paternalistic, autocratic and authoritarian science, to a participative, open and deeply democratic activity approach to science. This shift-in-the-making is particularly evident in the intellectual niche called artificial intelligence (AI) wherein the gauntlet has been thrown down by Rodney Brooks and his coworkers at the Mobile Robot Group at M.I.T. (Brooks, 1991). They pose a direct challenge to monolithic approach to intelligence characterised by traditional AI's approach of separate functional units for perception, modelling, inference engine, planning, and motor control, which in turn is representative of the conceptual edifices of top-down science. The work of both Kauffman and Langton have a similar quality of challenging cherished theoretical perspectives in the life sciences. Paradigm changes are often fuelled by challenges to authority which are picked up by the following generation of scientists. Certainly, we feel constrained by the dominant, linear, symbol-based, objectivist and rule-based models in both the physical and life sciences.

What is on offer from the behaviour-based artificial intelligence efforts of researchers like Brooks is an approach to the emergence of intelligence, through richly interacting networks which are in intimate touch with sensory and motor activities. Their mobile robots use only very local information, and instead of building a model of its world, it uses the best available model of the vagaries of the world, the world itself. This approach is called 'situated' and the intelligence of such devices is 'embodied' in its 'on-going participation and perception of the world'. 'Intelligence can only be determined by the total behaviour of the system and how that behaviour appears in relation to the environment.' What is being articulated by Brooks is that intelligence emerges both from the rich interactions amongst the components within a system, and the rich interactions without in its environment. It is emergence blended with holism.

The components are not very complex, especially when compared to top-down approaches to intelligence. Brooks mentions a 'density dependence' in terms of the number of components and resources which they consume, that could determine the global behaviour of the system. We hypothesise that characterisations of the rich connectivity is more likely to capture this dependence, and that in contexts established by environmental interactions, that phase changes in connectivity may be behind the acquisition of skills by his mobots. The system architecture which he has used is a fixed topology of augmented finite state machines. Based upon our work with random digraphs, we further hypothesise that if these systems could alter their own topology, that increased performance would emerge.

Hogeweg points out (1993) regarding the highly intelligent behaviour of the apes, '... the group structure acts as a substrate for intelligent behaviour, rather than being a result of it.' A similar thought is that society does not antedate the roles of individuals, but rather emerges from the rich interaction between them. Hence, complex systems can be seen to be upholding a strong role for the individual while examining the emergence of high-level behaviours globally for the ensemble of individuals. It is taking both a global and local perspective to whole systems.

The author has personally observed emergent processes in the acquisition of reading, and mathematical skills by his 'home-schooled' children. They were encouraged to explore and relate in a non-didactic and open environment without any formal teaching being applied, but where relationship skills were emphasised. In this manner, they acquired reading and mathematical ability common to their age group. The acquisition of many skills, including speech, may go through a similar process, and each of us in turn may become conscious of the process of emergence of skills and wisdom in our own lives.

Now that the process of emergence has been underscored by complex systems research (as opposed to its theoretical descriptions in the general systems theory and self-organising systems of the 1960s), we believe that it will grow to be a fundamental paradigm shift in the intellectual and scientific realms. Moreover, we agree with Brooks who points out that what we do in our daily lives is not the result of top-down problem-solving and planning, but rather emerges from our habits and routines as they interact with a dynamic world around us. Hence, this paradigm shift is not merely a theoretical idea, but rather an approach which is actually more in alignment with our own actual realities, and independent of what our egos may think we are doing. We believe that this chapter has given strong indications that graphical connectivity may fill a deep and fundamental role in complex systems and the phenomenon of emergence. In fact, it should be possible to observe graphical evolution unfold over recent and future years, as more and more connections are made between the phenomena of complex systems, and our science and understanding in all areas. This self-similarity between theory and the process of its acceptance will be its own validation.

Bibliography

Bak, P., & Chen, K. (1991). Self-organized criticality. *Scientific American*, **265**, 26–33.

Bollobas, B. (1985). *Random Graphs*. Academic Press, London.

Bossomaier, T., Pipitone, J., & Stuart, G. (1993). Neural dynamics in biological visual information processing. *Pages 361–371 of:* Green, D., & Bossomaier, T. R. J. (eds), *Complex Systems: From Biology to Computation*. IOS Press, Amsterdam.

Brooks, R. (1991). Intelligence without reason. *Proc. of the International Joint Conference on Artificial Intelligence*. ACM Press.

Cheeseman, P., Kanefsky, B., & Taylor, W. (1991). *Where the Really Hard Problems Are*. ACM Press. Pages 331–337.

Codd, E. F. (1968). *Cellular automata*. Academic Press, New York.

Eigen, M., & Schuster, P. (1979). *The Hypercycle: A Principle of Natural Self-Organisation*. Springer-Verlag, New York.

Erdos, P., & Renyi, A. (1960). On the evolution of random graphs. *Mat. Kutato Int. Kozl.*, **5**, 17–61.

Gardiner, M. (1970). Mathematical games: The fantastic combinations of John Conway's new solitaire game 'life'. *Scientific American*, October, 112–117.

Gislason, S. (1991). *Nutritional Therapy*. Persona Publications, Vancouver.

Gould, S. J., & Eldredge, N. (1977). Punctuated equilibria: The tempo and mode of evolution reconsidered. *Paleobiology*, **3**, 115–151.

Green, D. G. (1993). Emergent behaviour in biological systems. *Pages 24–35 of:* Green, D.G., & Bossomaier, T. R. J. (eds), *Complex Systems – from Biology to Computation*. IOS Press, Amsterdam.

Hillis, W. D. (1991). Co-evolving parasites improve simulated evolution as an optimization procedure. *Pages 313–324 of:* Langton, C.G., Taylor, C., Farmer, J. D., & Rasmussen, S. (eds), *Artificial Life II*. Santa Fe Institute Studies in the Sciences of Complexity, vol. 10. Addison Wesley, Reading, Mass.

Hillis, W. D. (1992). Punctuated equilibrium due to epistasis in simulated populations. *Emergent Computation*.

Hogeweg, P. (1993). As large as life and twice as natural: bioinformatics and the artificial life paradigm. *Pages 2–11 of:* Green, D. G., & Bossomaier, T. R. J. (eds), *Complex Systems – from Biology to Computation*. IOS Press, Amsterdam.

Kauffman, S. A. (1984). Emergent properties in random complex automata. *Physica D*, **10**, 145–146.

Kauffman, S. A. (1989). Cambrian explosion and permian quiescence: Implications of rugged fitness landscapes. *Evolutionary Ecology*, **3**, 274–281.

Kauffman, S. A. (1991). Antichaos and adaptation. *Scientific American*, **265**, 64–70.

Langton, C. G. (1992). Life on the edge of chaos. Langton, C. G., Taylor, C., Farmer, J. D., & Rasmussen, S. (eds), *Artificial Life II*. Santa Fe Institute Studies in the Sciences of Complexity, vol. 10. Addison Wesley, Reading, Mass.

Levins, R. (1977). The search for the macroscopic in ecosystems. *Pages 213–222 of:* Innes, G. S. (ed), *New Directions in the Analysis of Ecological Systems II*. Simulation Councils, La Jolla.

Levy, S. (1992). *Artificial Life; The Quest for a New Creation*. Penguin Books, London.

Lewin, R. (1992). *Complexity, Life at the Edge of Chaos*. Collier Books, New York. Page 17.

Maturana, H.R., & Varela, F. (1980). *Autopoiesis and Cognition*. D. Riedel Publish. Comp., London.

Minsky, M. (1986). *The Society of Mind*. Simon and Schuster, New York.

Nagels, G. (1985). Space as a bucket of dust. *General Relativity and Gravitation*, **17**(6), 545–557.

Palmer, E. (1985). *Graphical Evolution*. John Wiley and Sons.

Ronald, S. R. (1992). *Strong Connection of Random Digraphs of Order N*. Simulation and analysis, C.I.S. Tech. Report. University of South Australia.

Seeley, D. (1966). *The Structure of Homogeneous Markov Chains*. M.Sc. Thesis, Dept. of Industrial Engineering, University of Toronto.

Seeley, D. (1971). *The Use of Simulated Evolution in the Pattern Detection of Electro-encephalograms.* Ph.D. Thesis, Dept. of Industrial Engineering, University of Toronto.

Seeley, D. (1973). An animation system for the exploration of cellular automata. *Proc. of the Canadian Conference on Man/Machine Communication.*

Seeley, D., & Ronald, S. R. (1993). The emergence of connectivity and fractal time in the evolution of random digraphs. *Pages 12–23 of:* Green, D., & Bossomaier, T. R. J. (eds), *Complex Systems – From Biology to Computation.* IOS Press, Amsterdam.

Varela, F., Thompson, E., & Rosch, E. (1991). *The Embodied Mind.* MIT Press, Cambridge, Mass.

Wolfram, S. (1984a). Computation theory of cellular automata. *Commun. Math. Phys.*, **96**(November), 15–57.

Wolfram, S. (1984b). Universality and complexity in cellular automata. *Physica D*, **10**(January), 1–35.

Wolfram, S. (1985). Twenty problems in the theory of cellular automata. *Physica Scripta*, **T9**, 170–183.

Wolfram, S. (1994). *Cellular Automata and Complexity: Collected Papers.* Addison-Wesley, Reading, Mass.

4
Artificial life: growing complex systems

ZORAN ALEKSIĆ

Nagoya Shoka Daigaku
Sagamine,
Nisshin-shi,
Aichi 470-0114,
JAPAN

4.1 Introduction

The possibility of constructing systems so complex and functional that they might be considered alive is now upon us. We are all witnesses of the explosive growth in technology, but are there limits to the complexity of the systems we can construct? Technology operates within the constraints of nature. Yet under the same constraints biological evolution *constructed*† living systems, the source of much admiration and wonder. For many years to come natural living systems with all their complexity and relative perfection will be the model for the construction of artefacts.

At one level living systems can be seen as arrangements of interacting atoms. However a knowledge of quantum electrodynamics does not help us to understand the behaviour of bacteria nor the replication of cells, even though it is clear that there is no magic force behind life. Living systems show organisation of their component parts at each hierarchical level. At the level of biochemistry living systems are organised macromolecules. The tissue of higher organisms consists of organised cells. Our knowledge of atoms, macromolecules, cells, tissues, organs, individual living systems and upwards in the hierarchy of living systems indicates that what we can only vaguely define and call 'life' is always characterised by the arrangement of parts.

If a caveman were able to see a skyscraper how would he describe it? Maybe he would see similarities to a cave (Lem, 1977). The scientist studying the elegant compositions of living systems often compares them with machines (Monod, 1972) for want of a better analogy. A machine, especially a computing machine, is another apparently complex entity composed of simple parts. Both machines and living systems are composite. In principle there is nothing we know that suggests artificial systems cannot be at least as complex as living systems. We, ourselves, are examples of complex organisation. How hard is it to achieve such organisation? What type of organisation is a living system? How does nature organise parts? What are the useful parts? How long can it take to organise life-like complex systems? Let us look at some of these questions.

The growing number of researchers considering the above questions

---

† An important theme in this chapter and an important theme in artificial life in general, is the lack of any sort of central planning or control. It is sometimes difficult to avoid a teleological turn of phrase. But to say that nature does this or does that should not imply anything more than the observed phenomena.

has led to a new field of study called artificial life ('alife' for short). Typical of the many definitions of the field is the following:

Artificial life is the study of man-made systems that exhibit behaviours characteristic of natural living systems (Langton, 1989a).

The objects of study in ALife are constructed (synthesised) systems. To construct systems, especially complex ones, it is usually accepted that the constructor needs a detailed knowledge of the system to be constructed. But can something really be learned by analysing constructed systems? That the answer to this question is probably yes has helped the recent surge of the interest in simulating life-like systems. The availability of fast and affordable computers has provided an opportunity for relatively easy experiments. On one side the results of computer experiments demonstrate that even with a simple arrangement of simple components apparently complex behaviour can be achieved. The generated behaviour can be so complex that knowledge of the governing rules of the system is practically useless for anticipating the resulting behaviour. On the other hand it is possible to inject evolutionary processes into a digital medium. Hopefully the investigation of the process of evolution in well controlled and easily accessible media will increase our understanding of the ways of *blind constructivism*, i.e. construction without detailed planning.

The goals of ALife are ambitious indeed and the problems facing it are complex. When we consider the enormity of the task it is clear that research has hardly started. The results so far are modest. There is no room for unfounded and irresponsible optimism, but likewise we should not dismiss the field out of hand as a kind of science fiction. Even if it should happen that the construction of complex, life-like systems is beyond practical realisation, knowledge collected in the process of searching for the solution will undoubtedly be valuable.

Even at this early stage, the approaches taken by ALife researchers are diverse. But here we will focus on just one aspect of ALife: results gained through the construction and investigation of simple mathematical structures. This is a child-like way of forming complex systems. It parallels the recreational attitude that has often characterised the process of modelling living systems (Sigmund, 1993). It contrasts with an engineering approach where the constructor knows the process of construction and has a predetermined goal of construction. The child-like approach to construction is a heuristically guided search though the space of mathematical systems (usually implemented directly or indirectly as digital processes). The structures that emerge show some similarities to living systems. Investigating the structures that emerge has led to many useful insights.

The main characteristic of toy models investigated in ALife is that important questions related to the system's properties are computationally intractable, or even undecidable. In the next section we will consider a very simple toy universe, the *Game of Life*. It teaches us that a lot can be achieved by the simple arrangement of a large number of interacting non-linear components. It also warns us not to forget that, as in natural science, knowledge of fundamental laws does not help much in understanding behaviour at higher levels. But intractability is no problem for the process of evolution and a lot of effort in ALife research is oriented towards understanding evolution and trying to harness its principles for the construction of artificial systems. In the third section we describe a modified procedure of Dawkins (1988) to show how complex behaviour can arise by gradual accumulation of small changes. We will also describe the simple world of Tierra (Ray, 1992) that exhibits surprisingly rich evolutionary behaviour. In the fourth section we will consider the relationship between the intractability of complex systems and criticality. In the fifth section we crudely sketch the wide spectrum of ALife research with the aim not to give a detail review of the field, but rather to orient the interested reader to the growing literature.

4.2 Simple arrangement – complex, life-like behaviour

Many natural phenomena can be described by simple laws. Such laws may be simple in form, yet very complex in their action (Feynman, 1991). For example the law of gravity is simple but to predict the exact orbits of stars in a globular cluster is beyond our ability. While sciences like physics can generate a description of complex phenomena in terms of simple laws, in the field of ALife we attempt to construct complex systems (phenomena) using simple arrangement of interacting components. A prototype of a simple system that exhibits very complex behaviour is Conway's *Game of Life*.

4.2.1 The Game of Life

The *Game of Life* is a simple mathematical system invented by J. H. Conway (1982). It is an infinite two dimensional toy universe,

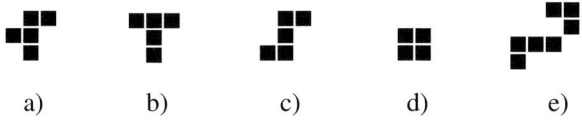

a)   b)   c)   d)   e)

Fig. 4.1. Some patterns observable in the *Game of Life universe*. Pattern (a) is known as an r-pentomino.

which we call here simply **C**. Parts of this universe can be easily embedded on the computer and observed on the display screen. A pixel of the screen corresponds to its smallest element and can be in one of two colours. To see the universe evolve all you have to do is to specify the initial state of the system. For example the pattern shown in fig. 4.1b can be an initial state. Except for a few black pixel/elements all other elements of the universe are white. After evolution starts you see a succession of new states. After several discrete time steps (generations) the pattern in fig. 4.1b evolves to a simple periodic pattern of period 2 known as *traffic lights*. The pattern shown in fig. 4.1c disappears after only three generations. The *r-pentomino* pattern shown in fig. 4.1a is growing. The state of the universe after 500 generations is shown in fig. 4.2. Some elements of this state are static patterns (*still life*) or simple periodic patterns. In other areas you can see some activity. It is hard to describe the dynamics of **C** verbally. The best first encounter with the *Game of Life* should be to observe its evolution on a large screen (for example 1000 × 800 pixels) when the updating is at least a few steps per second.

After seeing today's video games rich in colour and sound, the *Game of Life* may not appear so spectacular. To appreciate its power one has to contrast the simplicity of its governing rules with the apparent complexity of the evolution of the universe. The *Game of Life* can be thought of as a regular infinite two-dimensional array of simple elements/cells. Each cell can exist in either of two possible states: dead or alive (we represent a dead cell as white and a live cell as black). The states of cells are updated simultaneously in discrete time steps. The state of one cell in the next generation depends only on the states of its eight neighbouring cells. A living cell will survive only if it has two or three living neighbouring cells. A dead cell will return to life only if it has exactly three living neighbours. These two simple local rules fully determine the evolution of the universe from a given initial state.

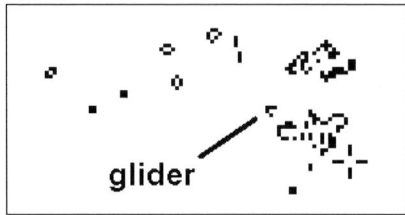

Fig. 4.2. The pattern of life universe after a number of generations evolving from the r-pentomino pattern.

Despite the simplicity of its rules the evolution of **C** is essentially unpredictable. The only way to say what will be the state of the universe after some number of generations given the initial state is to compute its evolution explicitly. There is no shortcut. It is provable that evolution in the *Game of Life* is unpredictable. The proof is an indirect one: we show that *life* is capable of universal computation.

For computer science what we usually think of as a computer is an instance of the *universal computer*. If something can be computed (like some mathematical functions) then a universal computer can compute it. All computers that are universal computers are *equivalent*. Put simply they are all capable of computing any computable functions. Of course for many practical purposes there are big differences in the efficiency between different universal computers. If it takes a universal computer A one hour to finish some task and a universal computer B is say 1000 times slower in performing the same task then you may not care at all about their equivalence. However when considering the theoretical capabilities of computers (or any entity that is capable of universal computation) the idea of equivalence is important. Once you prove something for a simple toy universal computer it is true for any universal computer. For example an important theorem about universal computers says that there is no general procedure for deciding whether a given program will ever stop (or halt). To see the proof of this undecidability theorem of the halting problem and find more on the theory of universal computation refer to standard textbooks on computation.

One simple question concerning the future states of the *Game of Life* is to predict whether or not all the activity in the **C** will ever stop given an arbitrary initial state. Conway showed that it is possible to construct

Artificial life: growing complex systems 97

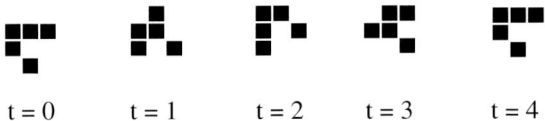

Fig. 4.3. After four generations the spatio-temporal pattern known as the 'glider' appears the same moved one cell up and one cell left.

a universal computer in **C**. When the initial states are appropriately selected the evolution of the *Game of Life* can be described as a process of universal computation. Whether any program (activity) executed on the *Game of Life* universal computer will ever stop is an undecidable question.

Understanding the behaviour of *gliders* is essential for computer engineering in the *Game of Life*. The *glider* is an important *elementary particle* in **C**. One can observe several gliders that appear in the evolution from the *r-pentomino* pattern. One of them is the marked pattern in fig. 4.2. After four generations the glider has moved one cell diagonally (fig. 4.3). Gliders can be easily detected because of their shape invariance. They can be considered as *physical* objects of **C** and the interaction of gliders or the interaction between gliders and other objects can be investigated. In the collision of two gliders both can disappear or only one can disappear and another turns back from whence it came. The most spectacular is the collision of 13 appropriately positioned gliders in which the *glider gun* appears. The *glider gun* is a spatio-temporal pattern that periodically fires a new glider. The collision of two gliders can also produce patterns called *eaters* (fig. 4.1e) and *block* (fig. 4.1d), two simple static patterns that together with the gun and glider form the four basic building blocks for a computer. In the universal computer Conway has constructed in **C**, gliders are the carriers of signals. Conway has shown how to construct crossing wires, logical gates and other parts of the computer, thus leading to the construction of a universal computer, capable of emulating any other computer. The possibility of such a construction within the *Game of Life* implies that it is capable of supporting arbitrarily complex computation.

Starting from the universal computer so constructed, it can be shown that the embedded computer can self-replicate (Sigmund, 1993). The command *copy* is one of the simple actions of any computer and the problem was to show how a program could copy itself. That Conway showed the possibility of self-replication of universal computers does not mean that only computers can live. It just means that there is in principle no problem for the replication of an arbitrary complex spatio-temporal pattern in **C**. Let us now speculate a little: imagine an infinitely large **C** evolving from a random initial state. Somewhere in its vast space there will be a relatively stable spatio-temporal pattern that can replicate. The replicas do not have to be exact. The only condition is that they can further replicate. The replicating patterns should be relatively stable so that at least some of them can create a few more or less similar copies of themselves before they disappear. The succession of patterns that preserve the capability of replication will slowly start to populate **C**. Replicated patterns that take longer to disappear can create more replicas. In the evolving **C** new types of phenomena can emerge, the functional equivalent of biological evolution. We used the term evolution to describe the succession of states in **C**. In observing the successions of replicating patterns we can talk about the patterns' evolution. The class of replicating patterns that populate the space of life better than other classes of patterns can be considered as better fitted entities at this stage of **C**. In principle the *Game of Life* may support *biological* evolution of complex spatio-temporal patterns.

4.2.2 Emergent complex behaviour

The events in the *Game of Life* can be described at different levels. At the level of generating the patterns we are talking about there is a homogeneous array of cells and local nearest neighbour interaction between cells: at this micro level we can talk about the propagation of excitation between cells, where we consider live cells as excited; at another level we can discriminate gliders and other simple objects. Shifting focus from the micro level to coarser spatial and temporal scales we may be able to discriminate recurrent patterns. Similarly, in the process of constructing a computer we start with logical gates, move to instruction sets and ultimately to the interaction of embedded universal computers. Somewhat like our own universe **C** can support different hierarchical levels of organisation. At each hierarchical level we see interacting components that can be further decomposed. At any level we can find recurrent patterns whose regularities can be described at that level without considering the lower generating level.

Of course, every possible pattern observed in the evolution of the *Game of Life* is ultimately the result of a local transition function. However, the observed patterns are defined with respect to the observed spatio-temporal regularities in the domain of observation/level of description. For example, a glider is characterised by its recurrent form. It is specified by the set of relations between the group of active cells independently of their position in space or time. The characterisation of the glider is independent of the generative mechanism because of its relational specification. Or let us take another example. In the *Game of Life*'s universal computer the logical gates are an appropriately arranged set of gliders. Logical gates in today's computers are constructed using semiconductors. In the first calculating machines the logical gates were mechanical. Later they were replaced first by vacuum tubes, then by integrated circuits. As long as the given entity at a level of the system organisation exhibits functional organisation the generative mechanism does not matter very much.

The regular patterns that can be discriminated in domains other than the generating domain are examples of emergent phenomena (see chapter 2). The main characteristic of an emergent phenomenon is that *it cannot exist and thus cannot be interacted with or observed at lower generating levels* (Baas, 1994). It exists as the result of the aggregate behaviour of more elementary parts. In some cases the emergent phenomena can be deduced or computed knowing the generating level, but the most interesting emergence is that one which is computationally intractable or even undecidable. The connection of interesting emergent phenomena at different organisational levels is not a trivial one.

For example, because the evolution of **C** is unpredictable, it is not straightforward to deduce the behaviour at higher levels from a given initial state. Yes, it is always possible to run the system and say what is happening *but there is no simple theory, there is no set of shortcut rules eliminating the need for direct computations.* The space of possible initial states of **C** is infinite. The computational effort necessary to find solutions grows exponentially with the size of the considered area of **C**. Without the effective order in this space any encounter of the regular behaviour at a higher level gives the appearance of surprise, hard discovered regularity. The same difficulties are encountered when trying the inverse approach: given the behaviour at the higher level is there an initial state leading to such behaviour? Again there is no shortcut. One has to resort to direct computation. Thus even a system that is easy to construct, like the *Game of Life*, is computationally intractable for general questions about large scale areas.

The *Game of Life* is a good example of a complex system. On the face of it, the *Game of Life* is an ideal reductionist system. We know the rules exactly but we have to talk about gliders or self-replicating computers to gain some semblance of understanding of its behaviour. If we were asked to deduce the rules of the *Game of Life* by observing its large-scale behaviour, we might try some level-by-level, top-down search procedure. But the reductionist approach does not work: neither an analytic one (from the top level to the elementary one) nor a naively synthetic one (from the bottom level directly to the top level). To understand the behaviour we have to deal with the emergent regularities at different levels of complexity (Anderson, 1972).

The notion of emergent complex behaviour looks like a kind of miracle to some people. If we consider the nature of the world we inhabit, emergent phenomena are inevitable. All natural objects and processes we can observe and interact with are composite entities, they are aggregations of components (molecules, atoms, ...). We are so accustomed to describing the things we find in nature with a small number of parameters that we easily forget their composite nature. We should not forget that most of the successes of natural science are connected with the explanation of *obvious* properties of phenomena in terms of the functioning of the component parts. For example, we take the concept of temperature as something simple but remember how difficult it was to reach our current understanding, to penetrate the illusory surface of the phenomena. We might even say that all good scientific results are the ones that show how a given phenomenon emerges from the functioning on the generative level. If this were not the case science would be a trivial activity. From this perspective the activities in ALife studies or complex system theory are not something strange. These activities connected with the attempts to construct complex phenomena can be seen as a normal development of science.

At this stage the understanding of emergent complex behaviour is rather intuitive. It is not even clear what should be called emergent behaviour. That is probably the reason why there are different opposing ideas. For example, for Cariani (1992) formal systems like the *Game of Life* are not capable of exhibiting emergent behaviour therefore according to his perspective the *Game of Life* is the wrong example. In principle, yes, there is nothing new through the evolution of a formal system. Everything in the future is already implicit in the initial state and the rules of evolution. It is unfortunate that the word 'emergence' has attracted a somewhat mystic flavour. In some usages it is a buzzword replacing the word novelty or creativity. Here we use it in a rather mundane way to denote relatively stable and regular structures that appear during the aggregation or formation of components. For the links between undecidability and emergence see Langton (Langton, 1989a). For more recent attempts to specify emergence formally see Baas (1994), Crutchfield (1994) and Darley (1994).

The *Game of Life* and other systems described in this book show how easy it can be to generate complex behaviour. For a constructor the emergent complex phenomena are not something just interesting to observe. The emergent regularities can be used for further construction. Emergent properties of some system A can be functionally important in the operation of the coupled system B. The composition of systems A and B can be more complex than the complexity of each system taken alone.

It appears that the possibilities for the simple construction of complex systems are greater than is naively thought. However, from the constructor's point of view the computational intractability of complex systems is a problem. Yes it can be easy to construct a complex system but is the search for the proper organisation an NP-complete problem? As we have seen, the problem of constructing a complex system is not solved, it is merely moved into another domain. Maybe it can be easier to solve the problem in the new domain where a complex system can emerge. The biosphere is full of entities whose very existence is proof of the possibility of constructing systems that are computationally intractable. In a way intractability precludes understanding. But evolution teaches us that it is not necessary to understand something, to have a blueprint or recipe for its construction.

4.3 Evolving complex systems

The theory of evolution is concerned with *the process of the accumulation of useful changes*. The process appears relatively simple if the organisation and functioning of the evolving entities is ignored. It is a kind of chain reaction. The entities that evolve can make copies of themselves. But these copies are imperfect: they contain small differences from the original. These replicas are in turn replicated with additional errors. The rate of replication of a given entity is a function of its relationship with its environment (i.e. the other entities with which it can interact and the non-living local universe). The process of evolution consists in the accumulation of small differences (errors) that enable increased numbers of copies to be made during the entity's existence. Within its environment a population of entities will evolve so that they are better adapted to survive and reproduce (Maynard Smith, 1989). Starting from simple systems, by gradual step-by-step transformations, the process of evolution can give rise to very complex structures. Any population, in which the entities have the properties of replication, variation and heredity, can evolve regardless of their actual structure.

A chicken is more complex than its genetic code. Any living organism is the result of a process of self-organisation which starts from the information contained in its genes. Discovering the nature of that process is one of the main interests in ALife research. The relationship between a 'genotype' (i.e. the underlying genetic code) and the resulting 'phenotype' (i.e. the full-grown organism) is similar to the way in which a small number of parameters can serve to specify complex behaviour in an organism. For instance, we can consider the observed behaviour of the *Game of Life* as a phenotype that is the expression of the underlying genotype. In this case the genotype consists of the parameters specifying the local transition rules plus the initial conditions (Langton, 1989a). An evolutionary approach to constructing ALife systems is to search for systems that exhibit emergent behaviour and are capable of adaptation. The aim is to learn how to navigate through the infinite space of possibilities and in the process accumulate knowledge of how to construct complex systems. Note that the process has two aspects. In one sense we use evolution as a search method, looking only at the products. However, we also want to learn as much as we can about adaptive systems in general. We want to discover what kinds of organisation and structures are adaptive.

4.3.1 Interactive evolution

As we saw above, a complex system can emerge from a simple one by the accumulation of small, useful changes. To see the power of cumulative selection, let us look at the process of interactive evolution developed by Dawkins (1988). In his simple computer model Dawkins used computer images as the organisms to be evolved. The model not only has an educational value but has recently been developed into a powerful method for computer graphics by Sims (1991).

The interactive evolution of graphical images works as follows. We take the set of parameters of a graphical procedure plus an initial state as the genotype of an evolving 'creature'. To each genotype there corresponds a graphical pattern, the creature's phenotype. The process of drawing the pattern on the computer screen can be thought of as the development or expression of the genes. Starting from an arbitrary genotype the program generates several new sets of parameters by mutation. The patterns corresponding to these mutated genotypes are drawn on the screen. The observer then selects the most interesting patterns. In the simplest case only one pattern/entity can be selected. The game continues in this fashion, repeating mutation and selection, until some target pattern is achieved or until some interesting pattern turns up.

Dawkins' procedure is the recursive drawing of a branching tree. For simplicity we will here use as the entity to be evolved the two-dimensional iterative map (4.1):

$$\begin{aligned} x_{n+1} &= by_n + F(x_n) \\ y_{n+1} &= -x_n + F(x_{n+1}), \\ F(x) &= ax + (1-a)\frac{2x^2}{1+x^2}. \end{aligned} \quad (4.1)$$

Starting with an initial state (a pair of numbers x_0, y_0) we can generate a sequence of pairs of numbers (x_n, y_n). Each pair in the sequence can be graphically represented as a point in the plane. For some parameters of the map the resulting graphs appear aesthetically complex. For example the pattern in fig. 4.5 has an attractive wing-like shape. The map (4.1) actually derives from research into the orbits of elementary particles in an accelerator but Gumowski and Mira noticed its intriguing potentiality as a generator of interesting pictures (Lauwerier, 1991). Here, we can again see that the observed pattern and the generating mechanism are not trivially related.

The parameters a and b are real numbers, which could be represented by points on the plane. To each such point there corresponds some graphical pattern. In principle there is an infinite number of patterns because the parameters are real. To start the interactive evolution we can randomly generate several pairs of parameters a and b and draw on the screen the corresponding patterns. We select one of these patterns. For example we could select the pattern shown in fig. 4.4a. The simplest way of mutating both parameters is to adjust each parameter by a small random value. By the process of interactive evolution we are moving through the space of possible patterns. The moves are determined by subjective judgment. By repeating this process we will slowly drift in the plane. From the pattern in fig. 4.4a we can reach patterns shown in fig. 4.4b and c. In this case patterns are close in the parameter space

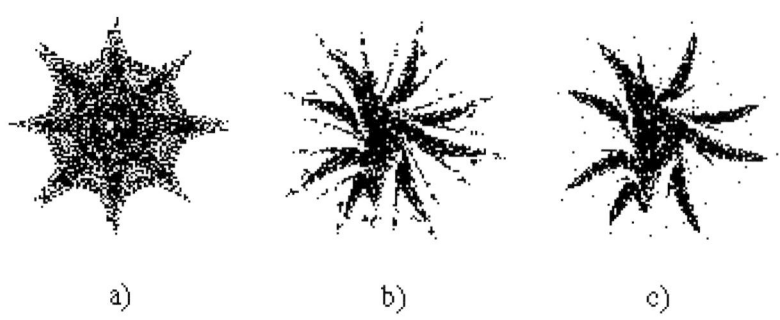

Fig. 4.4. Some patterns generated by the map (4.1): (a) $a = -0.7$, $b = 1.0$, (b) $a = -0.744$, $b = 0.972$, (c) $a = -0.741$, $b = 0.974$.

Fig. 4.5. The pattern generated by the map (4.1), $a = -0.8$, $b = 0.95$.

and are visually similar. Drifting further in the parameter space we can encounter the pattern shown in fig. 4.5.

Although we are talking about complex systems we have so far avoided saying exactly when something is complex. In the context of ALife the term 'complex' intuitively means 'similar in some respect to living systems'. Let us be more precise about the complexity in the case of the map (4.1): in the case of the map (4.1) we can talk about the complexity of the genotype, the development procedure and the complexity of generated pattern. As the complexity of the genotype we can simply take the number of parameters. The development process is a program that takes as an input the genotype and the initial point, and generate a picture on the screen. Its complexity can be taken as the complexity of the corresponding computer programs. For example the algorithmic complexity would be the length of its program. In our case there is a correlation between the size of genome and the algorithmic complexity. It is much harder to define the complexity of the generated pattern. That is exactly what the observer-breeder selects.

From one perspective the above example of interactive evolution is merely an aesthetically guided search in parameter space. Do the discovered creatures really emerge? Is it really a process of the growth of complexity? The complexity of both the genotype and the development procedure is unchanged. What is changing is the complexity of the generated patterns. The dimension of the parameter space is fixed and small, but it is a space of infinite size. Further, the maps like (4.1) live in very complex spaces. They are chaotic, so as in all chaotic systems, the properties of the system may change dramatically with a small change of parameters. The boundaries between different areas in parameter space are fractal boundaries. Considering the limited information capacity of the observer-selector even the drift through finite low-dimensional spaces may present the appearance of novelty and surprise.

4.3.2 Neutral evolution and the growth of complexity

To permit greater complexity to emerge when selecting the patterns above, we should first allow both the genotype and the development procedure to become more complex. Not only should we be able to move within the parameter space, we should be able to change its structure as well. Pattern selection and the generation of patterns operate in different domains, and we can exploit this difference to evolve the complexity of the parameter space that might lead to the increase in the complexity of the pattern.

The selector does not see the parameters of the patterns, the number of parameters or their values, nor even the method of generating the patterns. As long as two patterns are similar the different sets of parameters or different generating procedures are equivalent. Mutations that are indistinguishable at the level of selection are called neutral mutations in evolution theory (Kimura, 1983). We can transform the map (4.1) into many different maps without changing the phenotype. For

Fig. 4.6. The pattern generated by the map (4.2), $a = -0.8$, $b = 0.95$, $c = -0.16$, $d = 0.23$, $e = 0.21$.

example we could try to evolve a pattern existing in three dimensional space. The iterative map

$$\begin{aligned} x_{n+1} &= by_n + F(x_n) + cz_n \\ y_{n+1} &= -x_n + F(x_{n+1}), \\ z_{n+1} &= dy_n + ez_n^2. \end{aligned} \qquad (4.2)$$

can generate a sequence of triplets of numbers. When parameters c, d, and e are equal to zero the map (4.2) is equal to the map (4.1). We have a few new parameters, that is the size of the genotype is increased. We also changed the development procedure. Still, it is an iterative map but its form is different. We changed the map (4.1) dramatically but with the selected values we are really doing nothing at the phenotype level. However, now we have a bigger playground in which to try to get something more complex (Sigmund, 1993).

Figure 4.6 shows one of the patterns that could be encountered in the process of interactive drift starting from the set of parameters of the pattern in fig. 4.5. Through the process of neutral mutation we got one possible mechanism of increasing the complexity of the map. By neutrally changing the size of genotype and the process of phenotype expression we have the possibility for a drift in the parameter space of practically infinite dimension†. The drift of a finite observer/selector through infinite parameter space without known simple order means that many encountered patterns can certainly be considered as a discovery.

Neutral genotypes have indistinguishable (at the level of selection) phenotypes. Locally at the time of a transformation of the genotype (4.1) to a neutral form (4.2) there is nothing interesting. For selection it does not matter at that stage of evolution. But when evolution is seen as a long dynamic process the neutral transformation opens the space for structural changes in the evolving organisms. Neutral transformations provide for changes and the increases in the complexity of the evolving structures without disrupting the functioning of the organism. It is a way of introducing small changes in the creatures' structures that initially causes only small changes in the phenotype. However the small neutral changes may later unfold into very different creatures.

Let us try to be more formal when talking about neutral evolution. Different genotypes $\{G_i, G_j, G_k, \ldots\}$ can be expressed as the same phenotype P_m. Here when we say a phenotype we mean the traits of the emerging organisation that are relevant to selection. Ultimately, that means the traits that matter for the survival and replication within the given environment. Under the same conditions the genotypes $\{G_i, G_j, G_k, \ldots\}$ have the same fitness. The transition between the genotypes $\{G_i, G_j, G_k, \ldots\}$ does not result in any change of the fitness. Now, suppose that the genotype G_i can be transformed as $G_i \to G_{i+1}$ and the genotype G_j is transformed as $G_j \to G_{j+1}$. The transformation $G_i \to G_j$ is possible and it is neutral. We assume that the phenotype P_n of G_{i+1} has higher fitness than the phenotype P_o of G_{j+1}. When the transformation $G_j \to G_{i+1}$ is not possible, neutral evolution can open the path to increased fitness.

Up until now we have considered the selective force as something static. But both the non-living environment and other entities are constantly changing. At one moment of time (and space) the genotypes $\{G_i, G_j\}$ can have the same phenotype $P_i^t = P_j^t$. By the change in the environment the genotypes $\{G_i, G_j\}$ can be seen as having different phenotypes P_i^{t+1}, P_j^{t+1}. The phenotype P_j^{t+1} can be fitter than the phenotype P_i^{t+1}. Again we can see that the initially neutral (uninteresting) transition $G_i \to G_j$ may later lead to increasing fitness.

† The practical limitations of the dimension of the parameter space is the capacity of the system used for the embedding of evolving entities.

Interactive evolution can produce impressive results. Its power lies in the pattern recognition capabilities of the human observer. It is easy for us to quickly judge the interest of a picture. If the criteria for a more interesting or more desired structure could be expressed as a computable function the whole process could be automated. In that case it is an instance of a genetic algorithm as described in chapter 2. We could impose something like a fitness function that measures the desirability of structures. Instead of the drift of a small number of entities we can have evolution within the large population of entities. Usually the fitness function is itself the feature to be optimised. Using analogies with natural systems, the whole process can be described as an optimisation strategy.

4.3.3 Inoculating the process of evolution in Tierra

We know that evolution is capable of constructing very complex systems but the conditions under which complexity will grow is an open and important question in the field of ALife. In the previous subsection we deliberately chose an example in which there was the possibility of growth in complexity and we (the selector) searched for more complex patterns. However, there is no need for the chain of transformations to go in the direction of increased complexity. Let us repeat, what selection sees is the phenotype. As long as the system can survive and make replicas it does not matter what is happening with the complexity of the system. Fitter and complex are not synonymous. For some recent consideration of the growth of complexity consider (Arthur, 1994, Maynard Smith, 1994).

In the course of the evolution of natural living systems the selective pressure is nothing fixed, nothing imposed externally, that reflects a purpose. The cause of the variability of the selective force is not only the variability of a non-organic environment but to a large degree the simultaneous changes in the structure of interacting entities. When the growth of a population of different interacting living systems is correlated the systems *coevolve*. The more complex the coevolving entities become, the more complex their interaction. Being more complex, evolving beneficial emergent traits may mean having more chance to survive and reproduce. In some cases becoming more complex could be the only way to stay in the game. Coevolution plays a fundamental role in driving the growth of complexity seen in the biosphere (Dawkins, 1988, Maynard Smith, 1989).

One of the systems simple enough to be played with but complex enough that questions concerning coevolution can be investigated is the Tierra system created by Thomas Ray (1992). In the *Game of Life* simple rules can provide a space for the emergence of replicating patterns. Self-replicating entities are the starting point in the evolution of Tierra world.

Tierra is a digital world populated by digital 'creatures'. The physics and chemistry of this world are the Tierran machine language and the operating system that together control the activity of the system. Tierran creatures are active entities whose behaviour is determined by their program running on their own virtual processor simulated on either a sequential or parallel machine. A creature's behaviour could also be influenced by the parts of code they can find around it. Each virtual processor has an associated sets of registers and there is a corresponding area in the shared memory space where the program for a processor is written. The Tierra world consists of a fixed area of computer memory. This constrains the number of creatures. The CPU time of actual processor/processors of Tierra is allocated to creatures in the whole system either equally or proportionally to the creatures' program size. The main activity of successful creatures is to exist and replicate. The genotype code of Tierran creatures is their program code. This code is expressed by its execution by creatures' unchangeable virtual processor. Note that genotype and phenotype are different. There is a possibility for the emergence of complex behaviour of simple program code. It is also possible to have creatures with very different program codes but equivalent efficiency in replication.

The machine language of the Tierra system is a small set of 32 single word instructions. The main difference between the Tierran and other machine languages is *addressing by template.* For example in a standard machine language to move the pointer from the current position to the memory address say A4F1 one could say JMP A4F1 (jump to the memory address A4F1). In the Tierran machine language the area to where the instruction pointer can jump must be marked by a template. A template is a sequence of non-operational instructions NOP_0 and NOP_1. Instructions NOP_0 and NOP_1 have the role of marks for zero and one respectively. A sequence of instructions in Tierran JMP, NOP_0, NOP_0, NOP_1 means move the pointer to the position of the nearest occurrence of the template NOP_1, NOP_1, NOP_0 within a given range. The search for the templates is linear forward and/or backward away from the current pointer position. If the template is not found within the specified range the search is abandoned and the next instruction is executed. The templates or markers are expressed as non-operational instructions, not as direct binary data (a sequence of binary digits). In this way it is secured that the only content of the memory are instructions. The small instruction set and addressing by template mean that after mutation there is still a reasonable chance that the changed program will be operational. An additional property of the Tierran language not common for standard machine languages is the set of instructions MAL and DIVIDE. The instruction MAL allocates memory to a creature where the replica of the program will be written. The execution of DIVIDE gives birth to a new Tierran in this allocated area. There are several different machine languages developed for the Tierra world. The most recent one for parallel machines is expected to enable the development of multicellular creatures (Ray, 1994).

ALife in Tierra begins by the activation of a single creature. The ancestral creature, like the one written by Ray, self-replicates until it is terminated. The initial ancestor was 80 instructions long (Ray, 1992). At the beginning the creature finds its beginning and end marked by templates and calculates the size of its program. These templates can be seen as the membranes of the creature's cell. The cellularity of Tierran creatures is also enhanced by allowing a creatures to write only within the cell boundaries or within the area allocated to the daughter (after the execution of MAL and before the execution of DIVIDE). After the memory for the daughter creature is allocated all instructions are copied. When the replication is over the mother creature starts a new process (life of the daughter) using the DIVIDE instruction and at that moment loses access to the daughter memory space. When the total memory space of Tierra is almost filled with the creatures' genomes the operating system terminates a few of the oldest creatures. The speed of aging can be increased or decreased with respect to how well some instructions are performed.

Slowly, with the beginnings of ALife in Tierra the creatures become diverse. Randomly some bits of the memory in which the creatures' programs exist are flipped. The bits in the instructions can also be flipped during the process of copying. An additional source of noise in the system is the occasional erroneous execution of numerical instructions. Notwithstanding the simplicity of the Tierra world (as long as you feel comfortable with a machine language) there is surprisingly rich evolutionary behaviour. The replicas of the ancestor populate the whole memory space and the differences in the replication rate of mutated replicas comes to matter. Some creatures replicate better. We can say that fitter creatures replicate better but notice that there is no explicit fitness function. The efficiency of replication is a function of the evolving creature's genome. *The fitness function is implicit, it is not imposed externally, it emerges within the population of evolving creatures.* Within the constraints of Tierra creatures can evolve in an arbitrary, unprogrammed way. Tierra exhibits open-ended evolution relative, of course, to the context of the Tierran world.

The ancestor does not interact with other creatures. However the possibility for such interaction is implicit in Tierra and quickly (within the first few million executed instructions) during evolution the creatures *discover* a way to interact. They cannot write in the memory space of each other but they can read and execute other creatures' instructions. Through the mechanism of addressing by templates, mutations produce changes in the programs that cause some creatures to read the programs of nearby creatures. The possibility of interaction opens up the space for the appearance of parasites. A parasite does not have its own copy procedure, but uses the copy code of a neighbour. The part of the ancestor code for copying is marked by a template. One type of parasite loses its own copy part. When the copy procedure is necessary the parasite searches for the copy template. If there is a neighbouring creature within the range of search parasite can use this code. Parasites thus have shorter code then ordinary independent creatures. When the allocated processor time to each creature is fixed shorter code means more replicas can be created. As long as they live in a community of self-sufficient creatures, parasites are efficient.

Parasites do not directly harm their hosts, but they do compete for scarce resources. Thus some creatures develop resistance to parasites and, in turn, parasites evolve a way to circumvent the immunity, thus creating an *arms race*. This can get really vicious. Hyper-parasites manage to take control over parasites: when a parasite tries to use their copy procedure their registers' content is changed and they start to replicate the hosts' program. The emergent eco system of Tierra is diverse. Some creatures tend to decrease their size. Some organisms grow large and apparently complex, even longer than 1000 instructions. The growth of the genome is one way of exhibiting more complexity. Another is the appearance of groups of mutually beneficial creatures. The creatures become so coupled that the resulting group behaves as a complex unit. There are creatures called social hyper-parasites. They can replicate only within a group. The existence of each other supports the existence of a group. A group of social hyper-parasites can be considered as an emergent autopoietic organisation (McMullin, 1992).

Tierra indicates that evolving processes can proceed efficiently in constructed media (Ray, 1994). Investigation of Tierra and other systems with open-ended evolution indicates that some evolving entities can become more complex (Holland, 1994). The initial results are modest. How modest? It depends. If you are interested in immediate practical applications almost negligible. If you are interested in having a system that can give you some clues about the growth of complexity then quite considerable. It seems that intractability of complex systems is not an unsurpassable obstacle. It was not a problem for bio-evolution and it seems that a functionally similar process to bio-evolution can proceed in artificial media.

4.4 Around the critical state

The examples we used are intended to illustrate that some complex systems need not be the result of the complex construction of a rational, purposeful creator. Under certain conditions complex systems/complex behaviour may be the natural states of the aggregation of interacting parts. It appears that like the *Game of Life* many other ALife toy models exhibit ordered patterns at several hierarchical levels with the possible interactions over all scales. Many researches believe that interesting life-like behaviour is connected with critical states.

In chapter 2 you saw that the behaviour of many different systems can be represented using directed graphs. In chapter 3 Seeley shows that the directed graphs exhibit critical behaviour for a certain connectivity and argues that the critical area is the most interesting area from the point of view of complexity. In this section we will consider the question of criticality from the perspective of some of our ALife examples. It appears that both the *Game of Life* and Tierran world operate in the critical state.

4.4.1 Self-organised criticality

Inspecting large areas of **C** one can notice that some areas of it are dominated by static or simple periodic patterns. Simultaneously in different parts there is sustained activity. Occasionally one can see an almost diminishing small excitation arriving at an area where there is no change causing a sudden burst of activity. One way to measure the effects of perturbations is to investigate finite segments of **C** (finite array of cells with periodic or absorbing boundary conditions) (Bak et al., 1989, Alstrøm & Leão, 1994). The finite segments evolve from a random initial state to a state, akin to an equilibrium, that is a mixture of static patterns and simple periodic patterns. When such an equilibrium state is slightly perturbed (making one dead cell alive) after a number of generations the system settle into a new equilibrium state again. The distribution of the number of generations T the system takes to settle down is proportional to T^{-b} with $b \approx 1.6$. The distribution of the total number of changes s (births and deaths) caused by the perturbation is also a power-law relationship and it is proportional to $s^{-\theta}$ with $\theta \approx 1.4$. The activity in **C** does not decay or explode exponentially. Changes in **C** are correlated in time and space. The clustering of activity in the *Game of Life* has fractal geometry. The estimated fractal dimension is $D \approx 1.7$.

The power-law dependencies exhibited in the *Game of Life* are the signature of critical behaviour. In equilibrium physics criticality is the exceptional case. To achieve criticality one has to carefully tune some parameter to a unique value. But for the *Game of Life* and some large dissipative systems the critical scale-free spatial and temporal behaviour naturally self-organise. The critical state of these systems has a large basin of attraction: for a wide range of initial conditions the system will evolve into this critical state. The tendency of systems to organise themselves into a critical state is described as *self-organised criticality* (Bak, 1994).

The Tierra world is a self-contained universe like *Game of Life*. It appears that also Tierra operates in critical state like **C**. The variation and the distribution of species as well as the distribution of sizes of extinction in the Tierra world exhibit a power-law (Kauffman, 1993).

Systems in a critical state exhibit a hierarchical nature with interactions over all scales. The hypothesis has been made that the complexity and criticality are the same (Bak, 1994). Systems that operate near or at the critical state might be able to support evolutionary processes (Langton, 1992, Kauffman, 1993, Bak, 1994). At the moment the evidence for these hypothesis is rather vague. Bak and Seppen recently proposed a simple model of an ecology of interacting species that self-organises itself into a critical state (Bak & Sneppen, 1994).

4.4.2 Complex behaviour at the edge of chaos

The idea of self-organised criticality is connected with large systems, but we can also consider systems as members of some large family. For example, in the case of the *Game of Life* we can ask about the properties of other cellular automata.

The rule for computing the successive states of **C** is analogous to computing the orbit of a given differential equation. To understand more about the solution of the equation one has to compute and analyse orbits for a set of different initial conditions. This is very useful, but more general understanding can be gained when studying not only a given partial differential equation but a family of equations of which the given one is an instance. For example for a given form of equations one can analyse how the behaviour of the solutions varies with changes in the value of a given parameter. Similarly we could analyse the behaviour of a family of cellular automata of which the *Game of Life* automaton is a member. When cellular automata are seen as dynamic systems the rules of the automata are analogous to differential equations.

More formally, cellular automata (CA) are spatially extended discrete dynamic systems, which can be thought of as a regular array of simple identical processing units (or cells). Each cell takes on k possible values. The global evolution of cellular automata is determined by the local interaction between neighbouring cells. The simplest cellular automata are one-dimensional automata with nearest-neighbour interaction. The value a of the cells at each position i is updated in discrete time steps according to a fixed rule **f**

$$a_i^{t+1} = \mathbf{f}(a_{i-1}^t, a_i^t, a_{i+1}^t). \tag{4.3}$$

The *Game of Life* is a member of automata whose local transition function can be written as

$$a_{i,j}^{t+1} = \mathbf{f}(a_{i-1,j+1}^t, a_{i,j+1}^t, a_{i+1,j+1}^t, a_{i-1,j}^t, a_{i,j}^t, a_{i+1,j}^t,$$
$$a_{i-1,j-1}^t, a_{i,j-1}^t, a_{i+1,j-1}^t). \tag{4.4}$$

where each state a of the cell at the position i and j and the time t can take only two values. The local rule **f** is a Boolean function of nine binary variables. It can be represented as a look-up table that specifies how the value of the cell is updated for a given pattern of activities of the neighbouring cell. For example one entry in the look-up table for a *Game of Life* could look like $001000001 \to 0$. This entry is one instance of the rule *a cell will die if the number of live neighbouring cells*

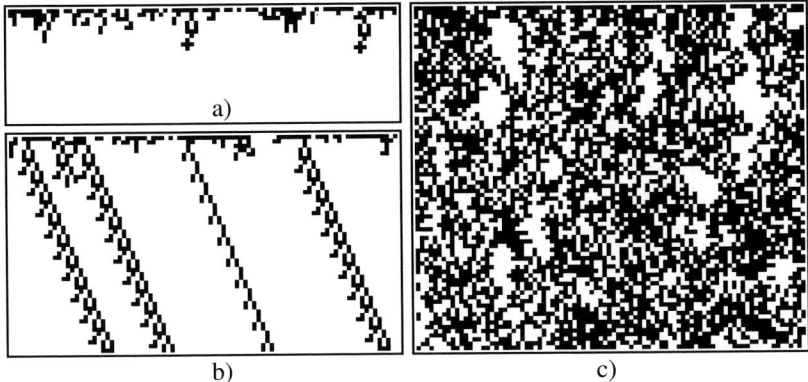

Fig. 4.7. Examples of the behaviour of one-dimensional cellular automata with respect to the parameter λ. White squares represent cells with value 0; black squares cells with value greater than 0. Each row is the state of the automaton at time t. The initial state is the first row. (a) small λ; (b) $\lambda < \lambda_c$; b) $\lambda > \lambda_c$

is not 2 or 3. The right single digit specifies the value of the cell at the next generation. The pattern on the left side specifies the corresponding condition. The right-most digit of the left pattern is the state of the cell at the previous moment of time. Other digits correspond to the values of neighbouring cell. For a *Life*-like cellular automaton the number of entries in the look-up table is 2^9. When the number of possible values the cell can take is k and the number of cells that affect the updating is n, the size of look-up table corresponding to the local transition function is k^n. The number of different possible functions $\mathbf{f}$ is k^{k^n}. Even for small k and n this number is so large that understanding the space of possible transition functions is practically impossible without discovering some kind of order.

The indication of some kind of order in the space of simple cellular automata was discovered by Wolfram (1986) who took a smaller subclass of simpler CA, and found that CA behaviour can be approximately categorised into four main classes which we shall discuss below. However his classification scheme still does not connect the properties of a CA rule with the corresponding automata behaviour. A better way of understanding the rule space of CA is to find a concise way to describe the space of transition functions such that there is a lawful relationship between the description of a transition function and corresponding behaviour of the automata. One possible parametrisation scheme was

Fig. 4.8. The behaviour of the cellular automaton at the edge of chaos.

discovered by Langton (1992). Let us call one value of the automaton cell a quiescent value. For a given transition function Langton defines the order parameter, λ, as the percentage of all entries in a look-up table which map to a non-quiescent value.

Let us illustrate the ordering of one-dimensional automata with respect to the λ parameter by considering the automata where the cell can take eight values, interacting only with the left and right neighbouring cell. We take 0 as the quiescent value. The number of entries in the look-up table of these automata and Conway's rule is the same. For a given value of the λ parameter we pick up arbitrarily one of several corresponding transition functions. The initial state of the automaton is a random state where values of the cells were generated using a uniform distribution. In fig. 4.7a for small λ the patterns quickly evolve to the quiescent state. The CA behaviour that evolves to a homogeneous static global pattern is classified by Wolfram as Class I. With an increase of λ the behaviour can be described as simple or periodic patterns (fig. 4.7b), which corresponds to Wolfram's Class II. For λ closer to 1 the behaviour exhibits deterministic chaos, class III behaviour (fig. 4.7c). For around some critical value of the parameter λ the behaviour exhibit interesting complex behaviour – something we saw in the behaviour of the *Game of Life* and what Wolfram calls class IV behaviour (fig. 4.8). The small (measure zero) area around λ_c is known as the *edge of chaos*.

The quantitative measures of the behaviour of cellular automata undergo changes around the critical value. When some measured values are plotted against the λ parameter the distribution of points appears to be bi-modal, which indicates a kind of transition phenomenon. Around the critical value λ_c the mutual information is high while the entropy and the spreading rate of difference patterns (Li et al., 1990) undergo a transition from low values in the domain of $\lambda < \lambda_c$ to higher values for $\lambda > \lambda_c$. The parameter λ is not a perfect parametrisation but considering its simplicity it is doing well. The λ parameter is a scalar value. Each value λ corresponds to some number of different rules. Ideally, all transition functions with the same parameter λ should exhibit similar behaviour. This desired behaviour is expected in large enough spaces of cellular automata (Wootters & Langton, 1990).

Not only do some systems that exhibit complex life-like behaviour exist near the edge of chaos but also there is an indication that successful evolving systems may evolve toward the edge of chaos region. For example, Kauffman (1993) studied the adaptive behaviour of game-playing Boolean networks (see chapter 2). The results shows that networks that are successful in playing the mismatch game adapt toward the ordered regime near the edge of chaos. Kauffman also studied coevolving systems. His model of coupled NK fitness landscapes exhibited coevolution to the edge of chaos. The results can be intuitively understood. To accomplish some task it is necessary to have enough capabilities. To solve more complex tasks the solving system needs more capabilities. At or around the edge of chaos the evolving systems are maximally complex.

What is common for both notions of the edge of chaos and self-organised criticality† is critical behaviour. The fact that the system operates at or near the critical state means that the smallest possible perturbations can cause drastic changes over large areas of the system. Small perturbations can cause changes very far within the system's space and time, practically infinitely far. Often the effect of perturbations would be localised and it would last for a short time. In the critical state the system is interconnected at all scales. Treating large systems in the critical state means that computational intractability is likely to occur.

† Analysing the exact relation between the two notions is beyond the scope of this chapter. The difference is about the 'role' of parameter tuning. From the perspective of self-organised criticality there is no external parameter tuning to bring the system to the critical state. The existence of a kind of a control parameter is essential for the idea of the edge of chaos which is a kind of second order phase transition. Both ideas are context dependent. Without mentioning a given context self-organised criticality appears as a magic property. The λ parameter is obviously useful as a way of parameterising a given system, but still it has to be shown to have an operational function for a concrete system, that it is not imposed by the observer for the sake of concise description.

4.5 The diversity of artificial life research

Even the few examples we have shown in this chapter indicate the diversity of the ALife field of research. The very diversity of the field mirrors the nature of the problems ALife is trying to solve. Actually, the question is not whether the diversity is something negative, a sign of an immature field, but the real problem is what is the proper methodology or approach to adopt in the face of difficult diverse problems.

Trying to classify research in ALife is neither an easy nor a rewarding task. For example, Taylor (1992) classified projects according to the classes of models. Some classes he discusses are computer viruses, evolution of digital processes, auto-catalytic networks, biomorphs and ontologically realistic processes, robotics, cellular automata, and artificial nucleotides. Of course the list is not comprehensive and there is overlap among the different classes. Another classification can be made according to the problem the research is addressing. For example there are classes of projects concerning the origin of life, evolutionary dynamics, and the development of artificial organisms.

ALife is a young research field that is still in a formative phase. The aim of this chapter was not to provide a comprehensive review of the ALife. We have presented a few, according to our perspective, key ideas of the field. The interested reader will find a broader perspective on the field in the journal *Artificial Life*. The proceedings of the ALife conferences, initiated by Langton (1989b, 1992, 1994), are also an important record of discoveries about artificial life. Let us now locate the examples presented in the chapter within the broader context of ALife research.

Cellular automata are historically one of the first systems developed to study some questions concerning ALife. Von Neumann used CA to investigate whether it is possible to achieve the growth of complexity in a purely deterministic system (McMullin, 1992). The *Game of Life* is one of the simplest CA that illustrate emergent behaviour. The most attractive feature of the *Game of Life* is the possibility of an elegant constructive proof of undecidability and computational intractability. For a recent use of CA in ALife to study the importance of spatial structure in evolution see Boerlijst and Hogeweg (1992). Studying ALife with CA is only one of their applications. Studying CA is a large field on its own (Toffoli & Margolus, 1987).

Interactive maps like the Mira–Gumowski map (4.1) are probably the simplest way of generating morphologically complex patterns. The morphology of natural living systems is the most obvious sign of the complexity of organisms. It is the process of development where the observable complexity of the living systems comes in to being. As Kauffman and Taylor (1993, 1992) point out, understanding the connection between development and evolution is very important. At the moment the models of open-ended evolution are mainly concentrated on RNA-like systems. The difference between genes and expressed phenotypes is small. The main reason for such approaches is the computational difficulties of generating more complex phenotypes. One of the formal systems capable of generating highly complex patterns (Prusinkiewicz, 1994) are L-systems, discussed in chapter 2.

The Tierran world is an example of a research approach where one studies the life of digital processes. Explorations of digital media to support artificial life are very popular. It is relatively easy to perform computer experiments. While Ray tries to understand the evolution of digital cellular organisms within the Tierran world and eventually evolve complex multicellular creatures, Rasmussen *et al.* (1992) study the possibilities of programmable matter to support functional self-organisation and cooperative dynamics. This research is complementary to the research on auto-catalytic networks (Bagley & Farmer, 1992, Bagley *et al.*, 1992) and it may throw some light on pre-biotic evolution. Both the Tierra system and the systems investigated by Rasmussen *et al.* (1992) use the Von Neumann computer architecture. Fontana and others (1992, 1994) study processes of evolution of organisation using Church's computational model. The importance of the organisation in living systems was understood a long time ago (Maturana & Varela, 1980) but the appearance of mathematical analysis was slow to come. The λ calculus and other mathematically minded computational models may be more useful in learning more about the organisation of life-like complexity.

Investigating and playing with mathematical systems is only one approach to ALife. In his sceptical futurological work Lem (1977) also investigated the possibilities for ALife. Of course in 1964 the term 'artificial life' did not exist. Lem coined the term *pantokreatika* for the theory of creating any possible complex systems, even those not yet realised by nature. For Lem pantokreatika should be based on the general theory of natural and mathematical systems. Lem saw the path to the point in time when the power of constructionism will be able to rival nature's capability for construction as a dual road. On one branch of the road there is old good science. Natural sciences will gain greater understanding of complex natural systems, both organic and inorganic.

New understanding will cause more development in technology. New materials and primitive parts will be available. On the other branch there will be an increasing understanding of formal systems – mathematical structures that exhibit complex behaviour. The real chance for ALife may appear after the results from both branches fully merge.

Research on systems like cellular automata, L-systems or the Tierra world could be seen as experimental mathematics investigating properties of formal systems. One application of the investigated systems is the modelling of natural phenomena. At this stage of research the concern for correspondence with something in nature is usually secondary. Yes, the structures to be investigated are selected heuristically so that they have some resemblance to living systems, but the correspondence is only in the domain of subjective observation. It is hard to relate in a quantitative way ALife systems and natural phenomena. There is nothing wrong with the heuristic approach to ALife. Like mathematics ALife is a generator of complex structures which can be used for modelling phenomena but at the moment it is reasonable to try to understand the complex behaviour of would-be models first. Only understood structures can be good models. The signs that some constructed complex systems have a potentiality to model complex natural phenomena are good reasons enough to warrant their further investigation.

Not only can the mathematical structures investigated in ALife be very useful for modelling natural phenomena but an even more exiting possibility is that promising structures may become *materialisable* (Lem, 1977). Systems currently investigated are those that can be straightforwardly implemented in computational systems. Digital media are not the only options for inoculating or investigating the processes of artificial life. Simply at this stage of development it is easier to construct and investigate digital worlds, but using purely digital processes for the construction of ALife may limit the achievable growth of complexity. The processes in natural living systems can be seen as computational processes (see chapter 2) but there is a big difference in computational power available to bio-evolution and to us as constructors. A constructor trying to arrange component parts so as to achieve a system exhibiting life-like behaviour is competing with nature who has vast number of parallel processors (Bagley & Farmer, 1992). We have difficulties in directly controlling computational potential of natural processes. But as we saw in the *Game of Life* what is a component part is somewhat arbitrary. There is no need to go to the ultimate level (that is equal to starting from scratch). We can use already available components provided within nature and try to arrange them to achieve life-like behaviour.

The straightforward approach is to use already complex macromolecules or natural replicators. Indeed it is already possible to achieve the evolution of synthetic molecules (Schuster, 1994). The research on nanotechnology may open the way to building silicon or other material replicators. If we learn enough about self-organisation and evolution in natural media we may be able to inoculate areas of particular media with simple organised entities that will be able to not only survive and perform some function but also be capable of exhibiting growth of complexity.

4.6 Conclusions

Given what we know today it is very unlikely that ALife is not possible. When do we decide that something is living? What is a definition of being alive? To some people (Dawkins, 1988, Maynard Smith, 1994) any entity that is capable of evolving is alive. In this sense the creatures in Tierra would be seen as alive. For others being alive implies a specific autonomous organisation, as in Maturela & Varela (1980). According to this definition we have to wait a bit longer to see ALife forms but the problem is of technological nature only. We avoided giving any definition of what is life. We are opting for a humble constructor approach. Often when one is talking about the construction of ALife one is actually talking about the construction of systems that exhibit complex spatio-temporal patterns somewhat similar in some aspects to natural life, whatever space and time in a given model means. Usually the life processes are in the background, they are implicitly known. Some similarity with living systems is a heuristic guide of the search/drift through the infinite space of (formal) models. It is one measure of interest of the model. Farmer talks about such an approach as *looking at stars with peripheral vision where there is higher sensitivity to twinkling light*†. To underline the need for an indirect approach and respect for a given medium Ray (1994) talks about Zen and the art of creation.

The toy models of ALife are mathematical objects and like mathematics ALife can be seen as a generator of potentially useful structures. However, ALife is still lacking some attractive characteristics of mathematics. There is an elegant proof of the unpredictability in the *Game of Life* and here and there interesting results but often ALife papers are no more than what Holland calls ... *simply points in the space of pos-*

† quoted in Waldrop (1993).

sibilities (Holland, 1992), featuring statements such as *dependencies on the parameter have not yet been systematically investigated*. Research in ALife is not only the quest for novel principles but also the quest for novel methodology. The nature of the problems facing ALife make conventional mathematics only partially helpful. Researchers have to rely extensively on computer experiments. In searching or waiting for a new methodology, new tools, a new calculus of complex systems, the solace is the existence of a large gap between what we know about living systems and the processes of complex construction on one hand and what can be learnt about it on the other. The gap is so big that even the easiest approach of computer experiments promises important knowledge.

The toy models of ALife should at least be interesting toys. Their behaviour is very often aesthetically interesting to observe. In some cases this could be an aim in its own right. Playing with (studying) these models can be seen as a kind of elaborate computer-assisted thought experiment (Holland, 1994). In the near future it is through easy to control computer experiments that insight is most likely to come. As before in the history of science the interesting mathematical systems discovered in ALife can be used for modelling. However, the most exciting prospect is that eventually discovered interesting structures may be embodiable not only in computational systems as today but in a different artificial natural medium. While today the questions on the reality of the Platonic world belongs to philosophy (Penrose, 1989), the embodiment of mathematical structures may be engineering questions in the future (Lem, 1977).

At present the products of ALife research only whet the constructor's appetite rather than satisfy it. By now we know that potentially interesting systems should exhibit emergent phenomena. It seems that the rich repertoire of emergent phenomena is a necessary condition for systems to achieve growth of complexity. Under *selection pressure* a good evolving organisation can move in the direction of developing further currently beneficial emergent traits. What is the appropriate organisation of entities that efficiently evolve? This is an important open questions. Almost as a rule interesting discovered organisation in ALife provides new insights but it also creates many new questions. Even the simple *Game of Life* is not yet a well understood system. Can we use such types of worlds to learn more about connections between different emergent hierarchical levels? How can we learn about large scale behaviour in life-like systems? What can happen in the $10^{12} \times 10^{12}$ cells large area of **C** during 10^{12} time steps? With this question we face directly the computational intractability of the system. Are we missing important phenomena through not being able to consider large sizes. What if the interesting emerging complex creatures are of prohibitively large sizes? For example, say in Tierra we set the system for the evolution of large creatures, of a few thousand instructions. In many cases it will not be easy to understand such an entity. Can we find an indirect way to use artificial, imperfectly understood, complex systems?

It seems that there is no problem in principle in inoculating a kind of directed evolutionary process and getting complexity for free. However, natural evolution is not, from a human perspective, a fast process. If you consider molecular processes as computation, then evolution is computationally very demanding. Natural evolution works well (in comparison with our capabilities) given enough time and resources. But can we wait even 100 years to get something? Evolution seen as the search in a living systems parameter space can be seen as a user of a rather simple heuristic to govern the search. Can artificial evolutionary processes be speeded up? Investigated ALife models usually exhibit hierarchical structure. Simon (1992) argues that complex systems evolve much more rapidly if their organisation is hierarchical. Not much research has been done in this direction.

ALife is not short of interesting, open questions. Searching for the answers we will gain new knowledge. We may not find answers to all questions but our perspective will be changed. Many of the outstanding questions will lose their meaning and new questions will emerge.

'Frankenstein, move over ... ' is written on the cover page of one popular book on ALife. Unfortunately, in this days when we can all feel the omnipotence of market laws it is easy to tell spectacular stories to make a quick profit. There is a danger of inflation of values and criteria. On the surface ALife may give the appearance of a video-game but it is (could be) a hard research program that demands serious work. Yes in ALife there is need for a lot of guessing, appreciation of beauty and simplicity but the last thing you can find there are slimy creatures offering you eternal life or the possibility for cyber-sex. The excitement of ALife is in the power of simple mathematical structures, in the potentialities for the organisation of (engineering with) macromolecules. Put simply, ALife is nothing special or radically new. It is only a potentially new way of doing science.

Bibliography

Alstrøm, P., & Leão, J. (1994). Self-organised Criticality in the 'game of Life'. *Phys. Rev. E*, **49**, 2507.

Anderson, P. (1972). More is different. *Science*, **177**, 393–396.
Arthur, B. W. (1994). On the Evolution of Complexity. *Page 65 of:* Cowan, G. A., Pines, D., & Meltzer, D. (eds), *Complexity: Metaphors, Models, and Reality.*
Baas, N. (1994). Emergence, Hierarchies, and Hyperstructures. *Page 515 of:* Langton, C. G. (ed), *Artificial Life III*. Addison Wesley, Reading, Mass.
Bagley, B. W., & Farmer, J. D. (1992). Spontaneous Emergence of a Metabolism. *Page 93 of:* Langton, C. G., Taylor, C., Farmer, J. D., & Rasmussen, S. (eds), *Artificial Life II*. Addison Wesley, Reading, Mass.
Bagley, R. J, Farmer, J. D., & Fontana, W. (1992). Evolution of Metabolism. *Page 141 of:* Langton, C. G., Taylor, C., Farmer, J. D., & Rasmussen, S. (eds), *Artificial Life II*. Addison Wesley, Reading, Mass.
Bak, P. (1994). Self-organised Criticality: A Holistic View of Nature. *Page 477 of:* Cowan, G. A., Pines, D., & Meltzer, D. (eds), *Complexity: Metaphors, Models, and Reality.*
Bak, P., & Sneppen, K. (1994). Punctuated equilibrium and criticality in a simple model of evolution. *Phys. Rev. Let.*, **71**(24), 4083.
Bak, P., Chen, K., & Creutz, M. (1989). Self-organised criticality in the 'game of life'. *Nature*, **342**, 780–782.
Berlekamp, E. R., Conway, J. H., & Guy, R. K. (1982). *Winning Ways for Your Mathematical Plays*. Academic Press, New York.
Boerlijst, M., & Hogeweg, P. (1992). Self-Structuring and Selection: Spiral Waves as a Substrate for Prebiotic Evolution. *Page 255 of:* Langton, C. G., Taylor, C., Farmer, J. D., & Rasmussen, S. (eds), *Artificial Life II*. Addison Wesley, Reading, Mass.
Cariani, P. (1992). Emergence and Artificial Life. *Page 775 of:* Langton, C. G., Taylor, C., Farmer, J. D., & Rasmussen, S. (eds), *Artificial Life II*. Addison Wesley, Reading, Mass.
Crutchfield, J. P. (1994). The calculi of emergence: Computation, dynamics, and induction. *Physica D*, **75**, 11–54.
Darley, V. (1994). Emergent Phenomena and Complexity. *Pages 411–416 of:* .Brooks, R. A, & Maes, P. (eds), *Artificial Life IV*. Addison Wesley, Reading, Mass.
Dawkins, R. (1988). *The Blind Watchmaker*. Penguin, London.
Feynman, R. (1991). *The Character of Physical Law*. Penguin, London.
Fontana, W. (1992). Algorithmic Chemistry. *Page 159 of:* Langton, C. G., Taylor, C., Farmer, J. D., & Rasmussen, S. (eds), *Artificial Life II*. Addison Wesley, Reading, Mass.
Fontana, W., Wagner, G., & Buss, L. W. (1994). Beyond digital naturalism. *Artificial Life*, **1**, 211.
Holland, J. H. (1994). Echoing Emergence: Objectives, Rough Definitions and Speculations for Echo-Class Models. *Page 309 of:* Cowan, G. A., Pines, D., & Meltzer, D. (eds), *Complexity: Metaphors, Models, and Reality.*
Holland, J.H. (1992). *Adaptation in Natural and Artificial Systems*, (2nd.ed). MIT Press, Cambridge, Mass.
Kauffman, S. A. (1993). *The Origins of Order*. Oxford University Press, Oxford.

Kimura, M. (1983). *The Neutral Theory of Molecular Evolution.* Cambridge University Press, Cambridge.

Langton, C. G. (1992). Life at the Edge of Chaos. *Page 41 of:* Langton, C. G., Taylor, C., Farmer, J. D., & Rasmussen, S. (eds), *Artificial Life II.* Addison Wesley, Reading, Mass.

Langton, C. G. (ed). (1994). *Artificial Life III.* Addison-Wesley, Reading, Mass.

Langton, C. G., Taylor, C., Farmer, J. D., & Rasmussen, S. (eds). (1992). *Artificial Life II.* Addison-Wesley, Reading, Mass.

Langton, C.G. (1989a). Artificial Life. *Page 41 of:* Langton, C.G. (ed), *Artificial Life.* Addison Wesley, Reading, Mass.

Langton, C.G. (ed). (1989b). *Artificial Life.* Addison-Wesley, Reading, MA.

Lauwerier, H. (1991). *Fractals.* Princeton Univ. Press, Princeton.

Lem, S. (1977). *Summa Technologiae.* Nolit, Beograd.

Li, W., Packard, N. H., & Langton, C. G. (1990). Transition phenomena in cellular automata rule space. *Physica D*, **45**, 77.

Maturana, H., & Varela, F. (1980). *Autopoiesis and Cognition.* D. Riedel Publish. Co., London.

Maynard Smith, J. (1989). *Evolutionary Genetics.* Oxford University Press, Oxford.

Maynard Smith, J. (1994). The Major Transitions in Evolution. *Page 457 of:* Cowan, G. A., Pines, D., & Meltzer, D. (eds), *Complexity: Metaphors, Models, and Reality.*

McMullin, F. V. (1992). *Artificial Knowledge: An Evolutionary Approach.* Ph.D. thesis, The National University of Ireland, Dublin.

Monod, J. (1972). *Chance and Necessity.* Vintage Books, New York.

Penrose, R. (1989). *The Emperors's New Mind.* Vintage, London.

Prusinkiewicz, P. (1994). Visual models of morphogenesis. *Artificial Life*, **1**, 61–74.

Rasmussen, S., Knudsen, C., & Feldberg, R. (1992). Dynamics of Programmable Matter. *Page 211 of:* Langton, C. G., Taylor, C., Farmer, J. D., & Rasmussen, S. (eds), *Artificial Life II.* Addison Wesley, Reading, Mass.

Ray, T. (1992). An Approach to the Synthesis of Life. *Page 371 of:* Langton, C. G., Taylor, C., Farmer, J. D., & Rasmussen, S. (eds), *Artificial Life II.* Addison Wesley, Reading, Mass.

Ray, T. (1994). An evolutionary approach to synthetic biology: Zen and the art of creating life. *Artificial Life*, **1**, 179–209.

Schuster, P. (1994). Extended molecular biology. *Artificial Life*, **1**, 39–60.

Sigmund, K. (1993). *Games of Life.* Oxford University Press, Oxford.

Simon, H. A. (1992). *The Sciences of the Artificial*, 2nd ed. MIT Press, Cambridge, Mass.

Sims, K. (1991). Artificial evolution for computer graphics. *Computer Graphics*, **25**, 319.

Taylor, C. E. (1992). Fleshing Out Artificial Life. *Page 25 of:* Langton, C. G., Taylor, C., Farmer, J. D., & Rasmussen, S. (eds), *Artificial Life II.* Addison Wesley, Reading, Mass.

Toffoli, T., & Margolus, M. (1987). *The Cellular Automata Machines.* MIT Press, Cambridge, Mass.

Waldrop, M. M. (1993). *Complexity, The Emerging Science at the Edge of Order and Chaos.* Viking, London.

Wolfram, S. (1986). *Theory and Applications of Cellular Automata.* World Scientific, Singapore.

Wootters, W.T., & Langton, C.G. (1990). Is there a sharp phase transition for deterministic cellular automata? *Physica D*, **45**, 95.

5
Deterministic and random fractals

JOHN E. HUTCHINSON

Department of Mathematics,
School of Mathematical Sciences,
Australian National University,
Canberra, ACT 0200, AUSTRALIA

5.1 Introduction

In this chapter I will discuss a mathematical framework for deterministic and random (or non-deterministic) fractals. The approach we take is via *scaling laws* and *scaling operators*, the latter also being known as *iterated function systems*. This gives a very attractive theory with many applications, particularly to computer graphics and to image and data compression. Various applications of fractals are discussed elsewhere in this book.

The essential ideas will first be presented by means of a number of standard examples. Readers who wish to obtain a relatively informal overview of the material can accordingly restrict themselves to §5.2 and §5.3, and to the less formal parts of the later sections. §5.4, §5.5 and §5.6 develop much of the mathematics behind these ideas. I have tried to keep the mathematics self-contained, and in particular have attempted to motivate and develop from first principles the relevant notions of metric spaces, measure theory, and probability theory. The later sections may perhaps serve as a brief introduction to some aspects of these subjects.

It is perhaps worth mentioning here one point that sometimes causes confusion. A 'mathematical' fractal in a certain precise sense looks the same at all scales; i.e. when examined under a microscope *at no matter what magnification* it will appear similar to the original object. On the other hand a 'physical' fractal will display this 'self-similarity' for only a range of magnifications or scales. The mathematical object will of course only be an accurate model within this particular range.

Examples of non-integer dimensional sets with scaling properties have long been known to mathematicians. It was Mandelbrot who introduced the term *fractal* and who in a series of papers and books (see (Mandelbrot, 1982) and the references there) developed the connections between these ideas and a range of phenomena in the physical, biological and social sciences.

In Hutchinson (1981) we showed that to each scaling operator there corresponds a unique fractal set (or measure) in a natural manner. Approximations via the scaling operator, fractal coding, and dimension and density properties of fractals were also developed. The terminology 'iterated function system' was introduced later by Barnsley and Demko (1985). In earlier papers, Moran (1946) proved the dimension results for (fractal) sets satisfying an 'open set condition' and Williams (1971) developed properties of iterates of contraction maps.

Applications to computer graphics were considered in Diaconis and Shahshahani (1984) in Barnsley and Demko (1985) and in Barnsley *et al.* (1986). The Markov process approach to generating fractals has been developed by Diaconis and Shahshahani (1984), Barnsley and Demko (1985), and others. See also Kolata (1984) for a popular view.

The results for random fractal sets are due to Falconer (1986), Graf (1987), and Mauldin and Williams (1986), and for random fractal measures are due to Mauldin and Williams (1986) and Arbeiter (1991). Zähle (1988) and Patzschke and Zähle (1990) developed an 'axiomatic' approach to random fractals and also established connections with the ideas developed by Falconer, Graf, Mauldin and Williams and Arbeiter. In Hutchinson and Rüschendorf (1998b) we develop a new approach, see also Hutchinson and Rüschendorf (1998a).

General references at an introductory level are Barnsley (1988), Falconer (1986, 1985), Peitgen *et al.* (1991), and Bélair and Duboc (1991) (in particular, the article by Vrscray).

5.2 Some important examples

We begin with an informal discussion of some of the main ideas. Our intention is to develop the reader's intuition. The relevant notions will be defined later in a more precise manner.

5.2.1 The Koch curve

A *fractal set* is a set K in $\mathbb{R}^n$† with certain *scaling properties*. The *Koch curve* or fractal is an example. Fig. 5.1 shows certain sets in a sequence $K^{(1)}, K^{(2)}, \ldots, K^{(j)}, \ldots$ which approximates the Koch curve K. The set K is the limit of this sequence, but in practice we can only draw an approximation.

From fig. 5.2,

$$K = K_1 \cup K_2 \cup K_3 \cup K_4. \tag{5.1}$$

The important point here is that each K_i is congruent to a scaled version of K. It is also clear from the diagram that the scaling factor is 1/3. More precisely, and significantly from our point of view, there are maps

$$S_i : \mathbb{R}^2 \to \mathbb{R}^2 \quad i = 1, \ldots, 4,$$

† $\mathbb{R}^n$ is n-dimensional Euclidean space. The important cases are the line $\mathbb{R} = \mathbb{R}^1$, the plane $\mathbb{R}^2$, and three-dimensional space $\mathbb{R}^3$.

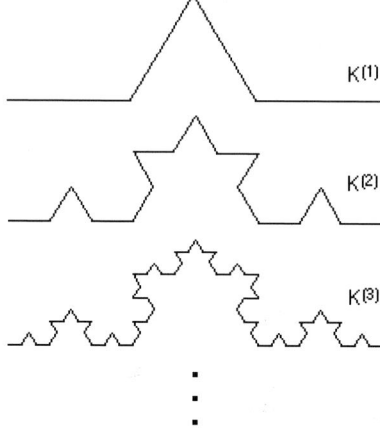

Fig. 5.1. Approximating the Koch fractal.

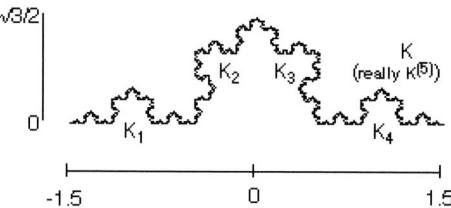

Fig. 5.2. Symmetry in the Koch fractal.

where each S_i is a composition of a translation, rotation, and scaling (with scaling factor $1/3$), such that (5.1) can be written

$$K = S_1(K) \cup S_2(K) \cup S_3(K) \cup S_4(K). \tag{5.2}$$

If we look again at the Koch curve in fig. 5.2, we see that we can also write it as the union of *two* sets (the left and right hand sides), each of which is obtained from K by means of a translation, rotation, *reflection* and scaling (in this case by $1/\sqrt{3}$, as is easy to calculate). Denoting the corresponding maps by T_i we have

$$K = T_1(K) \cup T_2(K). \tag{5.3}$$

Deterministic and random fractals 131

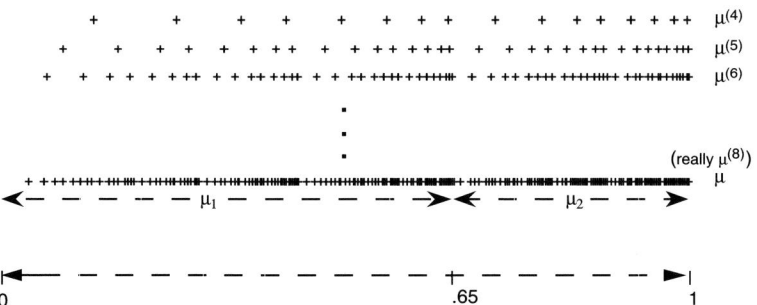

Fig. 5.3. Fractal measures.

5.2.2 A fractal measure on the line and the Koch measure

You should think of a *measure* μ in $\mathbb{R}^n$ as a 'mass-distribution' and the measure $\mu(E)$ of a subset E of $\mathbb{R}^n$ as the total mass in that set. Simple examples are a unit mass concentrated at a point in $\mathbb{R}^n$ or a unit mass uniformly distributed along a curve in $\mathbb{R}^n$ having finite length. The total mass (i.e. the measure of $\mathbb{R}^n$) is called the *mass* of the measure.

A *fractal measure* μ is a measure in $\mathbb{R}^n$ with certain *scaling properties*. In fig. 5.3 we have sketched various members in a sequence of approximations $\mu^{(1)}, \mu^{(2)}, \ldots, \mu^{(j)}, \ldots$ to a fractal measure μ on the interval $I = [0,1] \subset \mathbb{R}$ (as noted in the previous footnote, $\mathbb{R}$ is the real line). Each cross represents a point mass of a certain magnitude. The number of point masses describing $\mu^{(j)}$ is denoted by $N^{(j)}$ and each such point mass is given equal mass (or magnitude) $1/N^{(j)}$. If E is a set then the measure $\mu^{(j)}(E)$ of E is defined by

$$\begin{aligned}\mu^{(j)}(E) &= \text{total mass of all } j\text{th-level point masses in } E \\ &= (\text{total number of } j\text{th-level point masses in } E) \, / \, N^{(j)}\end{aligned}$$

(notice that $\mu^{(j)}$ has unit mass for every j). If j is large, then $\mu^{(j)}(E)$ is a very good approximation to $\mu(E)$ (at least for any 'reasonable' set E and in particular for any interval).

In this example we see from the diagram that

$$\mu = \mu_1 + \mu_2, \tag{5.4}$$

where μ_1 is 'supported' (cf. §5.5.1) on the interval $[0, 0.65]$ and μ_2 is 'supported' on $[0.65, 1]$. The important point is that μ_1 can be obtained from μ by first 'rescaling' by the factor 0.65 and then 'reweighting' by the factor 0.5. A similar remark applies to μ_2 except that the rescaling factor is 0.35 and the reweighting factor is again 0.5. More precisely, and again significantly from our point of view, there are linear maps

$$S_i : \mathbb{R} \to \mathbb{R} \quad i = 1, 2 ,$$

where each S_i is a composition of a translation and scaling (with scaling factors 0.65 and 0.35), and weighting factors

$$\rho_1, \ \rho_2$$

(here both equal to 0.5) such that (5.4) can be written

$$\mu = \rho_1 S_1(\mu) + \rho_2 S_2(\mu). \ \dagger \qquad (5.5)$$

In the present example the *support*‡ of the fractal measure is the entire interval $[0, 1]$. In many cases the support will be a more interesting fractal set. A fractal measure carries more information than its supporting set. *We can often usefully think of a fractal measure as a fractal set together with a grey scale, or weighting, at each point of the set.*

Another fractal measure, the *Koch measure*, can be constructed in a manner similar to the construction of the Koch curve. Thus (see fig. 5.1) we take $\mu^{(1)}$ to be a measure uniformly distributed along $K^{(1)}$ and with total mass one, so that the mass of each of the four line segments is $1/4$. Similarly $\mu^{(2)}$ is uniformly distributed along $K^{(2)}$ again with total mass one, so that the mass of each of the 16 line segments is now $1/16$. Likewise for $\mu^{(j)}$ with $j > 2$. Then μ is the limit, in a natural sense that can be made precise, of the sequence $\mu^{(1)}, \mu^{(2)}, \ldots, \mu^{(j)}, \ldots$. In this case

$$\mu = \rho_1 S_1(\mu) + \rho_2 S_2(\mu) + \rho_3 S_3(\mu) + \rho_4 S_4(\mu) \qquad (5.6)$$

where each S_i is a composition of a translation, rotation and scaling (with scaling factor $1/3$) and the scaling factors ρ_i all equal $1/4$.

A fractal set or measure is often simply called a *fractal*. We also sometimes instead use the terminology *deterministic fractal* (*set, measure*) to distinguish the present notions from the *random* (or *statistical* or *non-deterministic*) versions which follow.

† Think of $S_1(\mu)$ as the measure (mass-distribution) obtained by 'pushing' μ forward with the map S_1. Similarly for $S_2(\mu)$. The measures $S_1(\mu)$ and $S_2(\mu)$ both have total mass one, as does μ, and so if we want equality to hold in (5.5) it is necessary to reweight by factors ρ_1 and ρ_2 whose sum is one.
‡ See §5.5.1 for the definition.

Deterministic and random fractals 133

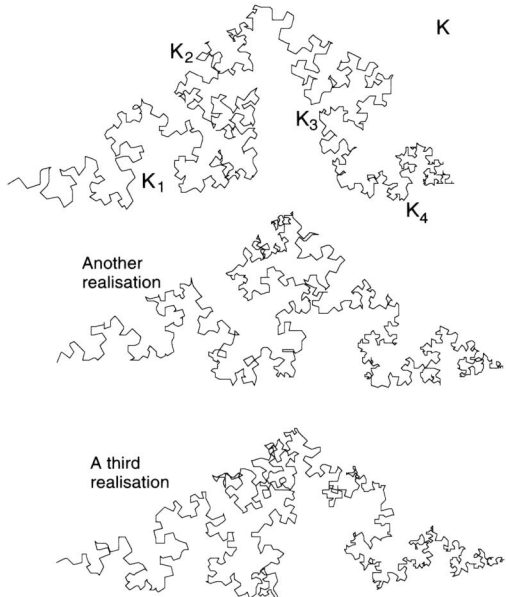

Fig. 5.4. A random Koch fractal.

5.2.3 The random Koch curve

A *random fractal set* can be thought of as a set generated according to some probability distribution $\mathcal{K}$ on sets, analogous to the manner in which a random number is generated by a probability distribution on numbers. The probability distribution $\mathcal{K}$ will be required to have a certain type of *scaling property*. When we sketch a random fractal set we are in fact sketching a particular realisation of $\mathcal{K}$.

What is most important is the probability distribution $\mathcal{K}$ itself, rather than any particular realisation. In fact, it is more precise to call $\mathcal{K}$ the random fractal set, and to distinguish this from particular *realisations* of $\mathcal{K}$.†

In fig. 5.4 we give three realisations of the same random fractal, the *random Koch curve*.

---

† It is perhaps worth remarking that a particular realisation of a random number gives little, or no, idea of the underlying probability distribution. In contrast, because of scaling properties, even a single realisation of a random fractal gives a lot of information about the underlying distribution.

For each realisation K of $\mathcal{K}$ we have
$$K = K_1 \cup K_2 \cup K_3 \cup K_4,$$
where each K_i looks, 'statistically' or 'on the average', like a rescaled version of K.

More precisely, and analogously to the deterministic case, we write
$$K \simeq S_1(K^{(1)}) \cup S_2(K^{(2)}) \cup S_3(K^{(3)}) \cup S_4(K^{(4)}). \tag{5.7}$$

Here $K^{(1)}, \ldots, K^{(4)}$ are chosen independently and at random via $\mathcal{K}$† and the single 4-tuple of maps (S_1, S_2, S_3, S_4) is chosen independently of the $K^{(i)}$ and at random via some probability distribution $\mathcal{S}$. Then the previous equation says that the probability distribution $\mathcal{K}$ on compact sets K is the same as the probability distribution on compact sets given by the right side of (5.7). In other words, (5.7) indicates equality in the sense of probability distributions.

Other physical examples are given by *Brownian motion*, the irregular oscillatory movement of microscopic particles in a limpid fluid.

5.2.4 The random Koch measure

A *random fractal measure* is a probability distribution $\mathcal{M}$ on measures which has a certain type of *scaling property*. An example can be constructed parallel to the construction of the random Koch curve. Imagine a measure of unit mass distributed 'uniformly' along the various realisations of the Koch curve in fig. 5.4. This will give three realisations μ of the same *random Koch measure* $\mathcal{M}$.

The relevant property is
$$\mu \simeq \rho_1 S_1(\mu^{(1)}) + \rho_2 S_2(\mu^{(2)}) + \rho_3 S_3(\mu^{(3)}) + \rho_4 S_4(\mu^{(4)}). \tag{5.8}$$

Here $\mu^{(1)}, \ldots, \mu^{(4)}$ are chosen independently and at random via $\mathcal{M}$ and the single 8-tuple of maps and weights $\big((S_1, \ldots, S_4), (\rho_1, \ldots, \rho_4)\big)$ is chosen independently of the $\mu^{(i)}$ via some probability distribution $\mathcal{S}$. The previous equation indicates equality in the sense of probability distributions.

5.3 Discussion of main properties

We discuss some of the main properties of fractals as they apply to the examples from the previous section.

† Do not confuse $K^{(1)}, \ldots, K^{(4)}$ with the $K_1, \ldots, K_4$ in fig. 5.4.

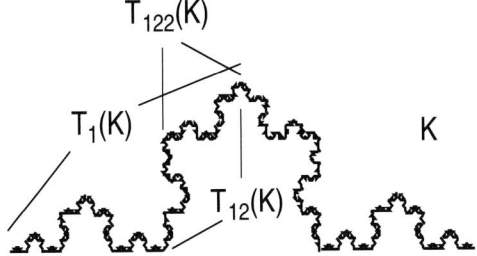

Fig. 5.5. Scaling for the Koch fractal.

5.3.1 Scaling laws and coding

From (5.3) we see that K is the union of two sets, each of which is a scaled version of K itself (fig. 5.5). We can interpret (5.3) as saying that K satisfies the *scaling law* (determined by) $\mathbf{T} = (T_1, T_2)$. Similarly, (5.2) can be interpreted as saying that K satisfies the scaling law $\mathbf{S}$ where $\mathbf{S} = (S_1, S_2, S_3, S_4)$.

If we replace each occurrence of K on the right side of (5.3) by $T_1(K) \cup T_2(K)$ we obtain

$$\begin{aligned} K &= T_1\Big(T_1(K) \cup T_2(K)\Big) \cup T_2\Big(T_1(K) \cup T_2(K)\Big) \\ &= T_{11}(K) \cup T_{12}(K) \cup T_{21}(K) \cup T_{22}(K), \end{aligned} \quad (5.9)$$

where $T_{ij} = T_i \circ T_j$. Then (5.9) is in fact just the decomposition (5.2), even in the same order. Repeating the construction, we obtain

$$K = \bigcup_{1 \leq \sigma_1, \sigma_2, \sigma_3 \leq 2} T_{\sigma_1 \sigma_2 \sigma_3}(K),$$

where $T_{\sigma_1 \sigma_2 \sigma_3} = T_{\sigma_1} \circ T_{\sigma_2} \circ T_{\sigma_3}$. Thus K is the union of eight sets, each of which is a rescaled version of K, the rescaling factor being $(1/\sqrt{3})^3$.

The previous construction can be iterated k times, writing K as the union of 2^k subsets each of which is a rescaled version of K with rescaling factor $(1/\sqrt{3})^k$:

$$K = \bigcup_{1 \leq \sigma_1, \ldots, \sigma_k \leq 2} T_{\sigma_1 \ldots \sigma_k}(K), \quad (5.10)$$

where $T_{\sigma_1 \ldots \sigma_k} = T_{\sigma_1} \circ \cdots \circ T_{\sigma_k}$. For each infinite sequence $\sigma = \sigma_1, \sigma_2, \ldots, \sigma_k, \ldots$ the corresponding sequence

$$K \supseteq T_{\sigma_1}(K) \supseteq T_{\sigma_1 \sigma_2}(K) \supseteq \cdots \supseteq T_{\sigma_1 \ldots \sigma_k}(K) \supseteq \cdots \quad (5.11)$$

is a decreasing sequence of sets whose intersection is a single point. This point is assigned the *code* or *address* $\sigma = \sigma_1\sigma_2\ldots\sigma_k\ldots$. Every point in K has such a representation.

Similar remarks apply to fractal measures. Thus equation (5.5) can be interpreted as saying that μ satisfies the *scaling law* **S** where **S** $= \Big((S_1, S_2), (\rho_1, \rho_2)\Big)$.

We can iterate (5.5) and obtain

$$\begin{aligned}\mu &= \rho_1 S_1\Big(\rho_1 S_1(\mu) + \rho_2 S_2(\mu)\Big) + \rho_2 S_2\Big(\rho_1 S_1(\mu) + \rho_2 S_2(\mu)\Big) \\ &= \rho_1{}^2 S_{11}(\mu) + \rho_1\rho_2 S_{12}(\mu) + \rho_2\rho_1 S_{21}(\mu) + \rho_2{}^2 S_{22}(\mu).\end{aligned}$$

Repeating this construction, we obtain

$$\begin{aligned}\mu &= \sum_{1 \le \sigma_1, \sigma_2, \sigma_3 \le 2} \rho_{\sigma_1\sigma_2\sigma_3} S_{\sigma_1\sigma_2\sigma_3}(\mu) \\ &= \sum_{1 \le \sigma_1, \ldots, \sigma_k \le 2} \rho_{\sigma_1\ldots\sigma_k} S_{\sigma_1\ldots\sigma_k}(\mu),\end{aligned}$$

where $\rho_{\sigma_1\ldots\sigma_k} = \rho_{\sigma_1} \cdot \ldots \cdot \rho_{\sigma_k}$ and $S_{\sigma_1\ldots\sigma_k} = S_{\sigma_1} \circ \cdots \circ S_{\sigma_k}$.

Analogous ideas apply in the random case. Thus in §§5.2.3 and 5.2.4 the probability distributions $\mathcal{S}$ on the (S_1, S_2, S_3, S_4) and on the $\Big((S_1, \ldots, S_4), (\rho_1, \ldots, \rho_4)\Big)$, can be considered as *scaling laws* satisfied by the random Koch curve and the random Koch measure respectively.

5.3.2 Existence and uniqueness of fractals

A basic fact is that the scaling law satisfied by the Koch curve in fact characterises the Koch curve. This is a consequence of a general result which has had interesting applications and which we now discuss.

We first need a little notation. A map $S: \mathbb{R}^n \to \mathbb{R}^n$ is said to be a *contraction map* if there exists some r satisfying $0 \le r < 1$ such that

$$|S(x_1) - S(x_2)| \le r|x_1 - x_2| \quad \text{for all } x_1, x_2 \in \mathbb{R}^n. \tag{5.12}$$

The number r is called a *contraction ratio* or *Lipschitz constant* for S.† The maps S_i and T_i in §5.2 are all contraction maps of a particularly simple kind in that *equality* holds in (5.12) with the appropriate choice of r.

---

† Thus the distance between the *images* of any two points in space is less than the distance between the two points by a factor $r < 1$, where r is independent of the two chosen points.

If $\mathbf{S} = (S_1, \ldots, S_N)$ (where $N \geq 2$) is an N-tuple of contraction maps as above and K is a compact subset (cf. §5.4.2) of $\mathbb{R}^n$, we say that K satisfies the scaling law $\mathbf{S}$ if

$$K = S_1(K) \cup \ldots \cup S_N(K). \tag{5.13}$$

Then we have the rather surprising result (theorem 5.2) that *to each scaling law there exists exactly one compact set (fractal) satisfying that law*. Note however that different scaling laws may give the same fractal. For example, both $\mathbf{S} = (S_1, S_2, S_3, S_4)$ and $\mathbf{T} = (T_1, T_2)$ give the Koch curve; see § 5.2.1.

Analogous results apply to scaling laws for fractal measures, and to scaling laws for random fractal sets and random fractal measures, cf. theorems 5.6, 5.10 and 5.14.

5.3.3 Approximating fractals

5.3.3.1 Deterministic approximations

Corresponding to the scaling law $\mathbf{S} = (S_1, \ldots, S_N)$ there is a *scaling operator*, also denoted by $\mathbf{S}$, such that for any compact set A the compact set $\mathbf{S}(A)$ is defined by

$$\mathbf{S}(A) = S_1(A) \cup \ldots \cup S_N(A).$$

Thus the fact (5.13) that K satisfies the scaling law $\mathbf{S}$ can be written

$$K = \mathbf{S}(K).$$

Beginning with any compact set A, sometimes called a *seed*, the scaling operator $\mathbf{S}$ can be iterated to obtain a sequence of sets

$$\mathbf{S}^1(A) = \mathbf{S}(A), \ \mathbf{S}^2(A) = \mathbf{S}(\mathbf{S}(A)), \ \mathbf{S}^3(A) = \mathbf{S}(\mathbf{S}(\mathbf{S}(A))), \ldots \tag{5.14}$$

then (theorem 5.2) the sequence $\mathbf{S}^1(A), \mathbf{S}^2(A), \ldots, \mathbf{S}^k(A), \ldots$ *converges to the unique compact set corresponding to the scaling law* $\mathbf{S}$. This is a constructive procedure that allows one to construct the fractal corresponding to the given scaling law to within any prescribed degree of accuracy.

In fig. 5.1 the sets $K^{(k)}$ equal $\mathbf{S}^k(A)$, where A is a horizontal line segment.

In fig. 5.6 another sequence of approximations is shown to the Koch curve. In this case we use the scaling operator $\mathbf{T} = (T_1, T_2)$ as in (5.3). Moreover, A contains a single point x and so $\mathbf{T}^k(A)$ consists of all points of the form $T_{\sigma_1 \ldots \sigma_k}(x)$.

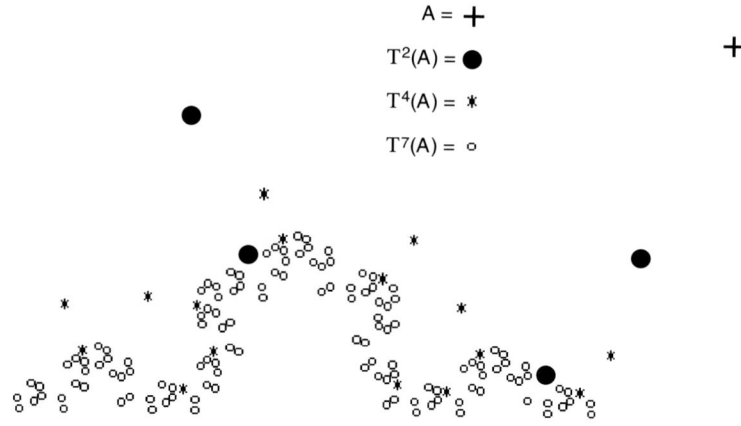

Fig. 5.6. Sequence of approximations to the Koch curve.

Analogous results apply to fractal measures. In this case a *scaling law* is a 2N-tuple

$$\mathbf{S} = \big((S_1,\ldots,S_N),(\rho_1,\ldots,\rho_N)\big)$$

where $N \geq 2$ is an integer, $S_1,\ldots,S_N \colon \mathbb{R}^n \to \mathbb{R}^n$ are contraction maps, and $\rho_1,\ldots,\rho_N$ are positive real numbers such that $\rho_1 + \cdots + \rho_N = 1$. Then we say μ satisfies the scaling law $\mathbf{S}$ if

$$\mu = \rho_1 S_1(\mu) + \cdots + \rho_N S_N(\mu). \tag{5.15}$$

The result (theorem 5.6) is that *there is exactly one unit mass fractal which satisfies the scaling law* $\mathbf{S}$.

There is a corresponding *scaling operator* also denoted by $\mathbf{S}$ and defined by

$$\mathbf{S}(\nu) = \rho_1 S_1(\nu) + \cdots + \rho_N S_N(\nu),$$

where ν is any compactly supported unit mass measure in $\mathbb{R}^n$. Thus the fact (5.15) that μ satisfies the scaling law $\mathbf{S}$ can be conveniently written

$$\mu = \mathbf{S}(\mu).$$

Beginning from a (finite mass) measure ν we can iterate the scaling operator $\mathbf{S}$ to obtain a sequence of measures

$$\mathbf{S}^1(\nu) = \mathbf{S}(\nu),\ \mathbf{S}^2(\nu) = \mathbf{S}(\mathbf{S}(\nu)),\ \mathbf{S}^3(\nu) = \mathbf{S}(\mathbf{S}(\mathbf{S}(\nu))),\ldots. \tag{5.16}$$

Then (theorem 5.6 again) *for any unit mass measure ν the sequence*

$S^1(\nu)$, $S^2(\nu)$, ..., $S^k(\nu)$, ... *converges to the unique unit mass measure corresponding to the scaling law* **S**.

Once again, analogous results to the preceding apply to the random case.

5.3.3.2 Random approximations

In (5.6) we saw that the Koch curve satisfies

$$\mu = \frac{1}{4}S_1(\mu) + \frac{1}{4}S_2(\mu) + \frac{1}{4}S_3(\mu) + \frac{1}{4}S_4(\mu),$$

where the S_i are certain maps $S_i : \mathbb{R}^2 \to \mathbb{R}^2$.

This leads to the following procedure, a *random* method to construct the *deterministic* Koch curve and measure.

(0) Begin with an arbitrary point (*seed*) $x_0 \in \mathbb{R}^2$.
(1) Choose S_{σ_1} from $(S_1, \ldots, S_4)$, where the probability of choice of any particular S_i is $1/4$. Let $x_1 = S_{\sigma_1}(x_0)$.
(2) Choose S_{σ_2} from $(S_1, \ldots, S_4)$ independently of S_{σ_1}, where again the probability of choosing any particular S_i is $1/4$. Let $x_2 = S_{\sigma_2}(x_1)$.

...

(k) Choose S_{σ_k} similarly and independently of the $S_{\sigma_1}, \ldots, S_{\sigma_{k-1}}$. Let $x_k = S_{\sigma_k}(x_{k-1})$.

...

In this way an *orbit* of points

$$x_0, \; x_1 = S_{\sigma_1}(x_0), \; x_2 = S_{\sigma_2}(x_1), \ldots, \; x_k = S_{\sigma_k}(x_{k-1}), \ldots$$

is constructed. Then with probability one (cf. §5.6.1) the orbit will come arbitrarily close to every point in the Koch curve K. Moreover, see Elton (1987), if μ is the Koch measure then the measure $\mu(B)$ of any subset $B \subseteq \mathbb{R}^n$ is given by

$$\mu(B) = \lim_{k \to \infty} \frac{\mathcal{N}(k,B)}{k+1}, \dagger$$

where

$$\mathcal{N}(k,B) = \text{number of points in } \{x_0, x_1, x_2, \ldots, x_k\} \cap B.$$

Thus the *relative visitation frequency* of the finite orbit $x_0, x_1, x_2, \ldots, x_k$ in the set B is a good approximation to $\mu(B)$ for k large.

---

† Technical condition: one needs to assume that $\mu(\partial B) = 0$, where ∂B is the boundary of B in the usual topological sense.

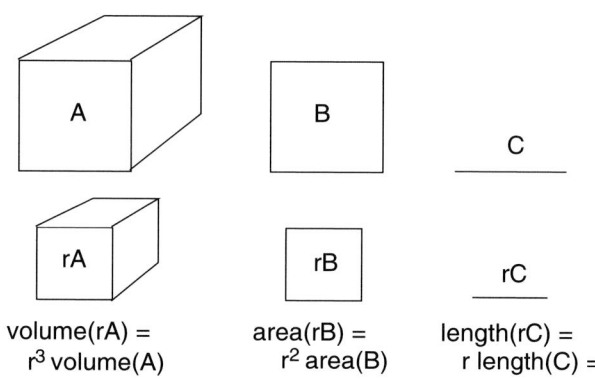

Fig. 5.7. Scaling properties.

In computer simulations using either the algorithm here or from the previous section, as successive points are plotted on the screen, increasingly accurate approximations to the Koch curve will be constructed. The density of points will indicate the distribution of mass according to the Koch measure. The Koch measure, as we have defined it, is in a certain sense distributed uniformly along the Koch curve, but in other cases such as the fractal measure on the line in §5.2.2, the mass is not uniformly distributed. The density of points constructed via appropriate modifications of either algorithm will closely approximate the mass distribution of the corresponding fractal measure.

5.3.4 Dimension of fractals

If S satisfies (5.12) with '$\leq$' there replaced by '$=$', we say S is a *similitude* with scaling factor r. There is a standard classical notion of 'D-dimensional' volume, $D\text{-vol}(E)$, for certain sets $E \subset \mathbb{R}^n$. For $D = 1, 2, 3$ this gives the length, area, and usual volume, respectively. A standard property is

$$D\text{-vol}(S(E)) = r^D D\text{-vol}(E)$$

if S is a similitude with scaling factor r (see fig. 5.7).

Now suppose

$$E = S_1(E) \cup \ldots \cup S_N(E)$$

where $S_1, \ldots, S_N$ are similitudes with scaling factors $r_1, \ldots, r_N$, and assume the $S_i(E)$ intersect each other on sets of lower dimension than D.† Then

$$D\text{-vol}(E) = r_1{}^D\, D\text{-vol}(E) + \cdots + r_N{}^D D\text{-vol}(E).$$

Thus we obtain the relation

$$r_1{}^D + \cdots + r_N{}^D = 1, \tag{5.17}$$

assuming $0 < D\text{-vol}(E) < \infty$.

Motivated by this, from (5.2) we expect that the 'dimension' D of the Koch curve K should satisfy

$$4\left(\frac{1}{3}\right)^D = 1. \tag{5.18}$$

This gives $D = \log 4/\log 3 = 1.2615\ldots$ for the dimension of the Koch curve. If we use (5.3) we obtain

$$2\left(\frac{1}{\sqrt{3}}\right)^D = 1,$$

which leads to the same value for D.

There is a notion of dimension, the *Hausdorff dimension*, which assigns to any subset of $\mathbb{R}^n$ a real number in the range $[0, n]$. Then (theorem 5.4) *if the scaling property* **S** *satisfies a certain 'Open Set Condition', the corresponding fractal* K *will have Hausdorff dimension* D *given by* (5.17)†. In particular, the Hausdorff dimension of the Koch curve is indeed $\log 4/\log 3$. Analogous results hold for random fractal sets (§5.6).

Note that the dimension of the Koch curve is greater than one, and in particular the Koch curve will have 'infinite length' in the sense that its one dimensional measure is infinite (cf. §5.4.6). Any realisation (more precisely, with probability one as discussed in §5.6) of the random Koch curve will also have dimension greater than one.

A coastline can often be modelled by a kind of random fractal, cf. (Mandelbrot, 1982). The mathematical model will have dimension greater than one and hence infinite length, so in this sense we say that the coastline itself has dimension greater than one and has infinite length.

† More precisely, the D-dimensional volume of the intersections is zero. In particular the intersections may be empty.
† Note that if we define $f(x) = r_1{}^x + \cdots + r_N{}^x$ then $f(0) = N$, $f(x) \to 0$ as $x \to \infty$, and $f(x)$ is strictly decreasing for $x \in \mathbb{R}$. It follows that there is indeed a unique value of D for which (5.17) is satisfied.

5.4 Fractal sets

In this and subsequent sections we develop the important ideas involved in the proofs of the main properties of fractals as discussed in §5.3. The mathematical level is accordingly a little higher, but we have attempted to keep the details to a minimum and hope that the ideas will be clear even to those to whom much of the background is new.

5.4.1 Contraction maps

Euclidean n-space is denoted by $\mathbb{R}^n$. We have already introduced the notion of a *contraction map* on $\mathbb{R}^n$ and the corresponding *contraction ratio*. Recall also that S is a *similitude* with scaling factor r if S satisfies (5.12) with '$\leq$' there replaced by '$=$'.

5.4.2 Metric spaces

A *metric space* (X, d) is a set X together with a notion of *distance* $d(a, b)$ between any two members $a, b \in X$ which satisfies the following properties:

(i) $d(a, b) \geq 0$ and $d(a, b) = 0$ precisely when $a = b$,
(ii) $d(a, b) \leq d(a, c) + d(c, b)$,
(iii) $d(a, b) = d(b, a)$.

One also calls d a *metric*.

A simple example of a metric space is where $X = \mathbb{R}^n$ (e.g. $\mathbb{R}^2$) and $d(a, b)$ is the usual distance between the two points a and b. Another example is where $X = D$ is the *unit disc* consisting of points in the plane whose distance from the origin is *strictly* less than one, and $d(a, b)$ is again the usual distance between the points a and b (see the next diagram). We will soon meet some other very important, but more complicated examples.

A metric space (X, d) is *complete* if whenever a sequence $x_1, x_2, \ldots, x_k, \ldots$ of members of X has the property that

$$d(x_j, x_k) \to 0 \quad \text{as} \quad j, k \to \infty$$

then $d(x_k, x) \to 0$ for some $x \in X$. In this case one says x_k *converges to* x and writes $x_k \to x$ as $k \to \infty$.

The metric space $(\mathbb{R}^n, d)$ is complete but the metric space (D, d) discussed above is not complete. In fig. 5.8 we show the sequence of points

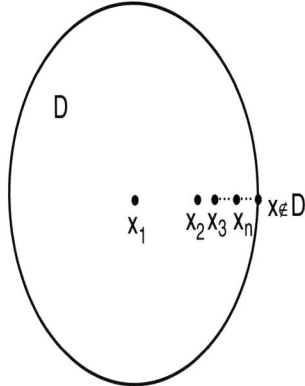

Fig. 5.8. Sequence of points with limit outside D.

$x_k = (1 - 1/k, 0)$. We see that

$$d(x_j, x_k) = \left| \frac{1}{j} - \frac{1}{k} \right| \to 0 \quad \text{as} \quad j, k \to \infty.$$

But the limit $x = (1, 0)$ of this sequence is not in D (it is on the boundary of D).

As in (5.12), a map $F: X \to X$ is said to be a *contraction map* if there exists some $r \in (0, 1)$ such that

$$d(F(x_1), F(x_2)) \leq r d(x_1, x_2) \quad \text{for all } x_1, x_2 \in X.$$

The *contraction mapping principle* asserts that: *if (X, d) is a <u>complete metric space</u> and $F: X \to X$ is a contraction map, then there exists exactly one $a \in X$, called the <u>fixed point</u> of F, such that $F(a) = a$. Moreover, for any $x \in X$, the sequence*

$$x, \; F^1(x) = F(x), \; F^2(x) = F(F(x)), F^3(x) = F(F(F(x))), \ldots$$

converges to a. Also, if $d(x, F(x)) \leq \varepsilon$ and r is a contraction ratio for F, then $d(x, a) \leq \varepsilon/(1 - r)$.

The point to the last claim is that if x is near $F(x)$ then in fact x is near the fixed point a. The proof of this claim is easy, since each application of F decreases the distance between points by the factor r. More precisely,

$$\begin{aligned} d(x, F^k(x)) \\ \leq \; & d(x, F(x)) + d(F(x), F^2(x)) + \cdots + d(F^{k-1}(x), F^k(x)) \end{aligned}$$

$$\leq d(x, F(x))(1 + r + \cdots + r^{k-1})$$
$$\leq \varepsilon \frac{1-r^k}{1-r}.$$

Now let $k \to \infty$ to see that $d(x,a) \leq \varepsilon/(1-r)$.

5.4.3 The metric space $(\mathcal{C}, d_\mathcal{H})$ of compact subsets of $\mathbb{R}^n$

A set $A \subset \mathbb{R}^n$ is *closed* if whenever $x_1, x_2, \ldots, x_k, \ldots$ is a sequence of points from A and $x_k \to x$ for some $x \in \mathbb{R}^n$, then $x \in A$ (in other words, A contains all its limit points). A set $A \subset \mathbb{R}^n$ is *bounded* if the distance between any two points in A is less than some fixed finite number b (where b depends of course on A, but not on the particular points in A).

A *compact* subset of $\mathbb{R}^n$ is a closed and bounded subset. The set of all such non-empty† compact subsets is denoted by $\mathcal{C}$. There is an important notion of distance between two members of $\mathcal{C}$, called the *Hausdorff distance* or *Hausdorff metric* and denoted by $d_\mathcal{H}$.

Before introducing the Hausdorff distance we need some preliminary definitions, see fig. 5.9.

(i) If $x \in \mathbb{R}^n$ and $A \in \mathcal{C}$ we define the distance $d(x, A)$ between the *point* x and the *set* A (fig. 5.9) by

$$d(x, A) = \min\{\, d(x,a) : a \in A \,\},‡$$

where $d(x,a)$ is the usual distance between points.

(ii) For a set $A \in \mathcal{C}$ and a real number $\varepsilon \geq 0$ define the ε-*enlargement* by

$$A_\varepsilon = \{\, x : d(x, A) \leq \varepsilon \,\}.$$

Thus $A = A_0$, and $A_{\varepsilon_1} \subseteq A_{\varepsilon_2}$ if $\varepsilon_1 \leq \varepsilon_2$.

(iii) For two sets $A, B \in \mathcal{C}$ define the *Hausdorff distance* between A and B (fig. 5.10, fig. 5.11) by

$$d_\mathcal{H}(A, B) = \min\{\varepsilon : A \subseteq B_\varepsilon \text{ and } B \subseteq A_\varepsilon \,\}.§$$

Then one can show that $(\mathcal{C}, d_\mathcal{H})$ *is a complete metric space.*

---

† For technical reasons. If A is the empty set, $d(x, A)$ will not have a finite value; see (i) below.
‡ Technical aside: one usually writes 'inf' for infimum, instead of 'min' for minimum, but this is equivalent in the present setting.
§ As in the previous footnote, one could equivalently write 'inf' instead of 'min'.

Deterministic and random fractals

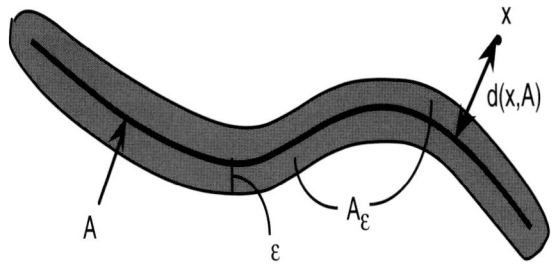

Fig. 5.9. Distance to a set.

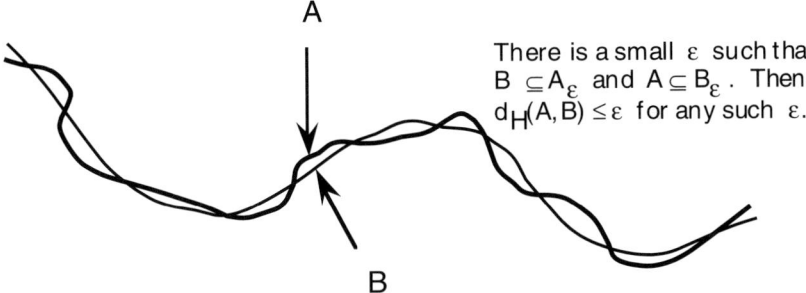

Fig. 5.10. Properties of the Hausdorff distance between sets.

Two important properties of the Hausdorff metric are

$$d_{\mathcal{H}}\left(\bigcup_{i=1}^{N} A_i, \bigcup_{i=1}^{N} B_i\right) \leq \max_{1 \leq i \leq N} d_{\mathcal{H}}(A_i, B_i); \quad (5.19)$$

$$d_{\mathcal{H}}(F(A), F(B)) \leq r d_{\mathcal{H}}(A, B), \quad (5.20)$$

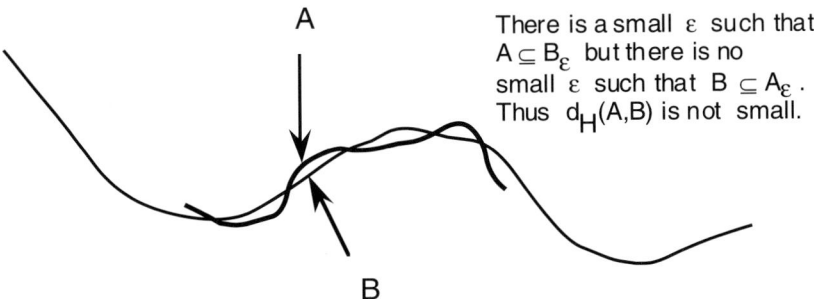

Fig. 5.11. Properties of the Hausdorff distance between sets.

where in the second inequality $F : \mathbb{R}^n \to \mathbb{R}^n$ is a contraction map with contraction ratio r.†

Finally, if $A \in \mathcal{C}$ then the *diameter* of A is defined by

$$\operatorname{diam}(A) = \max_{x,y \in A} d(x,y).$$

5.4.4 Existence, uniqueness and approximation

Definition 5.1 A *scaling law for sets* is an N-tuple **S** of contraction maps $(S_1, \ldots, S_N)$ defined on $\mathbb{R}^n$. The corresponding *scaling operator* **S** is defined on compact sets by $\mathbf{S}(A) = \bigcup_{i=1}^{N} S_i(A)$. We say K *satisfies the scaling law* **S** *if* $K = \mathbf{S}(K)$.

We can now prove the *existence and uniqueness result* for fractal *sets* discussed previously, see Hutchinson (1981, 3.2(1)).

Theorem 5.2 *There is a unique compact set K satisfying a given scaling law* **S**. *If A is any compact subset of $\mathbb{R}^n$ then $\mathbf{S}^k(A)$*† *converges to K in the Hausdorff metric as $k \to \infty$.*

Proof Let $\mathbf{S} = (S_1, \ldots, S_N)$ and let r be the maximum contraction ratio of the S_i's.

For any $A_1, A_2 \in \mathcal{C}$,

$$\begin{aligned} d_{\mathcal{H}}(\mathbf{S}(A_1), \mathbf{S}(A_2)) &= d_{\mathcal{H}}\left(\bigcup_{i=1}^{N} S_i(A_1), \bigcup_{i=1}^{N} S_i(A_2)\right) \\ &\leq \max_{1 \leq i \leq N} d_{\mathcal{H}}(S_i(A_1), S_i(A_2)) \\ &\leq r d_{\mathcal{H}}(A_1, A_2), \end{aligned}$$

by (5.19) and (5.20). It follows that the scaling operator $\mathbf{S} : \mathcal{C} \to \mathcal{C}$ is a contraction map.‡

The conclusion of the theorem now follows from the completeness of $(\mathcal{C}, d_{\mathcal{H}})$ and the contraction mapping principle. □

Remark: The set K satisfying the scaling law $\mathcal{S}$ is usually called a *fractal (set)*.

A simple corollary of the theorem is that if some set A satisfies

---
† One does not require $r \leq 1$ for this result.
† See (5.14) for notation.
‡ Do not confuse this with the fact that the S_i are contraction maps on $\mathbb{R}^n$.

$d_\mathcal{H}(A, \mathbf{S}(A)) \leq \varepsilon$ then $d_\mathcal{H}(A, K) \leq \varepsilon/(1-r)$, where K is the compact set satisfying the scaling law $\mathbf{S}$. This follows from the last part of the contraction mapping principle in §5.4.2. Barnsley calls this the *Collage Theorem* since it implies that if A can be covered by scaled copies of itself to within some small Hausdorff distance ε then there is a corresponding 'nearby' fractal K whose distance from A is at most $\varepsilon/(1-r)$.

The previous theorem also justifies the deterministic method for approximating fractals discussed in §5.3.3.1 in the context of the Koch curve.

5.4.5 Code space

Let K be the unique fractal set invariant under the scaling law $\mathbf{S} = (S_1, \ldots, S_N)$. Write

$$S_{\sigma_1 \ldots \sigma_k} = S_{\sigma_1} \circ \cdots \circ S_{\sigma_k}. \tag{5.21}$$

Then for any k, exactly as in (5.10) and (5.11), we can decompose K into 'smaller and smaller' pieces:

$$K = \bigcup_{1 \leq \sigma_1, \ldots, \sigma_k \leq N} S_{\sigma_1 \ldots \sigma_k}(K), \tag{5.22}$$

$$K \supseteq S_{\sigma_1}(K) \supseteq S_{\sigma_1 \sigma_2}(K) \supseteq \cdots \supseteq S_{\sigma_1 \ldots \sigma_k}(K) \supseteq \cdots. \tag{5.23}$$

If r is the maximum contraction ratio of $S_1, \ldots, S_N$, then it is easy to see that

$$\text{diam}\, S_{\sigma_1 \ldots \sigma_k}(K) \leq r^k \text{diam}\,(K).$$

Hence $\text{diam}\, S_{\sigma_1 \ldots \sigma_k}(K) \to 0$ as $k \to \infty$. Thus there is a unique point which belongs to every member of the sequence (5.23). We give this point the *code* or *address*

$$\sigma = \sigma_1 \sigma_2 \ldots \sigma_k \ldots$$

and denote the point by

$$\Pi(\sigma). \tag{5.24}$$

Every point in K has a code, but the code is not necessarily unique. For example, in fig. 5.5 in which the Koch curve is generated by two similitudes T_1 and T_2, the top-most point has the two addresses $1222\ldots$ and $2111\ldots$. On the other hand, if the generating maps $S_1, \ldots, S_N$ satisfy $S_i(K) \cap S_j(K) = \emptyset$ for $i \neq j$, then the N^k sets in (5.22) are also mutually disjoint and it follows that every point in K does have a unique address.

The set C_N of all codes σ with each $\sigma_k \in \{1, \ldots, N\}$ is called *code space*. It is a metric space with metric

$$\delta(\sigma, \tau) = \sum_{k=1}^{\infty} \frac{|\sigma_k - \tau_k|}{N^k}.$$

Then one has (Hutchinson, 1981, theorem 3.1):

Theorem 5.3 *For any $x \in \mathbb{R}^n$*

$$S_{\sigma_1 \ldots \sigma_k}(x) \to \Pi(\sigma) \in K \text{ as } k \to \infty.$$

Moreover, the map $\Pi : C_N \to K$ is a continuous map onto K.

Proof To prove the first result first note that if r is the maximum contraction ratio of the $S_1, \ldots, S_N$ then

$$d(S_{\sigma_1 \ldots \sigma_k}(x), S_{\sigma_1 \ldots \sigma_k}(K)) \leq r^k d(x, K) \to 0 \text{ as } k \to \infty.$$

Since

$$\Pi(\sigma) \in S_{\sigma_1 \ldots \sigma_k}(K)$$

and

$$\operatorname{diam} S_{\sigma_1 \ldots \sigma_k}(K) \to 0 \text{ as } k \to \infty$$

the first claim follows.

For the second claim suppose $\sigma, \tau \in C_N$, that $\sigma_i = \tau_i$ for $i = 1, \ldots, k$ and $\sigma_{k+1} \neq \tau_{k+1}$. Then

$$\delta(\sigma, \tau) \geq 1/N^k. \tag{5.25}$$

On the other hand

$$\Pi(\sigma), \Pi(\tau) \in S_{\sigma_1 \ldots \sigma_k}(K) = S_{\tau_1 \ldots \tau_k}(K),$$

and so

$$d(\Pi(\sigma), \Pi(\tau)) \leq \operatorname{diam} S_{\sigma_1 \ldots \sigma_k}(K) \leq r^k \operatorname{diam}(K). \tag{5.26}$$

From (5.25), $\delta(\sigma, \tau)$ is small implies k is large, and from (5.26) this implies $d(\Pi(\sigma), \Pi(\tau))$ is small. Thus Π is continuous. □

One can also give a direct proof of theorem 5.2 by arguing as in the previous proof, see (Hutchinson, 1981) theorem 3.1.

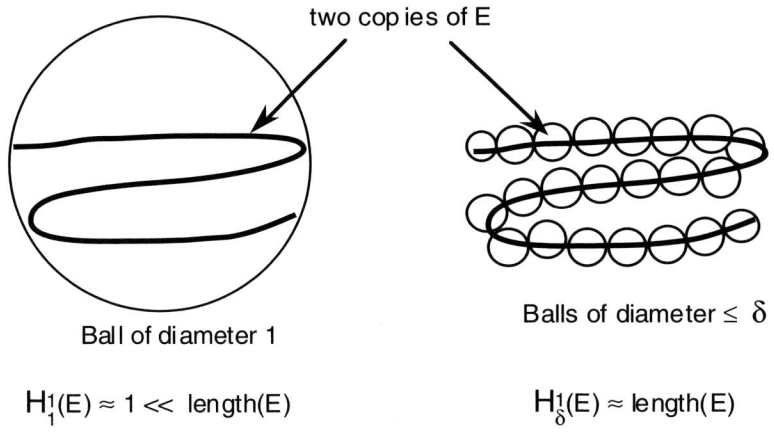

Fig. 5.12. Hausdorff measure.

5.4.6 Dimension

5.4.6.1 Hausdorff dimension

Assume $E \subseteq \mathbb{R}^n$ and $d \geq 0$. If $\delta > 0$ we define the *d-dimensional δ-approximating measure* by

$$\mathcal{H}^d_\delta(E) = \inf \sum_{i \geq 1} (\operatorname{diam} B_i)^d, \qquad (5.27)$$

where the infimum is taken over all (finite or infinite) sequences of balls† $(B_i)_{i \geq 1}$ such that (i) $\operatorname{diam} B_i \leq \delta$ and (ii) $E \subseteq \bigcup_{i \geq 1} B_i$.

If $\delta_1 \leq \delta_2$ then $\mathcal{H}^d_{\delta_1}(E) \geq \mathcal{H}^d_{\delta_2}(E)$ since there are *fewer* allowable families of balls for δ_1 than for δ_2. Thus $\mathcal{H}^d_\delta(E)$ is increasing as $\delta \to 0$. We define the *d-dimensional Hausdorff measure of E* (fig. 5.12) by

$$\mathcal{H}^d(E) = \lim_{\delta \to 0} \mathcal{H}^d_\delta(E). \qquad (5.28)$$

Clearly,

$$0 \leq \mathcal{H}^d(E) \leq +\infty.$$

It is not too hard to show that there exists a *unique* $\overline{d} \in [0, n]$ such that

$$\mathcal{H}^d(E) = \begin{cases} \infty & \text{if } d < \overline{d} \\ 0 & \text{if } d > \overline{d}. \end{cases} \qquad (5.29)$$

† Technical aside: One usually allows arbitrary subsets of $\mathbb{R}^n$. This will make no difference to the later definition of the Hausdorff dimension $\overline{d}$ but may change the value of $\mathcal{H}^{\overline{d}}(E)$.

This unique $\overline{d}$ is called the *Hausdorff dimension* of E.

It is possible to construct examples of sets E where $\mathcal{H}^{\overline{d}}(E) = a$ for any given a satisfying $0 \leq a \leq \infty$. But if $\mathcal{H}^d(E)$ is *finite and non-zero for some d* then $\overline{d} = d$. If $\mathcal{H}^d(E) = 0$ then $\overline{d} \leq d$, if $\mathcal{H}^d(E) = \infty$ then $\overline{d} \geq d$.

In case E is a smooth curve in $\mathbb{R}^2$ or $\mathbb{R}^3$ then $\overline{d} = 1$ and $\mathcal{H}^1(E)$ is the length of E. In case E is a smooth surface in $\mathbb{R}^3$ then $\overline{d} = 2$ and $\mathcal{H}^2(E)$ is the area of E. For the fractal sets we are considering, $\overline{d}$ will not usually be an integer.

5.4.6.2 Scaling dimension and the open set condition

Assume $\mathbf{S} = (S_1, \ldots, S_N)$ where the S_i are *similitudes* with scaling factors $r_1, \ldots, r_N$ respectively. Then the *scaling dimension* corresponding to $\mathbf{S}$ is defined to be the unique value of D such that

$$r_1^D + \cdots + r_N^D = 1, \tag{5.30}$$

cf.§5.3.4 and the final footnote there. Under certain conditions (theorem 5.4) this scaling dimension will equal the Hausdorff dimension of the associated fractal set.

For this reason, we say that $\mathbf{S}$ satisfies an *Open Set Condition* if there exists a non-empty open set $O \subseteq \mathbb{R}^n$ such that

(i) $\bigcup_{i=1}^{N} S_i(O) \subseteq O$,
(ii) $S_i(O)$ and $S_j(O)$ are disjoint if $i \neq j$.

The scaling operator $\mathbf{S} = (S_1, \ldots, S_4)$ for the Koch curve K satisfies the open set condition, where O is the interior of the large triangle in fig. 5.13. Notice, on the other hand, that the sets $S_1(K), \ldots, S_4(K)$ are not disjoint.

The following result showing the equality of scaling dimension and Hausdorff dimension if the open set condition is satisfied, is due to Moran (1946).

Theorem 5.4 *If $\mathbf{S}$ consists of similitudes and satisfies the open set condition then the scaling dimension D of $\mathbf{S}$ equals the Hausdorff dimension of the associated fractal set K. Moreover, $\mathcal{H}^D(K)$ is finite and non-zero.*

Proof We will prove the easier result that the Hausdorff dimension is less than or equal to the scaling dimension D. This does *not* require the open set condition, and so holds much more generally.

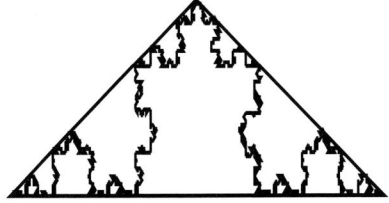

O is the interior of the large triangle; $S_1(O),..., S_4(O)$ are the interiors of the shaded triangles.

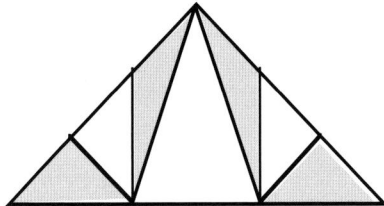

Fig. 5.13. Scaling dimension.

Using (5.30) first note that

$$\begin{aligned}
1 &= r_1^D + \cdots + r_N^D \\
&= r_1^D\left(r_1^D + \cdots + r_N^D\right) + \cdots + r_N^D\left(r_1^D + \cdots + r_N^D\right) \\
&= \sum_{i,j=1}^{N} r_i^D r_j^D.
\end{aligned}$$

Similarly,

$$\sum_{i_1,\ldots,i_k=1}^{N} r_{i_1}^D \cdot \ldots \cdot r_{i_k}^D = 1 \tag{5.31}$$

for any k.

Now choose a ball B, with diameter b (say), such that $K \subset B$. From (5.22),

$$K = \bigcup_{1 \leq \sigma_1,\ldots,\sigma_k \leq N} S_{\sigma_1\ldots\sigma_k}(K)$$

$$\subseteq \bigcup_{1 \leq \sigma_1, \ldots, \sigma_k \leq N} S_{\sigma_1 \ldots \sigma_k}(B). \tag{5.32}$$

Notice that $S_{\sigma_1 \ldots \sigma_k}(B)$ is a ball of diameter $r_1 \cdot \ldots \cdot r_k \cdot b$.

Take any $\delta > 0$. By choosing k sufficiently large we may assume $r_1 \cdot \ldots \cdot r_k \cdot b \leq \delta$ (if $r = \max\{r_1, \ldots, r_n\}$ is the maximum contraction ratio just choose k so $r^k b \leq \delta$). It follows from (5.32) and (5.31) that K is covered by a finite number of balls of diameter $\leq \delta$ such that

$$\begin{aligned}
\sum (\text{diameter of the balls})^D &= \sum_{i_1, \ldots, i_k = 1}^{N} (r_{i_1} \cdot \ldots \cdot r_{i_k} \cdot b)^D \\
&= b^D \sum_{i_1, \ldots, i_k = 1}^{N} r_{i_1}^D \cdot \ldots \cdot r_{i_k}^D \\
&= b^D.
\end{aligned}$$

Hence $\mathcal{H}^D_\delta(K) \leq b^D$ from (5.27), for all $\delta > 0$. But then $\mathcal{H}^D(K) \leq b^D$ from (5.28). It follows from the remarks following (5.29) that if $\overline{d}$ is the Hausdorff dimension of K then $\overline{d} \leq D$.

The proof that $\overline{d} \geq D$ (and hence that $\overline{d} = D$) is more involved. The idea is to use the open set condition to cover K by small balls that do not overlap 'too much'. We refer the reader to (Moran, 1946), or to (Hutchinson, 1981) theorem 5.3 where a result involving densities is also proved. The fact that $\mathcal{H}^D(K)$ is finite and non-zero is also a consequence of the argument. □

5.4.7 Parameter space

An *affine* map $S : \mathbb{R}^n \to \mathbb{R}^n$ is a composition of a linear map and a translation. Such an affine map is completely described by means of an $n \times n$ matrix and an n vector, and hence by a finite number of parameters. Thus a scaling operator $\mathbf{S} = (S_1, \ldots, S_N)$ of affine maps can be thought of as a point in some finite dimensional *parameter space* $\mathbb{R}^m$, cf. (Hutchinson, 1981) §5.5 and (Oppenheimer, 1980). Not all points in $\mathbb{R}^m$ will actually correspond to a family of *contraction* maps and so parameter space will in fact correspond to a certain (open) subset of $\mathbb{R}^m$. It is often natural to work with certain restricted families of scaling operators, in which case the dimension m may be considerably reduced.

It follows from theorem 5.2 and (5.3) that the Koch curve is uniquely characterised by the scaling law $\mathbf{T} = \{T_1, T_2\}$, which in turn corresponds to a point in $\mathbb{R}^m$ where $m = (2 \times 2 + 2) + (2 \times 2 + 2) = 12$. If T_1 is replaced

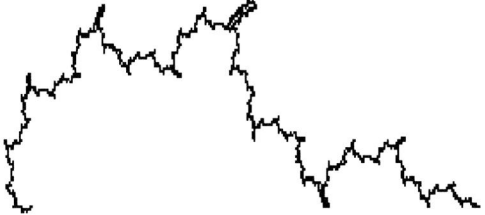

Fig. 5.14. The dragon fractal.

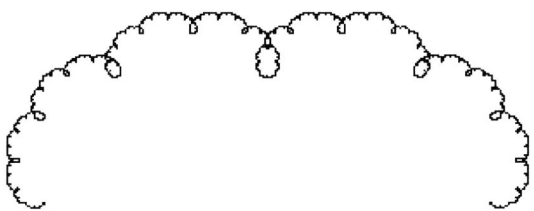

Fig. 5.15. The brain fractal.

by the same affine map *except* that no reflection is performed, and if T_2 is left unchanged, then the *Dragon* fractal (fig. 5.14) is obtained.

The same calculation as in (5.18) shows that the scaling dimension of the scaling law which gives the dragon is $\log 4/\log 3$. Presumably the Hausdorff dimension of the dragon is also $\log 4/\log 3$, but the open set condition does not apply here.

If T_1 and T_2 are as for the Koch curve, except that no reflection is performed in either case, then the *Brain* fractal (fig. 5.15) is obtained. The scaling dimension is again $\log 4/\log 3$, and the Hausdorff dimension is presumably the same.

If T_1 and T_2 are similar to the previous example except that now T_1 fixes $P = (-1.5, 0)$ and maps $Q = (1.5, 0)$ to $(-0.225, 0.9)$ instead of to $(0, \sqrt{3}/2)$, and T_1 fixes Q and maps P to $(0.225, 0.9)$ instead of to $(0, \sqrt{3}/2)$, then the *cloud* fractal (fig. 5.16) is obtained. We can only

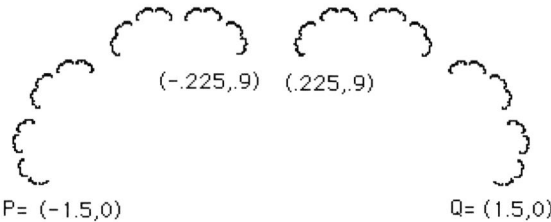

Fig. 5.16. Cloud fractals.

show an approximation, and in fact the actual fractal in this case will in a certain precise sense be 'totally disconnected'.

The contraction ratios are here each $r = 0.5202\ldots$ and so the dimension of the Clouds is $\log 2/\log(1/r) = 1.0606\ldots$.

5.5 Fractal measures

5.5.1 Some notions from measure theory

A *finite measure*† ν on $\mathbb{R}^n$ assigns to every set $E \subseteq \mathbb{R}^n$ a positive real number such that (i) $\nu(\emptyset) = 0$ where $\emptyset$ is the empty set, (ii) if $E_1 \subseteq E_2$ then $\nu(E_1) \leq \nu(E_2)$, and (iii) if $E \subseteq \bigcup_{i=1}^{\infty} E_i$ then $\nu(E) \leq \sum_{i=1}^{\infty} \nu(E_i)$. We also require that there be a class of subsets of $\mathbb{R}^n$, called the class of (ν-)measurable subsets, which includes the open and closed sets, is closed under complements and under finite and countably infinite intersections and unions, and satisfies $\nu(E) = \sum_{i=1}^{\infty} \nu(E_i)$ whenever $E = \bigcup_{i=1}^{\infty} E_i$ and the sets E_i are mutually disjoint and measurable. The *mass* of ν is $\nu(\mathbb{R}^n)$. If the mass of ν is one, we say that ν is a *unit mass* measure. The *integral* $\int f\, d\nu$ of various functions $f : \mathbb{R}^n \to \mathbb{R}$ is defined in such a way that, roughly speaking, '$\int f\, d\nu$ is a weighted sum of the values of f, where the weighting is done according to ν'.

The *support* $\mathrm{spt}(\nu)$ of ν is the intersection of all closed sets whose complement has ν measure zero; and so $\mathrm{spt}(\nu)$ is the *smallest* closed set whose complement has ν measure zero. If $S : \mathbb{R}^n \to \mathbb{R}^n$ then the finite measure $S(\nu)$ is defined by $(S(\nu))(E) = \nu(S^{-1}(E))$; think of $S(\nu)$ as the measure ν 'pushed forward' by the function S. If ρ is a positive number then the finite measure $\rho\nu$ is defined by $(\rho\nu)(E) = \rho \cdot \nu(E)$; think of the

† What we here call a finite measure is often called a *finite Radon outer measure*.

measure $\rho\nu$ as the measure ν reweighted by the factor ρ. Moreover, we have the important properties

$$\int f\, dS(\nu) = \int f \circ S\, d\nu \tag{5.33}$$

where $f \circ S$ is the usual composition of functions, and

$$\int f\, d(\rho\nu) = \rho \int f\, d\nu \tag{5.34}$$

where ρ is any positive number.

5.5.2 The metric space $(\mathcal{U}, d_{\mathcal{MK}})$ of measures on $\mathbb{R}^n$ with unit mass and compact support

The set of all *unit mass measures*† on $\mathbb{R}^n$ with compact support is denoted by $\mathcal{U}$.‡

For our purposes there is a useful notion of distance between two such measures, called the *Monge Kantorovitch distance*, which is defined by

$$d_{\mathcal{MK}}(\mu, \nu) = \sup\left\{ \left| \int f\, d\mu - \int f\, d\nu \right| \;\middle|\; f : \mathbb{R}^n \to \mathbb{R} \text{ and} \right.$$
$$|f(x_1) - f(x_2)| \leq |x_1 - x_2|$$
$$\left. \text{for all } x_1, x_2 \in \mathbb{R}^n \right\}. \tag{5.35}$$

As a simple example, suppose P and Q are two point in $\mathbb{R}^2$. Let δ_P and δ_Q be unit measures concentrated at P and Q respectively – thus for $E \subseteq \mathbb{R}^n$, $\delta_P(E) = 1$ if $P \in E$ and $\delta_P(E) = 0$ if $P \notin E$, and similarly for δ_Q and Q. Then with any f as in the definition of $d_{\mathcal{MK}}$, we see

$$\left| \int f\, d\delta_P - \int f\, d\delta_Q \right| = |f(P) - f(Q)| \leq |P - Q|.$$

Thus $d_{\mathcal{MK}}(\delta_P, \delta_Q) \leq |P - Q|$. On the other hand, it is easy to find a (in fact linear) function f as in the definition of $d_{\mathcal{MK}}$ such that $\left| \int f\, d\delta_P - \int f\, d\delta_Q \right| = |P - Q|$. It follows that $d_{\mathcal{MK}}(\delta_P, \delta_Q) = |P - Q|$, the usual distance between the *points* P and Q.

It can be shown that $(\mathcal{U}, d_{\mathcal{MK}})$ *is a complete metric space*.

---

† A unit mass measure μ is itself a probability distribution in a natural way, but we do not here think of μ in this way.
‡ If μ satisfies a scaling law as described in the next section, so does any multiple of μ. Thus there is no essential loss of generality in restricting to *unit mass* measures.

5.5.3 Existence, uniqueness and approximation

Definition 5.5 A $2N$-tuple $\mathbf{S} = \big((S_1,\ldots,S_N),(\rho_1,\ldots,\rho_N)\big)$ of contraction maps $S_i\colon \mathbb{R}^n \to \mathbb{R}^n$ and positive numbers ρ_i such that $\rho_1+\cdots+\rho_N = 1$ is called a *scaling law for measures*. The corresponding *scaling operator* $\mathbf{S}$ is defined on measures in $\mathcal{U}$ by $\mathbf{S}(\nu) = \sum_{i=1}^{N} \rho_i S_i(\nu)$. We say ν *satisfies the scaling law* $\mathbf{S}$ if $\mathbf{S}(\nu) = \nu$.

We now prove the *existence and uniqueness* result from (Hutchinson, 1981) theorem 4.4(1) for fractal *measures*.

Theorem 5.6 *There is a unique measure $\mu \in \mathcal{U}$ satisfying a given scaling law $\mathbf{S}$. If ν is any measure in $\mathcal{U}$ then $\mathbf{S}^k(\nu)$† converges to μ in the Monge Kantorovitch metric as $k \to \infty$. If K is the support of μ, then K is the unique compact set satisfying the corresponding scaling law $(S_1,\ldots,S_N)$ for sets.*

Proof We outline the main points.

The scaling operator $\mathbf{S}$ is a contraction map on the metric space $(\mathcal{U}, d_{\mathcal{MK}})$. The main point in showing this fact is that for $\nu_1, \nu_2 \in \mathcal{U}$, if f is as in (5.35) and r is the maximum contraction ratio of the S_i's, then

$$\left| \int f\, d\big(\mathbf{S}(\nu_1)\big) - \int f\, d\big(\mathbf{S}(\nu_2)\big) \right|$$
$$= \left| \sum_{i=1}^{N} \left[\int f\, d(\rho_i S_i(\nu_1)) - \int f\, d(\rho_i S_i(\nu_2)) \right] \right|$$
$$\leq \sum_{i=1}^{N} \left| \int f\, d(\rho_i S_i(\nu_1)) - \int f\, d(\rho_i S_i(\nu_2)) \right|$$
$$\leq \sum_{i=1}^{N} \left| \rho_i \int (f \circ S_i)\, d\nu_1 - \rho_i \int (f \circ S_i)\, d\nu_2 \right|$$
$$= \sum_{i=1}^{N} \left| \rho_i r \int r^{-1}(f \circ S_i)\, d\nu_1 - \rho_i r \int r^{-1}(f \circ S_i)\, d\nu_2 \right|$$
$$\leq \sum_{i=1}^{N} \rho_i r\, d_{\mathcal{MK}}(\nu_1,\nu_2)$$
$$\leq r\, d_{\mathcal{MK}}(\nu_1,\nu_2).$$

† See (5.16) for notation.

For the second inequality we have used (5.33) and (5.34). For the penultimate inequality we have used the fact that since f is as in (5.35) then

$$\left| r^{-1}(f \circ S_i)(x_1) - r^{-1}(f \circ S_i)(x_2) \right| \leq |x_1 - x_2|,$$

as can be readily checked, and hence (5.35) can now be applied with f there replaced by $r^{-1}(f \circ S_i)$.

It follows from the contraction mapping principle that $\mathbf{S}$ has a unique fixed point $\mu \in \mathcal{U}$, and if $\nu \in \mathcal{U}$ then $\mathbf{S}^k(\nu)$ converges to μ in the Hausdorff metric as $k \to \infty$. The first two claims in the theorem follow.

The final claim follows from elementary properties of the support of a measure. Namely,

$$\begin{aligned}\operatorname{spt} \mu &= \operatorname{spt} \mathbf{S}(\mu) = \operatorname{spt} \sum_{i=1}^{N} \rho_i S_i(\mu) \\ &= \bigcup_{i=1}^{N} \operatorname{spt} S_i(\mu) = \bigcup_{i=1}^{N} S_i(\operatorname{spt} \mu).\end{aligned}$$

Thus the compact set $\operatorname{spt} \mu$ satisfies the scaling law $(S_1, \ldots, S_N)$ and so is the unique such compact set by theorem 5.2. □

Remark: The measure μ satisfying the scaling law $\mathcal{S}$ is usually called a *fractal measure*.

The method from §5.3.3.2 generalises to any fractal measure, see (Elton, 1987). At the kth stage in the iteration, the map S_{σ_k} is chosen from $(S_1, \ldots, S_N)$, where the probability of choice of any particular S_i is p_i.

5.6 Random fractals

5.6.1 Probability theory

We begin with some *examples*:

(i) If a fair dice is thrown, the set of possible outcomes is $X = \{1, 2, \ldots, 6\}$. The associated probability distribution or measure P assigns to each of these numbers the equal measure (or probability) $1/6$. If $E = \{1\}$ then $P(E) = 1/6$, if $E = \{1, 5\}$ then $P(E) = 1/3$, and of course $P(X) = 1$.

(ii) Consider an experiment in which a dart is thrown at a dartboard, all points are 'equally likely' to be hit, and the dart never misses

the board. If we take the dart board to be the disc of radius R given by
$$B_R = \{(x,y) \in \mathbb{R}^2 : \sqrt{x^2 + y^2} \leq R\},$$
then the probability measure P is a unit mass uniformly distributed over B_R. Thus we write
$$P = \tfrac{1}{\pi R^2}\, dx\, dy|_{B_R} = \tfrac{1}{\pi R^2} \mathcal{L}^2|_{B_R},$$
by which we mean the usual Lebesgue (i.e. 'unit density') measure on $\mathbb{R}^2$ restricted to the disc B_R, and reweighted so the total mass (measure) for B_R is one. For any event E (which we identify with a subset of B_R) the probability of E (i.e. the probability that the dart lands in the set E) is given by
$$P(E) = \int_E \tfrac{1}{\pi R^2}\, dx\, dy = \tfrac{1}{\pi R^2} \mathcal{L}^2(E).$$
By $\int_E$ we mean integration over the set E.

(iii) Suppose a number x is selected according to the *normal distribution* with mean a and variance σ^2. Then the frequency function is
$$f(x) = \tfrac{1}{\sqrt{2\pi}\sigma} \exp\left(-\frac{(x-a)^2}{2\sigma^2}\right),$$
and the corresponding probability measure is denoted by
$$P = f(x)\, dx.$$
The probability of an event E, i.e. the probability that a number selected by this normal distribution lies in the set $E \subseteq \mathbb{R}$, is
$$P(E) = \int_E f(x)\, dx.$$

Motivated in part by these examples, one defines a *probability distribution* or *probability measure* P to be a unit mass measure on the set X of possible outcomes of an experiment. The definition of a finite measure is as in the first sentence of §5.5.1 with $\mathbb{R}^n$ there replaced by X, but we also now require that $P(X) = 1$ (so $0 \leq P(E) \leq 1$ if $E \subseteq X$).

If E is an *event*, i.e. E is a set of possible outcomes of the experiment and so $E \subseteq X$, then $P(E)$ is interpreted as the *probability of E occurring*. We might also think of $P(E)$ as the 'measure' of E via P.

One point that sometimes causes confusion is the following. To say that an event E *occurs with probability zero* does not mean that it cannot occur. For example, in (b) above the mathematical probability of hitting a pre-designated point on the board with an idealised dart whose tip is also a (mathematically ideal) point is zero. One sometimes says that E occurs *almost never*. Similarly, if an event *occurs with probability one* we say it occurs *almost surely*. For example, the (idealised) dart almost surely will *not* hit the origin.

In the previous three examples we have the situation where a *number*, or a point in $\mathbb{R}^2$, is 'selected' via some probability distribution (or measure) P. The set X of outcomes was identified with a subset of $\mathbb{R}$ or of $\mathbb{R}^2$.

We, however, will be interested in conceptually more difficult cases where the set X of outcomes is either the set $\mathcal{C}$ of compact subsets of $\mathbb{R}^n$ (as in the case of random fractal sets), or the set $\mathcal{U}$ of compactly supported unit mass measures on $\mathbb{R}^n$ (as in the case of random fractal measures), or certain sets of N-tuples $(S_1, \ldots, S_N)$ of contraction maps S_i on $\mathbb{R}^n$, or certain sets of $2N$-tuples $\big((S_1, \ldots, S_N), (\rho_1, \ldots, \rho_N)\big)$ where the S_i are again contraction maps and the 'weights' ρ_i are positive numbers.

5.6.2 Random fractal sets

We first make precise the ideas from §5.2.3.

Definition 5.7 A *random set* $\mathcal{E}$ is a set generated according to a probability distribution on the set $\mathcal{C}$ of compact subsets of $\mathbb{R}^n$.

More precisely, a random set $\mathcal{E}$ can be considered to be the probability distribution itself, and one distinguishes this from particular 'realisations' of $\mathcal{E}$. (This is a little different from the standard convention where a random set is a function, or mapping, from some underlying sample space to some collection of sets.)

Thus when we think of a random set we usually consider, somewhat imprecisely, various 'typical' realisations (i.e. outcomes) selected according to the underlying probability distribution $\mathcal{E}$, as in the diagrams in §5.2.3.

Definition 5.8 A *scaling law* for random sets is a probability distribution $\mathcal{S}$ on the set of N-tuples $(S_1, \ldots, S_N)$ of contraction maps $S_i : \mathbb{R}^n \to \mathbb{R}^n$, for some fixed $N \geq 2$. Moreover, we assume that for some fixed compact set $B \subset \mathbb{R}^n$, $S_i(B) \subseteq B$ for $1 \leq i \leq N$ whenever $(S_1, \ldots, S_N)$ is selected via $\mathcal{S}$.†

Example In the case of the Koch curve as described in (5.3), recall that $T_1(1.5, 0) = T_2(-1.5, 0) = (0, \sqrt{3}/2)$.

Suppose now that instead of the point $(0, \sqrt{3}/2)$ we select a point $(0, a)$ according to the uniform probability distribution on the vertical line segment joining the points $(0, \sqrt{3}/2 - 0.15)$ and $(0, \sqrt{3}/2 + 0.15)$, say. This induces in a natural way a probability distribution $\mathcal{T}$ on pairs of maps (T_1, T_2).†

Definition 5.9 Suppose compact sets $E^{(1)}, \ldots, E^{(N)}$ are each selected independently according to the probability distribution (i.e. random set) $\mathcal{E}$, and a single N-tuple $\mathbf{S} = (S_1, \ldots, S_N)$ of contraction maps is selected independently of the $E^{(i)}$ and according to the scaling law $\mathcal{S}$ (remember that $\mathcal{S}$ is a probability distribution)‡. Consider the compact set

$$\bigcup_{i=1}^{N} S_i(E^{(i)}). \tag{5.36}$$

The method of selection of the $E^{(i)}$ and of $\mathbf{S} = (S_1, \ldots, S_N)$ induces in this way a natural probability distribution on compact sets. This probability distribution on compact sets is denoted by

$$\mathcal{S}(\mathcal{E}).$$

Since $\mathcal{S}$ when applied to a random set $\mathcal{E}$ yields another random set $\mathcal{S}(\mathcal{E})$, we also say $\mathcal{S}$ is a *scaling operator*.

We say the random set $\mathcal{K}$ *satisfies the scaling law* $\mathcal{S}$ iff

$$\mathcal{K} = \mathcal{S}(\mathcal{K}).$$

The following result is due independently to Falconer (1986), Graf (1987) and Mauldin and Williams (1986). See also Hutchinson and Rüschendorf (1998b).

---

† Note that this last condition implies that if $(S_1, \ldots, S_N)$ is selected via $\mathcal{S}$ then the fixed point of each S_i is contained in B (to be precise, with probability one).

† More precisely, after selecting $(0, a)$ according to the prescribed uniform distribution, T_1 is obtained by reflecting in the x-axis, followed by a contraction and rotation about $(-1.5, 0)$ so that $(1.5, 0)$ is mapped to $(0, a)$. One similarly obtains T_2, but in this case the contraction and rotation are about the point $(1.5, 0)$, and $(-1.5, 0)$ is mapped to $(0, a)$.

‡ Note that the S_i are not selected 'individually'. The entire N-tuple is chosen according to $\mathcal{S}$.

Theorem 5.10 *For each scaling law $\mathcal{S}$ on random sets, there exists a unique random set $\mathcal{K}$ which satisfies $\mathcal{S}$. Moreover, if $\mathcal{E}$ is any random set then $\mathcal{S}^k \mathcal{E}$ converges§ to $\mathcal{K}$ as $k \to \infty$.*

Proof (*Ideas and remarks only*) Completely disregarding significant technical difficulties, we make the following comments.

The method in Falconer (1986) is to define a certain metric on the set of random sets and then to show that $\mathcal{S}$ is a contraction map with respect to this metric. This is analogous to the proof of theorem 5.2.

The method in Graf (1987) and Mauldin & Williams (1986) is to suitably randomise the construction in §5.4.5 and in the proof of theorem 5.3. Informally the idea is to build a (*random*) *construction tree* as follows:

(0) choose an N-tuple (of contraction maps) $\mathbf{S} = (S_1, \ldots, S_N)$ according to $\mathcal{S}$;

(1) for each $1 \leq i_1 \leq N$ choose N-tuples (of contraction maps) $\mathbf{S}^{i_1} = (S_1^{i_1}, \ldots, S_N^{i_1})$ according to $\mathcal{S}$, independently of one another and of the $\mathbf{S}$ chosen in (0);

(2) for $1 \leq i_1, i_2 \leq N$ choose N-tuples (of contraction maps) $\mathbf{S}^{i_1 i_2} = (S_1^{i_1 i_2}, \ldots, S_N^{i_1 i_2})$ according to $\mathcal{S}$, independently of one another and of the previously chosen N-tuples;

(3) for $1 \leq i_1, i_2, i_3 \leq N$ choose N-tuples $\mathbf{S}^{i_1 i_2 i_3} = (S_1^{i_1 i_2 i_3}, \ldots, S_N^{i_1 i_2 i_3})$ according to $\mathcal{S}$, independently of one another and of the previously chosen N-tuples;

...

Define

$$S_{\sigma_1 \ldots \sigma_k} = S_{\sigma_1} \circ S_{\sigma_2}^{\sigma_1} \circ S_{\sigma_3}^{\sigma_1 \sigma_2} \circ \cdots \circ S_{\sigma_k}^{\sigma_1 \ldots \sigma_{k-1}}. \tag{5.37}$$

The reasons for this definition are as follows:

(i) First note the correspondence with (5.21), with the difference that here there is no longer a single *fixed* N-tuple $\mathbf{S} = (S_1, \ldots, S_N)$. Analogously to (5.24), there is associated to each code $\sigma = \sigma_1 \sigma_2 \ldots \sigma_k \ldots$ a point

$$\Pi(\sigma) = \lim_{k \to \infty} S_{\sigma_1 \ldots \sigma_k}(x),$$

which is independent of x.

§ That is, the sequence $\mathcal{E}, \mathcal{S}(\mathcal{E}), \mathcal{S}^2(\mathcal{E}) = \mathcal{S}(\mathcal{S}(\mathcal{E})), \mathcal{S}^3(\mathcal{E}) = \mathcal{S}(\mathcal{S}^2(\mathcal{E})), \ldots$ converges in the technical sense of weak convergence of probability measures.

(ii) The set $\Pi(C_N)$† of all points $\Pi(\sigma)$ will form a compact set analogous to those in each of the three diagrams in §5.2.3‡. For a fixed $x \in \mathbb{R}^n$ and a fixed large k, the set of points of the form $S_{\sigma_1 \ldots \sigma_k}(x)$ will be a good approximation to $\Pi(C_N)$, analogous to the diagram in §5.3.3.1 for the deterministic case (see also the sentence preceding that diagram).

Since Π depends on the choices made in (0), (1), (2), ..., we actually have a *random* set $\Pi(C_N)$, or more precisely a probability distribution on the set of compact subsets of $\mathbb{R}^n$. This random set is, moreover, invariant under the scaling law $\mathcal{S}$. The point is that if $(S_1^*, \ldots, S_N^*)$ is chosen via $\mathcal{S}$ and independently of all the choices made in (0), (1), (2), ..., then $\Pi(C_N)$ and $\bigcup_{i=1}^N S_i^*(\Pi(C_N))$ will have the same probability distribution. □

5.6.3 Random fractal measures

We next make precise the ideas from §5.2.4.

Definition 5.11 A *random measure* $\mathcal{N}$ is a measure generated according to a probability distribution on the set $\mathcal{U}$ of compactly supported unit mass measures in $\mathbb{R}^n$.

More precisely, a random measure $\mathcal{N}$ can be considered to be the probability distribution itself, and one distinguishes this from particular 'realisations' of $\mathcal{N}$. (As in the case of random sets, this is a little different from the standard conventions.)

We think of a random measure as a measure 'selected according to' the distribution $\mathcal{N}$.

Definition 5.12 A *scaling law* for random measures is a probability distribution $\mathcal{S}$ on the set of $2N$-tuples $\big((S_1, \ldots, S_N), (\rho_1, \ldots, \rho_N)\big)$ of contraction maps $S_i : \mathbb{R}^n \to \mathbb{R}^n$, and of positive numbers $\rho_1, \ldots, \rho_N$

---

† The independence follows as in the deterministic case if we assume that for some fixed $r < 1$ the contraction maps in any selected N-tuple have contraction ratio less than or equal to r; cf. (5.12). But more general conditions are also possible.
† Recall from §5.4.5 that C_N is code space consisting of all codes $\sigma = \sigma_1 \sigma_2 \ldots \sigma_k \ldots$ where each $\sigma_i \in \{1, \ldots, N\}$
‡ More precisely, the diagrams in §5.2.3 are three different realisations of $\mathcal{T}^9 \mathcal{E}$, where $\mathcal{E}$ is (with probability one) the line segment joining the endpoints $P = (-1.5, 0)$ and $Q = (1.5, 0)$, and $\mathcal{T}$ is a probability distribution on pairs (T_1, T_2) similar to the one in the example near the beginning of this section.

satisfying $\rho_1 + \cdots + \rho_N = 1$, for some fixed $N \geq 2$.§ Moreover, we assume that for some fixed compact set B, $S_i(B) \subseteq B$ for $1 \leq i \leq N$ whenever $(S_1, \ldots, S_N)$ is selected via $\mathcal{S}$.

Example
Consider the example of the random Koch curve in the previous section. If the same probability distribution $\mathcal{T}$ on pairs of contraction maps (T_1, T_2) is again used, and if the weights (ρ_1, ρ_2) are each always taken to be $1/2$, then this induces a probability distribution $\mathcal{T}^*$ on 4-tuples $\Big((T_1, T_2), (\rho_1, \rho_2)\Big)$ which satisfies the previous definition with $N = 2$.

Definition 5.13 Suppose measures $\nu^{(1)}, \ldots, \nu^{(N)}$ are each selected independently according to the probability distribution (i.e. random measure) $\mathcal{N}$, and a single $2N$-tuple $\mathbf{S} = \Big((S_1, \ldots, S_N), (\rho_1, \ldots, \rho_N)\Big)$ is selected independently of the $\nu^{(i)}$ and according to the scaling law $\mathcal{S}$. Consider the measure

$$\sum_{i=1}^{N} \rho_i S_i(\nu^{(i)}). \tag{5.38}$$

The method of selection of the $\nu^{(i)}$ and of $\mathbf{S}$ induces in a natural way a probability distribution on measures (selecting in particular, measures of the form (5.38) with probability one). This probability distribution (i.e. random measure) is denoted by

$$\mathcal{S}(\mathcal{N}).$$

Since $\mathcal{S}$ when applied to a random measure $\mathcal{N}$ yields another random measure $\mathcal{S}(\mathcal{N})$, we also say $\mathcal{S}$ is a *scaling operator*.

We say the random measure $\mathcal{M}$ *satisfies the scaling law* $\mathcal{S}$ iff

$$\mathcal{M} = \mathcal{S}(\mathcal{M}).$$

The following result is due to Maudlin and Williams (1986) and Arbeiter (1991). See also Zähle (1988, 1990) for related ideas and Wicks (1993) for connections between these approaches.

Theorem 5.14 *For each scaling law $\mathcal{S}$ on random measures, there exists a unique random measure $\mathcal{M}$ which satisfies $\mathcal{S}$. Moreover, if $\mathcal{N}$ is any random measure then $\mathcal{S}^k \mathcal{N}$ converges† to $\mathcal{M}$ as $k \to \infty$.*

§ Weaker conditions are sometimes used, in particular that the *expected value* (or *average value*) of $\rho_1 + \cdots + \rho_N$ equals 1.
† In the technical sense of 'weak convergence of probability measures'.

Proof (*Some ideas only*) Given a scaling operator $\mathcal{S}$ for random measures, one can proceed parallel to the selection of the $S_j^{i_1\ldots i_p}$ in the proof of theorem 5.10 and simultaneously obtain weights $\rho_j^{i_1\ldots i_p}$. This leads to a natural (random) measure on code space C_N.

Namely, for each k and for each $\sigma_1, \ldots, \sigma_k$, the set of all codes of the form $\sigma_1 \ldots \sigma_k \ldots$ is given the measure

$$\rho_{\sigma_1\ldots\sigma_k} = \rho_{\sigma_1} \cdot \rho_{\sigma_2}^{\sigma_1} \cdot \rho_{\sigma_3}^{\sigma_1\sigma_2} \cdot \ldots \cdot \rho_{\sigma_k}^{\sigma_1\ldots\sigma_{k-1}}.$$

One can check that for fixed k this defines the measure of sets in a partition of C_N and that the total measure of these sets is one. As k increases this partition becomes 'finer'. One can show that there is a unique corresponding measure on C_N which will be denoted by Ψ†.

If Π is the map constructed in the proof of theorem 5.10 then $\Pi(\Psi)$ is a measure on $\mathbb{R}^n$. Since Π and Ψ depend on the choices in (0), (1), (2), ..., the measure $\Pi(\Psi)$ is in fact a *random* measure on $\mathbb{R}^n$. Moreover, it follows from the construction process that $\Pi(\Psi)$ satisfies the scaling law $\mathcal{S}$. □

Remark: In Hutchinson and Rüschendorf (1998b, 1998a) we give a simple proof of generalisations of the previous theorem using the contraction mapping principle.

Final Remark: Conditions under which the dimension of random fractals can be computed have been extensively investigated. See Arbeiter (1991),Falconer (1986), Graf (1987), Graf *et al.* (1988), Mauldin and Williams (1986) and Patzsche and Zähle (1990).

Acknowledgement

I would particularly like to thank Jacki Wicks for many valuable discussions we had during the writing of her undergraduate honours thesis (Wicks, 1993), and Andy Wood for various helpful suggestions concerning probabilistic notions.

Bibliography

Arbeiter, M. A. (1991). Random recursive constructions of self-similar fractal measures. the noncompact case. *Prob. Th. Rel. Fields*, **88**, 497–520.

---

† If we assume that the expected value of $\rho_1 + \cdots + \rho_N$ is one, rather than $\rho_1 + \cdots + \rho_N = 1$ with probability one, then the definition of the measure is technically more complicated.

Barnsley, M. F. (1988). *Fractals Everywhere.* Academic Press.

Barnsley, M. F., & Demko, S. (1985). Iterated function systems and the global construction of fractals. *Proc. Roy. Soc. London,* **A399**, 243–275.

Barnsley, M.F., Ervin, V., Hardin, D., , & Lancaster, J. (1986). Solution of an inverse problem for fractals and other sets. *Proc. Nat. Acad. Sci,* **83**, 1975–1977.

Bélair, J., & Dubuc, S. (eds). (1991). *Fractal Geometry and Analysis. Proceedings of the NATO Advanced Study Institute, Montréal, Canada (1989).* Kluwer.

Diaconis, P, & Shahshahani, M. (1984). *Products of Random Matrices and Computer Image Generation.* Preprint, Dept. of Stats. Stanford.

Elton, J. (1987). An ergodic theorem for iterated maps. *Ergod. Th. Dynam. Syst,* **7**, 481–488.

Falconer, K. J. (1985). *The Geometry of Fractal Sets.* Cambridge University Press.

Falconer, K. J. (1986). Random fractals. *Math. Proc. Camb. Philos. Soc,* **100**, 559–582.

Graf, S. (1987). Statistically self-similar fractals. *Prob. Th. Rel. Fields,* **74**, 357–392.

Graf, S., Mauldin, R. D., & Williams, S. C. (1988). The exact Hausdorff dimension in random recursive constructions. *Mem. Amer. Math. Soc,* **71**(381).

Hutchinson, J. E. (1981). Fractals and self-similarity. *Indiana. Univ. Math. J.,* **30**, 713–747.

Hutchinson, J. E., & Rüschendorf, L. (1998a). *Random Fractal and Probability Metrics.* Tech. rept. MRR98-048 < http://wwwmaths.anu.edu.au/research.reports/98mrr.html >. Australian National University.

Hutchinson, J. E., & Rüschendorf, L. (1998b). Random fractal measures via the contraction method. *Indiana. Univ. Math. J,* 471–487.

Kolata, G. (1984). Esoteric math has practical results. *Science,* **225**, 494–495.

Mandelbrot, B. B. (1982). *The Fractal Geometry of Nature.* W.H. Freeman, San Francisco.

Mauldin, R. D., & Williams, S. C. (1986). Random recursive constructions; asymptotic geometric and topological properties. *Trans. Amer. Math. Soc.,* **295**, 325–346.

Moran, P. A. P. (1946). Additive functions of intervals and Hausdorff measure. *Proc. Camb. Phil. Soc.,* **42**, 15–23.

Oppenheimer, P. (1980). *Constructing an Atlas of Self-Similar Sets,.* Undergraduate thesis, Princeton.

Patzschke, N., & Zähle, U. (1990). Self-similar random measures. iv — the recursive constuction model of Falconer, Graf, and Mauldin and Williams. *Math. Nachr.,* **149**, 285–302.

Peitgen, H-O., Jürgens, H., & Saupe, D. (1991). *Fractals for the Classroom.* Springer-Verlag, New York.

Wicks, J. (1993). *Deterministic and Random Self-Similar Fractals and Fractal Measures.* Undergraduate honours thesis, Aust. Nat. Univ.

Williams, R. F. (1971). Compositions of contractions. *Bol. Soc. Brasil Mat,,* **2**, 55–59.

Zähle, U. (1988). Self-similar random measures. i — notion, carrying Hausdorff dimension, and hyperbolic distribution. *Prob. Th. Rel. Fields,* **80**, 79–100.

6
Non-linear dynamics

DAVID E. STEWART[1] & ROBERT L. DEWAR[2]

[1] *School of Mathematical Sciences,*
[2] *Research School of Physical Sciences & Engineering,*
The Australian National University,
Canberra, ACT 0200, AUSTRALIA

6.1 Introduction

Since it impossible to do justice to the whole of non-linear dynamics and chaos in one chapter we shall give a broad-brush overview, but with emphasis on two aspects of the subject not normally given much attention in textbooks on dynamical systems – the emergence of low-degree-of-freedom dynamical systems as a description on a *macroscopic* scale of systems with large numbers of elements on a *microscopic* scale, and the *numerical analysis* of dynamical systems. These are related, as computational approaches can give much new insight in the field of complex systems. This chapter can be read on two levels – on the one hand we endeavour to give heuristic arguments and physical motivations so that the beginner should be able to get a feel for the subject, but on the other hand we also give a flavour of the rigorous mathematical approach and give references to the mathematical literature.

Note: References to sections in this chapter are indicated using the § symbol, words defined in the text are indicated by *italics* when they are first defined, and words defined in the glossary of this chapter are indicated in **bold** type when they first occur.

6.1.1 Chaos – the dynamical systems approach

The insight that has caused the great upsurge in interest in 'chaos theory' in recent years (for an introductory overview and reprints of original papers, see Hao (1990)) is that complex behaviour may arise from simple mathematical models. The models we have in mind are *dynamical systems* – sets of equations describing the deterministic evolution, with respect to time or a time-like parameter, of a number of dynamical variables forming a vector in a finite or infinite-dimensional *state* or *phase space* (§6.2). The time variable may be *continuous*, so that the dynamical system is a system, first order in time, of ordinary differential equations (ODEs) in the finite-dimensional case, or partial differential equations (PDEs) in the infinite-dimensional case. Alternatively the time variable may be *discrete*, so that the dynamical system is an iterated map.

The use of the word *chaos* is usually taken nowadays to imply (a) irregular behaviour not driven by externally imposed random noise[†]; and (b) that the description of the irregular behaviour can be reduced to a

---

[†] Chaos is sometimes called '(intrinsic) stochasticity', but the word 'stochastic' is better reserved for truly random phenomena.

low-dimensional dynamical system (even if the original physical system is infinite-dimensional). In this chapter we shall restrict consideration to 'pure' chaos in the sense of (a) above, but we caution that a full theory of complex systems requires dynamical systems theory to be extended by including 'noise' terms representing stochastic fluctuations (Haken, 1983, p. 315).

For mathematical texts on dynamical systems theory the reader is referred to such books as those of Guckenheimer and Holmes (1983), Devaney (1989) and Arrowsmith and Place (1990). Books on non-linear dynamics oriented more toward those who are not professional mathematicians are Arrowsmith and Place (1992), Lichtenberg and Lieberman (1992), Moon (1992) and Ott (1993).

In this chapter we shall be concerned primarily with dynamical systems with a finite (small) number of dimensions. Although such systems may exhibit complex, chaotic behaviour, they cannot in themselves be called complex systems. We also therefore give some consideration to systems with a large or infinite number of degrees of freedom and show how it can happen that behaviour in such complex systems can often be described by a reduced dynamical system with a small number of degrees of freedom. We take as our paradigm the development of organised collective motion in fluids, as this problem has been intensively studied by physicists, engineers and mathematicians for many years (though there is still much to understand).

A powerful influence on the spread of the field of non-linear dynamics beyond a handful of mathematicians capable of visualising complex behaviour in the mind's eye has been the advent of easy-to-use computers and computer graphics. This has allowed a much wider community of scientists to grasp the concepts of low-dimensional dynamical systems theory, and to apply them to practical problems. In turn the ability to perform computer experiments has stimulated many new theoretical developments.

Another area, within the fluid dynamics paradigm referred to above, in which computation has shed new light on old problems is in the simulation of liquids, gases and plasmas using 'particle-pushing' or *molecular dynamics* methods. In this approach supercomputers are used to advance the equations of motion of a large number of particles (§6.2.8), in an effort to model the actual real-world situation in which we know that the seemingly continuous behaviour of matter on a macroscopic scale is in fact the result of the motion of an enormous number, N say, of particles, where N is proportional to Avogadro's number, 6×10^{23} molecules per mole. No conceivable supercomputer can of course handle numbers of degrees of freedom of this order, but, by using cunning numerical tricks and by being careful with the interpretation of results, sufficient numbers of particles can often be followed to obtain physically relevant macroscopic information. However it is through the ability to obtain *microscopic* diagnostic information easily from a numerical simulation that this technique can lead to new insights into the nature of complex systems.

In this chapter we provide an overview of dynamical systems theory with an emphasis on *numerical analysis*. That is, we introduce some of the most important concepts in dynamical systems theory in a mathematical but constructive way, suitable for numerical implementation, and we also discuss error analysis (§6.3) and introduce the concept of shadowing (§6.9). This provides the mathematical underpinning for the computer experimentation in which we encourage the reader to engage while reading §6.2, or other dynamical systems texts. Two excellent interactive graphical computer programs which allow users access to the source code, so that they may provide their own dynamical systems in addition to the default ones, are the Unix X11 code dstool (Guckenheimer *et al.*, 1992) and the DOS/Unix X11 code dynamics by Yorke and Kostelich whose use is described by Nusse and Yorke (1993). Also very useful are general purpose algebra, computation and graphics packages like Mathematica (Wolfram, 1991) or Matlab (Moler *et al.*, 1990), which provide facilities for such things as two-dimensional arrow plots to depict vector fields.

In the next three subsections of this introduction we attempt to explain the paradox of how complex systems may sometimes be described by low-dimensional dynamical systems. This is of interest not only to understand the genesis of dynamical systems, but because we believe dynamical systems theory itself casts light on the nature of physical theory by introducing the concept of inertial manifolds (§6.8) and related techniques for obtaining a reduced description of the long-time behaviour of a system. The basis for such reductions has been called the *slaving principle* by Haken (1978, 1983), who has emphasised its key role in complex systems theory (called by him *synergetics*).

6.1.2 Conservative dynamical systems

The basic dynamical equations of physics are Hamiltonian and consequently conserve phase space volume (see classic texts like Landau and

Lifshitz (1971) and Goldstein (1980) or more modern ones discussing dynamics on **manifold**s, such as Arnol'd (1980) or Scheck (1990)). They also exhibit continuous symmetries such as translation invariance in space (giving rise to momentum conservation) and in time (giving rise to energy conservation), and discrete symmetries such as parity and time-reversal invariance (Sudarshan & Mukunda, 1974).

Because of the fundamental nature of Hamiltonian dynamical systems, their non-linear properties have been much studied in recent years (see the reprint collection of MacKay and Meiss (1987)). It is now realised that the emphasis in the classic texts on problems that could be solved by action-angle transformations (integrable systems) was misleading, in that a typical Hamiltonian will not have a phase space foliated by invariant tori, as is the case if an action-angle transformation exists, but will have at least some regions of chaotic motion. In these regions there exists a time scale (the inverse of the maximum Lyapunov exponent, see §6.6) over which small initial errors in position or velocity amplify exponentially, so that, on a time scale long compared with this Lyapunov scale, the motion is essentially unpredictable.

6.1.3 Dissipative dynamical systems

There are some large-scale physical systems, such as the solar system, which are well described as low-dimensional Hamiltonian dynamical systems. However many problems of physics (let alone chemistry or biology) involve far too many degrees of freedom for analysis from first principles as a Hamiltonian dynamical system to be practicable. Nevertheless, a low-dimensional dynamical system (typically *not* Hamiltonian) is often a valid physical description over some macroscopic range of length and time scales. That is, new physics appears when we refocus on larger scales than those of elementary particle physics or atomic physics where (quantum) Hamiltonian dynamics reigns supreme. The essentially emergent nature of physical theory on different scales has been emphasised, and contrasted with the traditional reductionist approach, by the distinguished physicists Anderson (1972) and Schweber (1993).

How may we understand this loss of the stupendous number of degrees of freedom in the microscopic description and their replacement with a relatively few non-trivial degrees of freedom on a macroscopic scale? The main mechanism by which simplicity can emerge from complexity is **irreversibility**. Since Boltzmann's work around the turn of the century (Lebowitz 1993a, 1993b) we have understood this as arising from the inherently coarse-grained nature of macroscopic observations, in which many microstates are projected onto one macrostate and a statistical or probabilistic description must be adopted, so that knowledge of the initial conditions is lost. The theory describing the evolution of probability densities or *distribution functions* in phase space, and the derivation from them of macroscopic transport coefficients, is called *non-equilibrium statistical mechanics*.

One might expect modern chaos theory to shed new light on the origin of irreversibility, and indeed an entropy (the Kolmogorov–Sinai entropy – see §6.6) which measures rate of information loss is used in dynamical systems theory. However the relation of this entropy rate to dissipation in physical systems is a surprisingly subtle and controversial question. In fact, on the basis of what is known in the rigorous mathematical literature (Lanford, 1981), Lebowitz (1993a) states that the essential features of irreversibility do not depend on 'the positivity of Lyapunov exponents, ergodicity or mixing'.

On the other hand, Evans *et al.* (1990, 1992) and found a clear and practical relation between Lyapunov exponents and viscosity for a 'thermostatted' model quasi-Hamiltonian system developed to allow molecular dynamics computations of stressed systems (e.g. shear flows) to be carried out without having to introduce boundaries to remove the heat generated. This thermostatted Hamiltonian is discussed in §6.2.8. The 'thermostat' is spatially distributed for computational convenience and has the remarkable effect of producing a strange attractor in the phase space of an otherwise Hamiltonian system. The relevance of this result to the more realistic case where heat is removed at the boundaries is a topic of hot debate (literally! – see pp. 327–343 of Mareschal and Holian (1992)). As stated by Cohen (Cohen, 1993) 'The relation between kinetic theory and the theory of dynamical systems has hardly been developed'. A first step in this direction starting from the low-dimensional dynamical system end is the study of transport in area-preserving maps (MacKay *et al.*, 1985), and Wiggins (1992) has begun to extend this into a theory for multi-dimensional Hamiltonian systems.

Another area in which our modern understanding of non-linear dynamics, combined with computation, raises fundamental issues in the field of statistical mechanics is the question of the compatibility between the Kolmogorov–Arnol'd–Moser theorem (KAM theorem – see §6.2.4) and the 'ergodic hypothesis' at the heart of equilibrium statistical mechanics. The ergodic hypothesis is the assumption that the state vector describing an isolated many-body system wanders in time over the entire 'energy-surface' – the set of points in phase space with the same energy. The KAM theorem, on the other hand, asserts that, under appropriate conditions, a finite fraction of orbits is confined to a more restricted region than the entire energy surface. It was the failure to confirm the ergodic hypothesis in a computational 'experiment' by Fermi, Pasta and Ulam (1965) which led eventually to the discovery of the theory of solitons. Current knowledge of the scaling with the number of particles of the chaotic region of phase space is reviewed by Lichtenberg and Lieberman (1992, §6.5).

Irreversibility allows the observable effects of the many-body, microscopic degrees of freedom to damp out on a short time scale. Just what level of macroscopic description is used depends on the problem. If the time and length scales of the physical experiment are comparable with the statistical relaxation scales (inverse collision frequencies and mean free paths) then we must use a *kinetic description.*

Kinetic theory works with PDEs describing the evolution of *single-particle distribution functions* – the probability of finding a particle species in its 6-dimensional phase space of position and velocity at time t. This is to be contrasted with the $2N$-dimensional phase space of the complete dynamical description of the system (assuming classical mechanics is an appropriate description at the microscopic level). Thus for kinetic theory to work we need to assume that all the information we need to know about the N-body system is contained in the single-particle distribution function. This is the hypothesis behind Bogoliubov's closure of the infinite hierarchy of equations for many-particle distribution functions (Montgomery & Tidman, 1964, Ch. 5) – after an initial transient all many-particle correlation functions are slaved to the single particle distribution function, so that a kinetic equation such as the Boltzmann equation can be derived. Another approach is the projection operator formalism (Zwanzig, 1964) (reprinted with many related articles in Oppenheim, Shuler and Weiss (1977)) for deriving the 'master equation" or generalised kinetic equation. Also Muncaster (Muncaster, 1983) has developed a general formalism for deriving coarse-grained theories from fine-grained theories and has discussed it in relation to the theory of attracting invariant manifolds (see §§6.5.6 and 6.8).

If the time and length scales of the physical experiment are long compared with the statistical relaxation scales the description may be reduced further using the Chapman–Enskog method described in Chapman and Cowling (1939) to derive *fluid equations.* These are still PDEs but are defined on the ordinary three dimensional space of position (plus time of course). If the boundaries of the system are sufficiently close to mutual thermodynamic equilibrium, the system will relax to a simple equilibrium state, but if the system is bordered by regions sufficiently far from mutual thermodynamic equilibrium, spontaneous symmetry breaking of the simple state can occur via one or more *bifurcations* (see §6.7.2), and new non-trivial degrees of freedom can thus emerge. These are driven by the input of energy at low entropy (e.g. heat from the hotter parts of the boundary) and its removal at a higher entropy (from the colder parts). The classic example of this is the onset of Rayleigh–Bénard convection rolls in a fluid between two plates when the temperature of the lower plate exceeds that of the upper by a critical amount (Swinney & Gollub, 1981), but the occurrence of life itself can also be thought of as a self-organising phenomenon driven by the temperature difference between the Sun and the Earth.

It should also be remarked that the low-entropy energy input need not be in the form of heat; for instance electrical energy can drive magnetic tearing instabilities which destroy the translational symmetry of a plasma current sheet (Parker *et al.*, 1990, Grauer, 1989, Wessen, 1993).

In dynamical systems terms the non-trivial degrees of freedom are coordinates on a low-dimensional 'inertial manifold' (Temam, 1988) which all orbits approach asymptotically at large times so that the attractor(s) (§6.4) of the system are embedded within it. The damped degrees of freedom have a remanent excitation, but only as a forced, adiabatic response to the motion on the invariant manifold – they are 'slaves' to the 'master' coordinates on the invariant manifold (Roberts, 1990). The slave modes in turn react non-linearly back on the master modes to produce a renormalised vector field in the reduced dynamical system of the master modes (§6.8). The master coordinates are also called 'order parameters' (Haken, 1978, Haken, 1983) or 'central modes' (Manneville, 1990).

In summary, we have put forward the view that self-organised dynamics is the result of a competition between the enslaving effect of irreversible dissipation and the tendency of low-entropy energy input to produce new master modes via symmetry-breaking bifurcations.

6.1.4 Spatio-temporal chaos

We have already remarked that PDEs first order in time are (infinite-dimensional) dynamical systems, so the study of systems distributed in space is included in dynamical systems theory. Furthermore the existence of a centre (§6.5.6) or inertial manifold (§6.8) can make the long-time behaviour describable by a low-dimensional dynamical system. This system may exhibit irregular behaviour in time, but not in space – *temporal chaos* – or may be irregular in both time and space – *spatio-temporal chaos*.

The field of spatio-temporal chaos is not as mature as that of temporal chaos, but includes many interesting aspects of complex system theory – pattern formation, lattice dynamics, cellular automata, Lagrangian chaos. An early general review is that of Crutchfield and Kaneko (1987), while a review of more recent work can be found in chapter 8 of Moon (1992). A much more complete treatment of spatio-temporal chaos in fluids is to be found in Manneville (1990).

6.1.5 I want dynamics – why study numerics?

The study of dynamical systems has been going on for at least a century now (starting with Poincaré (1892)). Most of this work has been done in terms of topology and geometry – much of it 'hard' analysis. On the other hand applications to any real situations suffer from the problem that analytical solutions of the governing differential or difference equations are either impossible to find or give little understanding about the existence or non-existence of chaos. Certainly computers have been used to study a great many phenomena arising from chaotic and other dynamical systems. This is the way the fractals that are used to decorate offices and bedroom walls are generated. Furthermore, these computations are not exact – they are almost always done using floating point arithmetic.

There are a number of issues that arise from these computations which are difficult to resolve and involve a great deal of sophisticated mathematics. The first point that should be made is that in dynamical systems, trajectories remain bounded, and yet diverge exponentially fast (until they are no longer close). Because of the finite precision of floating point arithmetic, numerically computed trajectories are constantly being perturbed. And these perturbations lead to exponentially growing errors in the solution. Very quickly the errors will swamp the trajectory itself, and the errors will be of about the same size as the features of the trajectory – thus there will be no significant digits in the numerically computed trajectories. Yet, the results that come out of simulations of the Lorenz equations or any other particular chaotic equations are remarkably independent of how the equations have been discretised, or how they have been implemented. Why is this? This question leads to some interesting and deep problems in analysis, both numerical and analytical.

One of the ideas that has come out of this work is the idea of *shadowing* (§6.9); that is, computed solutions of an iteration really satisfy $\|y_{n+1} - g(y_n)\| < \delta$ for some small $\delta > 0$. These 'δ pseudo-orbits' are 'ϵ-shadowed' by some exact solution x_n where $\|x_n - y_n\| < \epsilon$. Given a δ pseudo-orbit we want to find $\epsilon > 0$ for which there is an ϵ shadow. Sometimes this shadowing radius can be numerically computed for a particular trajectory.

Another way in which numerical analysis is of use is in computing and analysing objects that organise the long-term dynamics of the system. The objects might be equilibrium points, periodic trajectories, or more exotic objects like homoclinic orbits, and invariant tori. These structures often give a great deal of information about the structure of the dynamics. For example, *homoclinic orbits* are orbits that approach the same point both as time goes to $+\infty$ and as time goes to $-\infty$. Usually these give rise to complicated dynamics (chaos), a fact which was first discovered by Henri Poincaré (1892). *Invariant tori* often arise in Hamiltonian dynamics. These are tori for which trajectories that start on any **torus** remain on that torus for all time. Knowledge of the behaviour of the trajectories on the torus can be used to analyse the stability or otherwise of the dynamical system.

Bifurcations (§6.7.1) can also be analysed numerically, and branches of solutions can be followed using *continuation methods* (§6.7.3). These are mathematical techniques for tracking, as a parameter is varied, objects such as equilibrium points or periodic orbits from simple cases where the solutions are easy to find, to highly non-linear cases where solutions would otherwise be very difficult to locate. Continuation methods can also be used for calculating homoclinic and heteroclinic orbits (§6.7.4).

Numerical analysis can also be used for computing quantities of interest in the study of dynamical systems. These include the *dimensions* of fractal attractors (§6.4.2), *Lyapunov exponents* (§6.6) and the *entropy* (which gives information on the rate at which the system 'loses' information about where it started – see §6.6).

A book which covers many questions about numerical algorithms for non-linear dynamics and chaos is Parker and Chua's book (1989). It should be remembered that there is a great deal that can be said about, or used to study, dynamical systems and related computations. So there are many topics that go well beyond this chapter, or beyond any one textbook for that matter. The bibliography should provide you with at least a start in discovering the many questions and (incomplete) answers there are about chaotic dynamics.

6.2 Dynamical systems

Before we start our journey with dynamical systems we need some notation and jargon to describe the path we take and the views we see. To start with, the two types of dynamical systems we deal with are continuous time systems described by differential equations, and discrete time systems described by difference equations or iterations of maps. An **autonomous** continuous time system describing the state vector $x = x(t)$ evolving under the influence of the vector field $f(x)$ is described by

$$\dot{x} \equiv \frac{dx}{dt} = f(x)$$

and a discrete time system describing the iterates $x = x_i$ (the integer i being the 'time') under a mapping g is described by

$$x_{i+1} = g(x_i) \, ,$$

where in both cases x moves in an n-dimensional phase space, X (which can be the space $\mathbf{R}^n$ (see Glossary), or, more generally, an n-dimensional **differentiable manifold** (Arrowsmith & Place, 1990)).

A driven, non-autonomous system is defined the same way, but with f a function of t. Such a system can always be made autonomous by introducing a new dependent variable τ, say, such that $\dot{\tau} = 1$, with $\tau = 0$ at $t = 0$, thus increasing the dimension of the state space by one. Since the restriction to autonomous systems therefore entails no loss of generality, the formal theory of dynamical systems is usually developed under this restriction and we shall do likewise unless otherwise specified.

Continuous time systems define *flows*, which are functions

$$\varphi \colon X \times \mathbf{R} \to X$$

evolving initial positions x_0 to their values at time t: $\varphi(x_0, t) = x(t)$, with $\dot{x} = f(x)$ and $x(0) = x_0$. These flows have the following properties (assuming, as usual, the system to be autonomous): for all $x \in X$ and $t, s \in \mathbf{R}$,

$$\varphi(x, 0) = x, \qquad \varphi(\varphi(x, t), s) = \varphi(x, t + s) \, .$$

In some systems such as PDEs we might have to restrict the time argument of the flow to be non-negative (see §6.2.6). Such a restriction gives rise to *semiflows*; these have the above properties except that we restrict t, s to be ≥ 0.

In the discrete time case, if g is smooth and has an inverse function which is also smooth, then g is, by definition, a **diffeomorphism**; if it is not a diffeomorphism (or we don't know if it is or not), we use the more general term *map*. The function $\varphi(\cdot, t)$ for an ODE is always a diffeomorphism (but not for a PDE when φ is a semiflow).

Studying flows, rather than the differential equations which generate them, unifies the continuous time and discrete time cases. Indeed we can discover a discrete-time dynamical system in a continuous-time one by 'strobing' it periodically: simply take $g(x) = \varphi(x, T)$, where T is the sampling period. This is particularly natural in the case of a non-autonomous system with T-periodic time dependence. As argued above, an n-dimensional non-autonomous system is equivalent to an $(n+1)$-dimensional autonomous one, and we have just shown that it generates an n-dimensional discrete time system. This illustrates the close connection between n-dimensional discrete time systems and $(n+1)$-dimensional (autonomous) continuous time systems.

For notational convenience, we will use the same symbol φ for iterated diffeomorphisms and iterated maps, but will use integers (respectively, non-negative integers) as the 'time' argument: $\varphi(x, i+1) = g(\varphi(x, i))$, i integer.

Although we are concerned with non-linear systems, the truth is that most of the arsenal of mathematics only works against linear systems. Thus the best strategy is almost always to build up a knowledge of the non-linear system by finding related linear systems to study. This means that we shall often be *linearising* the dynamical system about an orbit, say $y(t)$, which is assumed known (e.g. it may be an equilibrium, or fixed point, $y = $ const). If ξ is a perturbation away from the known orbit, i.e. if $x = y + \xi$, then to linear order ξ obeys the equation found by Taylor expansion about y

$$\dot{\xi} = \nabla f(y)\, \xi$$

or

$$\xi_{n+1} = \nabla g(y)\, \xi_n \ ,$$

where we use the notation $\nabla f(x)$ for the *Jacobian matrix* whose element in the ith row and jth column is $\partial f_i / \partial x_j$, i and j being indices indicating the components of the vectors f and x, respectively. (Note that the order of the indices is reversed from what would be assumed in 3-tensor dyadic notation – in the above equations, ξ is contracted with the ∇ operator. Note also that the symbol D is often used instead of ∇.)

In the remainder of this section we give representative examples of the various types of dynamical system.

6.2.1 A one-dimensional map

Perhaps the best-known map is the 'discrete logistic map' on the unit interval $[0, 1]$. This is given by $x_{n+1} = \lambda x_n(1 - x_n)$ where $0 \leq \lambda \leq 4$. The behaviour of this map for various values of λ runs the range from having a single stable fixed point ($x = 0$ for $0 \leq \lambda \leq 1$) through a number of bifurcations, and eventually becomes chaotic by a cascade of period doubling bifurcations. For $\lambda > 1$ there are two fixed points: $x = 0$, which is unstable, and $x = 1 - 1/\lambda$, which is stable for $1 < \lambda \leq 3$. The first period doubling (flip) bifurcation occurs at $\lambda = 3$. The fixed point at $x = 1 - 1/\lambda$ becomes unstable. There follows a sequence of period doubling bifurcations which are progressively closer together. In fact if the kth period doubling bifurcation occurs at $\lambda = \mu_k$, then $(\mu_k - \mu_{k-1})/(\mu_{k+1} - \mu_k) \to 4.669201609 \cdots$ which is known as Feigenbaum's constant (Feigenbaum, 1978).

There is an infinite number of period doubling bifurcations in the region $\lambda < 3.569944 \cdots$. This sequence of period doubling bifurcations is called a period doubling cascade, and is typical of many dynamical systems. Immediately after the period doubling cascade is the onset of chaos (Devaney, 1989, Block & Coppel, 1992). This chaotic regime is interrupted by occasional windows where there is a stable periodic orbit. These periodic orbits also undergo a period doubling cascade to return to chaotic behaviour. Finally at $\lambda = 4$ the behaviour is chaotic and the attractor has expanded to fill the entire interval $[0, 1]$. In fact, the dynamical system can be exactly solved for $\lambda = 4$! The exact solution is $x_n = \sin^2(2^n \theta_0)$ for some θ_0. If θ_0/π is irrational, then the values x_n, $n = 0, 1, 2, \ldots$ fill the entire interval $[0, 1]$.

6.2.2 A one-dimensional diffeomorphism

Maps $g : S^1 \to S^1$ of the circle onto itself form an important area of study in dynamical systems theory because they form a prototype for the toroidal attractors and invariant surfaces occurring in higher dimensional systems. One of the most important properties of iterates of these maps is the average rate of rotation, or *rotation number*, $\nu = \rho(g)$, and one of the most important questions is 'under what circumstances can a change of variable (conjugacy) be found such that the map becomes the rigid rotation $R(\theta) = \theta + \nu$'. A standard example, whose numerical implementation is easy, is the Arnol'd circle map (Arrowsmith & Place, 1990, p. 248)

$$g(\theta) = \alpha + \theta + \epsilon \sin \theta .$$

For $|\epsilon| < 1$ this is a diffeomorphism, and chaotic behaviour does not occur in one-dimensional diffeomorphisms. However, for $|\epsilon| > 1$ it is 'just a map' because g no longer increases monotonically with θ. The transition point $\epsilon = 1$ is often studied as an example of critical behaviour (Lichtenberg & Lieberman, 1992, p.529), for which universal scaling constants can be found by renormalisation group methods (Delbourgo, 1992).

6.2.3 A dissipative two-dimensional diffeomorphism

The map $(x', y') = g(x, y)$:

$$\begin{aligned} x' &= y + 1 - ax^2 , \\ y' &= bx \end{aligned}$$

was introduced by Hénon (Hénon, 1976). Its Jacobian determinant is a constant, $\det \nabla g = -b \neq 0$, so the map can be inverted uniquely and is hence a diffeomorphism. For $|b| < 1$ the Hénon map contracts each element of phase space, and can therefore be used to model a dissipative system. The case $a = 1.4$, $b = 0.3$ was shown by Hénon to give rise to a fractal attractor.

6.2.4 An area-preserving diffeomorphism

The return map taking a Poincaré section of the energy surface of a two-degree of freedom autonomous Hamiltonian system back to itself, or the stroboscopic period-map of a periodically forced one-degree of freedom Hamiltonian system (§6.2.7) can be shown to preserve the area of an infinitesimal quadrilateral (Meiss, 1992). Since flows of finite-dimensional dynamical systems are always invertible, and smooth if the vector field is smooth, these area-preserving maps are in fact diffeomorphisms.

The most commonly studied area-preserving diffeomorphism is probably the Taylor–Chirikov or *standard map* $(x', y') = g(x, y)$, where

$$x' = x + y' \mod 1 ,$$

$$y' = y - \frac{k}{2\pi}\sin 2\pi x .$$

The first equation has been written in an implicit form purely for compactness – y' can be eliminated immediately in favour of x and y using the second equation. Conversely the reader can easily verify that the inverse map can be written down explicitly, verifying that it is indeed a diffeomorphism. Area preservation can be demonstrated by showing that the Jacobian determinant, $\det \nabla g$, is unity.

As defined above, the phase space is topologically cylindrical, with x the angle around the cylinder (in units of 2π radians). There is also a periodicity in the y-direction and other symmetries (Dewar & Meiss, 1992, Roberts & Quispel, 1992). When the non-linearity parameter k is zero, the map is *integrable* – the phase space is composed entirely of invariant circles $y = $ const encircling the cylinder (i.e. 'rotational' circles) and there is no chaos. As k is increased from 0, it is found that the circles for which the rotation number y was a rational fraction immediately evaporate into a complicated structure, known as an *island chain*, of periodic orbits surrounded by invariant, but not rotational, circles, chaotic regions, and island chains (forming an infinite hierarchy of islands within islands).

On the other hand, the celebrated KAM theorem by Kolmogorov (1954), Arnol'd (1963) and Moser (1962) predicts that, for sufficiently small k, most of the rotational invariant circles with irrational rotation numbers will survive, though deformed, as curves (still topologically rotational circles, or 1-tori) invariant under the map. These KAM curves act as barriers to chaotic diffusion in y, but the last of the original invariant circles are destroyed at $k = 0.971635406\cdots$ (Greene, 1979). These circles have rotation numbers of $\gamma^{-2} = 0.38196\cdots$ and $\gamma^{-1} = 1 - \gamma^{-2} = 0.61803\cdots$, where $\gamma \equiv (\sqrt{5}+1)/2$. The number $\gamma^{-1} \equiv (\sqrt{5}-1)/2$, known as the 'golden mean', is the number that is worst approximated by rationals, and thus gives rise to the KAM curves least affected by the resonances which lead to islands (Meiss, 1992).

In fig. 6.1 we show the phase space of the standard map at the critical non-linearity for globally connected chaos, $k = 0.971635406\cdots$, illustrating the islands arising from the unperturbed circles with rotation numbers 0/1, 1/1, 1/2, 1/3 and 2/3. Also shown are the two golden mean KAM curves mentioned above, constructed numerically by minimising the 'quadratic flux' functional of Dewar and Meiss (1992) using nine Fourier modes.

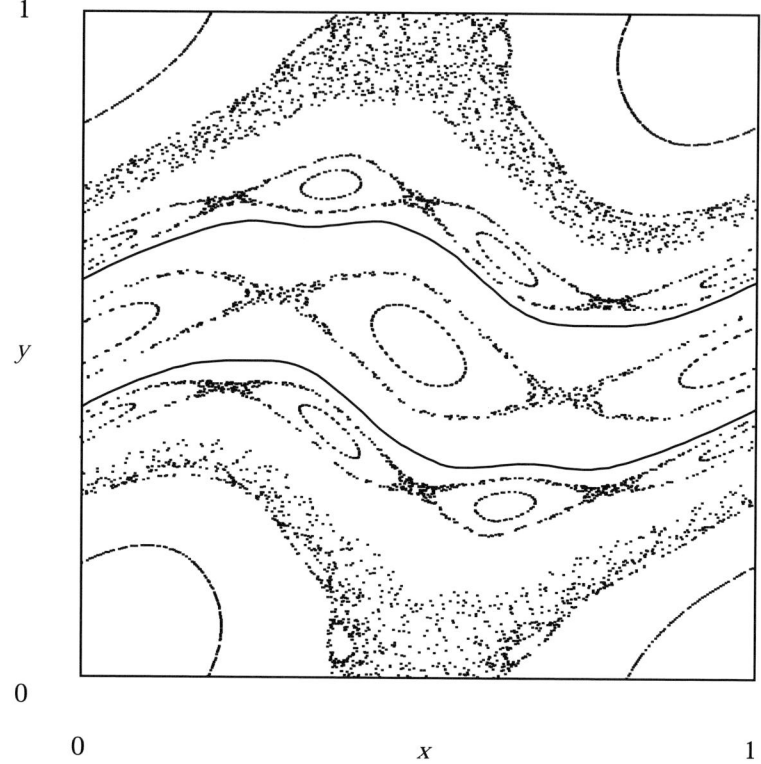

Fig. 6.1. Some representative orbit segments for the standard map, as described in the text.

6.2.5 A two-dimensional continuous-time dynamical system

One of the most studied two-dimensional continuous-time dynamical systems is the van der Pol oscillator. This was based on an electronic vacuum tube circuit devised in the 1920s and analysed by van der Pol (1927). It is given by the second order differential equation

$$\ddot{x} + \epsilon(x^2 - 1)\dot{x} + x = 0 \ .$$

This can be made into a two-dimensional system by setting $v \equiv \dot{x}$. The van der Pol equation then becomes the dynamical system

$$\dot{x} = v \ , \qquad \dot{v} = -\epsilon(x^2 - 1)v - x \ .$$

This is an example of an autonomous system with a **limit cycle** for $\epsilon \neq 0$. The only equilibrium point is the origin $(x, v) = (0, 0)$. For $\epsilon = 0$

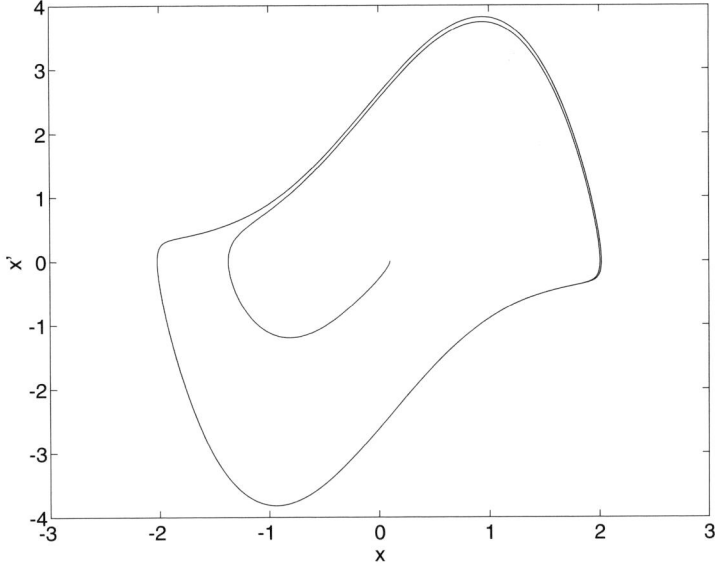

Fig. 6.2. Van der Pol oscillator phase plane portrait, $\epsilon = 2$.

the system reduces to simple harmonic motion, which has an infinite number of periodic orbits, each of which is a circle. For $\epsilon > 0$ a single periodic orbit exists (the limit cycle), which attracts trajectories; for $\epsilon < 0$ there is also a single periodic orbit, which repels trajectories.

Careful asymptotic analysis shows that for small ϵ, the periodic orbit is nearly circular with a radius close to two. For large ϵ the periodic orbit looks much more distorted (fig. 6.2).

6.2.6 An infinite-dimensional continuous-time dynamical system

We shall consider a simple linear PDE – the diffusion or *heat equation* with periodic boundary conditions:

$$\frac{\partial T}{\partial t} = \frac{\partial^2 T}{\partial x^2} ,$$

where $T(x,t)$ is the temperature at time t (in units such that the thermal conductivity is unity) and position x (in units such that the periodicity length is 2π). To see that this is a dynamical system, expand T in a Fourier basis

$$T = \sum_{m=-\infty}^{\infty} T_m(t) \exp imx$$

Then the infinite-dimensional vector $\mathbf{T}$, whose elements are T_m is the state vector and the dynamical system is

$$\dot{\mathbf{T}} = -\mathbf{diag}\,(m^2)\,\mathbf{T}\,,$$

where $\mathbf{diag}\,(\lambda_m)$ is the (infinite-dimensional) diagonal matrix with diagonal elements λ_m.

It is immediately seen that the flow map for this system is given by

$$\varphi(\mathbf{T}_0, t) = \mathbf{diag}(\exp -m^2 t)\,\mathbf{T}_0\,, \qquad t \geq 0\,.$$

Note that the *time-reversed* heat equation is extremely unstable, and will give infinite temperatures in finite time t_b unless its Fourier coefficients decay faster than $\exp(-m^2 t_b)$ for large $|m|$. This is a very strong restriction on allowable initial conditions, even for arbitrarily small but non-zero t_b. Since we do not wish to confine ourselves to such restricted function spaces, $\varphi(\cdot, t)$ is defined only for positive t – it is a semiflow.

6.2.7 One- and $1\frac{1}{2}$-degree of freedom Hamiltonian systems

A 1-degree of freedom mechanical system has only one *position* coordinate, q say. However the phase space is two-dimensional because, in order to make the equation of motion a first order ODE system we introduce a second dynamical variable, the *conjugate momentum*, p say, which is defined so that the system of equations (for a non-dissipative system) has the form

$$\dot{q} = \frac{\partial H}{\partial p}\,,$$

$$\dot{p} = -\frac{\partial H}{\partial q}\,,$$

where $H(q, p, t)$ is the *Hamiltonian*. Normally H is assumed independent of t, in which case it is a constant of the motion and no chaos can occur.

To indicate a *non*-autonomous system, the terminology $1\frac{1}{2}$-degree of freedom system is sometimes used, particularly when the time dependence is periodic. By destroying the constancy of H, the time dependence allows chaos to occur. Indeed the stroboscopic map given by the flow during one period is an area-preserving diffeomorphism, as discussed in §6.2.4.

The physical pendulum Hamiltonian

$$H = \frac{p_\theta^2}{2} - \omega_0^2 \cos\theta,$$

(where $\omega_0/2\pi$ is the frequency of small vibrations) is the prototype for developing a qualitative understanding of weakly non-linear Hamiltonian systems (Lichtenberg & Lieberman, 1992, p. 29).

6.2.8 N-body Hamiltonian systems

The n-degree of freedom Hamiltonian dynamical system has exactly the form given above for the 1-degree of freedom system, except that q and p are now n-vectors and the phase space is of dimension $2n$. Chaos can occur for $n \geq 1\frac{1}{2}$. In D spatial dimensions an N-body system has $n = ND$ degrees of freedom and a phase space of dimension $2ND$.

The Fermi–Pasta–Ulam experiment (Fermi et al., 1965) was an early attempt at studying an N-body Hamiltonian system computationally. This experiment attempted to simulate the approach to statistical equilibrium of a closed Hamiltonian system – a lattice of coupled oscillators.

To study self-organisation in complex systems, on the other hand, we need to simulate a *non*-equilibrium statistical mechanical problem, one in which the system is being driven by the input of low-entropy energy. Furthermore, to achieve a statistical steady state we need a mechanism for removing an equal amount of energy to that input (and a greater amount of entropy). An ingenious device has been developed for molecular dynamics simulations to achieve this removal in a distributed fashion, rather than at the boundaries of the computational domain (Evans & Morriss, 1990). This avoids 'wasting' large amounts of computation on boundary layer phenomena, at the expense of a certain artificiality in the dynamical system. For example, to simulate a shear flow with a mean velocity $\mathbf{u} = \gamma y \mathbf{e}_x$, where $\mathbf{e}_x$ is the unit vector in the x-direction, the dynamical system

$$\dot{\mathbf{q}}_i = \frac{\mathbf{p}_i}{m} + \gamma \mathbf{e}_x \mathbf{e}_y \cdot \mathbf{q}_i$$
$$\dot{\mathbf{p}}_i = \mathbf{F}_i - \gamma \mathbf{e}_x \mathbf{e}_y \cdot \mathbf{p}_i - \alpha \mathbf{p}_i$$

(Evans et al., 1990) has been used, where $\mathbf{q}_i$ is the position of the ith particle, $\mathbf{u} + \mathbf{p}_i/m$ is its velocity and $\mathbf{F}_i$ is the sum of the forces from the other particles. The last term, $-\alpha\mathbf{p}_i$, is the thermostatting term, required to remove the heat generated by the force maintaining the shear flow. It is this term which causes the phase space volume to contract and gives rise to the strange attractor referred to in §6.1.3 (Hoover, 1993).

6.3 Forward vs. backward error

One idea that is very important in numerical analysis is that of *backward error analysis*. The other kind of error analysis is a *forward error analysis*. The forward error of a calculation is the difference between what is computed and what would have been computed had there been no error. On the other hand, the backward error is the size in the perturbation in the *data* of a problem that would have given the computed result had the computations been done *exactly*.

To obtain estimates or bounds on the errors that are produced in the results of a computation, we need to combine backward error analysis with a perturbation theory for the problem under question. This can take the place of forward error analysis. And it is also often better than using forward error analysis. To illustrate this, suppose we consider solving a system of linear equations $Ax = b$ using Gaussian elimination; we might factor $A = LU$ and obtain large (forward) error bounds on the L and U factors, even though $\|A - LU\|$ might be quite small. Unless a great deal of information is included about the *structure* of the errors, forward error analysis can give gross overestimates of the errors resulting from a computation.

All-in-all, backward error analysis is better than forward error analysis for most classes of computational problems.

Not all problems are amenable to backward analysis. There is a number of significant classes of problem for which backward error analyses cannot be done. Often this is associated with a qualitative difference in the behaviour of the computational (perturbed) system and the original system.

Since computations with floating point numbers on computers almost always incur errors due to finite machine precision, lack of backward stability in the mathematical formulation of a problem indicates that simulations cannot be expected to give reasonable results for that model. On the other hand, in the real world, the differential equations themselves are just approximations, and there is also a great deal of 'noise' and other errors. If these unmodelled phenomena can invalidate the model, then the model is inappropriate and should be replaced by one that is backward stable.

6.3.1 Forward and backward error and dynamical systems

Consider a dynamical system

$$\dot{x} = f(x), \quad x(0) = x_0, \qquad \text{or} \qquad x_{n+1} = g(x_n).$$

Forward errors in dynamical systems correspond to the perturbations in the results due to small perturbations in the initial conditions and/or the differential or difference equation(s). Thus if the computed value is $y(T)$ or y_n then the forward error analysis relates $\|x(T) - y(T)\|$ or $\|x_n - y_n\|$ to $\|x_0 - y_0\|$ and the error per unit time or per iteration, $\|\eta\|$, defined by

$$\dot{y}(t) = f(y(t)) + \eta(t) \qquad \text{or} \qquad y_{n+1} = g(y_n) + \eta_n.$$

For stable systems the forward error remains small. For chaotic systems, the forward errors ($\|x_n - y_n\|$) in a trajectory grow exponentially fast. However, objects such as attractors have good backward error properties under certain conditions (such as attracting trajectories exponentially fast). Thus, even though individual trajectories may not be worthwhile objects to compute for chaotic systems, at least for large time intervals, the attractors that they generate are usually much more reliably computed.

For discrete time dynamical systems ($x_{n+1} = g(x_n)$), provided the function evaluation is done in a numerically stable way, the computed iterates y_n can easily be shown to satisfy

$$y_{n+1} = g(y_n) + \eta_n$$

where η_n is small, of the same order as *machine epsilon* or *unit roundoff* **u**. This quantity describes the precision of the floating point arithmetic used by a particular computer or software system. It is usually defined as the smallest number **u** such that the computed value '1+**u**' is different from 1. For single precision arithmetic this is usually around 10^{-6}, and for double precision using IEEE arithmetic is around 10^{-16}. Thus single precision usually has a precision of about 6 digits, while IEEE double precision arithmetic has a precision of about 16 digits. The exact value of **u** depends on the particular hardware and software used.

For ODEs the problem is a bit more complicated, as we need to use a time stepping scheme of some sort (e.g. Euler, implicit Euler, Runge–Kutta, or a multistep method). Even in exact arithmetic we have to approximate the solution of the ODE. The errors involved in going a small step of the integration scheme are called *local truncation errors*.

These errors are usually much larger than the roundoff errors; furthermore, they depend smoothly on the current iterate, unlike roundoff errors which can vary unpredictably.

It would be tempting to try to set up a backward error theory for **autonomous** ODEs assuming exact arithmetic. That is, to write a perturbation of $\varphi(\cdot, h)$ where h is the step size, as the flow generated by some perturbation of f. Unfortunately this is not possible if we are looking for autonomous perturbations. Partial results can be found which put most of the errors into an autonomous perturbation of the ODEs, but we cannot put *all* of the errors into such a perturbation. A discussion of this is given by Sanz–Serna (1992).

However, if we allow non-autonomous perturbations, then we can get a proper backward error theory for ODEs.

6.4 Attractors and dimensions

If one simulates a dynamical system then usually the system will settle down to wandering around a fairly well-defined (if complicated) set. Such sets are attractors. The formal definition of the attractor of a dynamical system is that a set $A \subset X$ is an *attractor* if

Definition 1

- *A is* invariant*; that is* $\varphi(A, t) = A$ *for all* t.
- *there is an open set V containing A such that for any open set U containing A, there is a t_0 where $t > t_0$ implies $\varphi(V, t) \subset U$.*

Terminology: the union of all such V is called the *basin of attraction* of A. The set A is a *global attractor* if the basin of attraction is the entire space. On the question of the existence of global attractors, we need only the condition that trajectories be bounded, or approach some compact set, to verify their existence.

6.4.1 Approximation of attractors

We do not expect numerically computed *trajectories* to be good approximations to the actual trajectories (at least with the initial conditions that we started with). However, it is easier to believe that the *attractors* we see coming out of a simulation are close to 'the real thing'. To study these questions more closely we need to have some machinery for understanding approximations of sets.

There are some standard metrics for sets, although different situations will require different metrics (just as different problems require different norms for function spaces). First there is a semi-norm: $\delta(A, B)$, which gives how far A is from being a subset of B. More formally,

$$\delta(A, B) \equiv \sup_{a \in A} d(a, B) \equiv \sup_{a \in A} \inf_{b \in B} d(a, b)$$

where $d(x, y)$ is the ordinary distance between the points x and y. (See the Glossary for **inf** and **sup**.) Note:

- $\delta(A, A) = 0$,
- $\delta(A, B) < +\infty$ if A is bounded & B non-empty
- if A and B are closed then $\delta(A, B) = 0 \implies A \subseteq B$
- $\delta(A, B) = \inf\{\eta > 0 \mid A \subset B + \eta B_0\}$

where B_0 is the (open) **unit ball** in X. We usually define $\delta(A, B) = +\infty$ if either A or B is empty.

Associated with this is the *Hausdorff* metric given by

$$d_H(A, B) = \max\{\delta(A, B), \delta(B, A)\}.$$

Set valued functions $A(\alpha)$ are said to be *upper semicontinuous* if

$$\lim_{\alpha \to \beta} \delta(A(\alpha), A(\beta)) = 0;$$

lower semicontinuous if

$$\lim_{\alpha \to \beta} \delta(A(\beta), A(\alpha)) = 0;$$

and *continuous* if

$$\lim_{\alpha \to \beta} d_H(A(\alpha), A(\beta)) = 0.$$

While we would like to get *continuity* of the attractors with respect to perturbations, the best that can be done in general is only upper semicontinuity.

Theorem 1 *Suppose that A is the global attractor of a dynamical system and that we have a numerical process with attractor A_h where $\|\varphi(x, 1) - \varphi_h(x, 1)\| \to 0$ as $h \downarrow 0$ and the A_h lie in a common compact set. Then $\delta(A_h, A) \to 0$ as $h \downarrow 0$.*

Furthermore, if $\|\varphi(x, 1) - \varphi_h(x, 1)\| = O(h^p)$ and A attracts trajectories exponentially fast, then $\delta(A_h, A) = O(h^p)$.

A proof can be found in, for example, Kloeden and Lorenz (1986), or Hale, Lin and Raugel (1988). The basic idea of the proof is described below. Consider a region U around A which is attracted to A under the unperturbed flow. Then for any region $A(\epsilon)$ containing all points within distance $\epsilon > 0$ of A, there is finite time T (depending on ϵ) after which all the points of U have been mapped into $A(\epsilon)$ by the unperturbed flow. Note that we choose ϵ small enough so that $A(2\epsilon) \subset U$. For h sufficiently small the perturbed flow $\varphi_h(x,T)$ is within ϵ of the unperturbed flow $\varphi(x,T)$ for all x in U, since the time T is independent of h. So after time T under the perturbed flow, the trajectories starting in U lie within $A(2\epsilon) \subset U$, and the same argument can be applied to show that for any x in U, $\varphi_h(x, 2T)$ is in $A(2\epsilon)$, and by induction, so is $\varphi_h(x, kT)$ for $k = 1, 2, 3, \ldots$.

To get the $O(h^p)$ estimate of the error, choose T to be large enough that points in a neighbourhood have the distance to the exact attractor halved. Then the same argument goes through, but this time it also gives an estimate of the value of ϵ. Full continuity results can be proven if we assume much stronger conditions. Raugel and Hale (1989) have proven lower semicontinuity as well for *gradient* systems.

One thing that should be pointed out: it is rather unfair to expect that a single trajectory should flesh out the entire global attractor. Of course, if an attractor has several disconnected parts and does not map points in one component into the other, then a single orbit can only give information about one component of the global attractor.

Sometimes though, numerical orbits fail badly to give sensible information about attractors. One simple example, which gives a warning to us all, is the tent map (fig. 6.3)

$$g\colon [0,1] \to [0,1], \qquad g(x) = \begin{cases} \alpha x & x \leq 1/2 \\ \alpha(1-x) & 1/2 \leq x \end{cases}.$$

It can be proven quite easily that for $2 \geq \alpha > 1$ this has chaotic orbits and that for $\alpha = 2$ the attractor of the map is the whole interval $[0, 1]$. But if $\alpha = 2$, the trajectories will quickly end up at zero. (You can try this yourself, to see that it happens.) But zero is not even an attracting fixed point. Why? The answer is that if you consider what is happening from the point of view of the machine's arithmetic, every iteration effectively shifts the bits of the mantissa left. (The operation of forming $(1-x)$ can be thought of basically as inverting the bits of the mantissa.) When all bits have been shifted off, the result is zero.

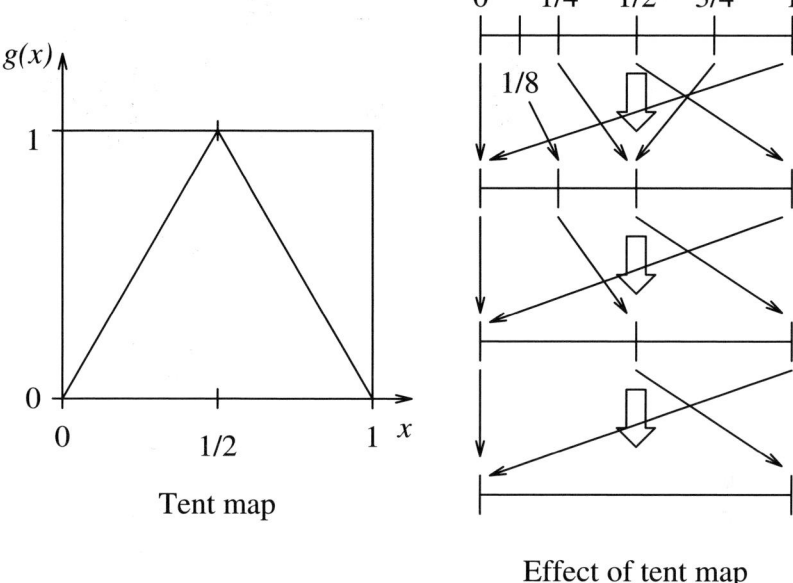

Fig. 6.3. Tent map for $\alpha = 2$ illustrating its effects on selected fractions of the form $k/2^n$.

By the way, one of the easiest ways to show that this map with $\alpha = 2$ is chaotic is precisely in terms of bit shifting (Devaney, 1989, pp. 48–52).

6.4.2 Fractals and approximating dimensions

Assuming that we have a decent approximation for the attractor we will often find that it is a rather complicated self-similar object. Such objects are often referred to as *fractals*; this name was invented by Benoit Mandelbrot because the sets have fractional (Hausdorff) dimension.

The notion of the Hausdorff dimension of a fractal is discussed in detail in §5.3.4 of J. Hutchinson's chapter 'Deterministic and random fractals' (chapter 5). It should be noted that the Hausdorff dimension is associated with a d-dimensional volume measure $\mathcal{H}^d(E)$ of a set E, as described in Hutchinson's article.

This is one of a number of definitions of dimension that are used in dynamical systems theory. From a theoretical perspective the Hausdorff dimension is very useful as it is based on a true measure. However, no-

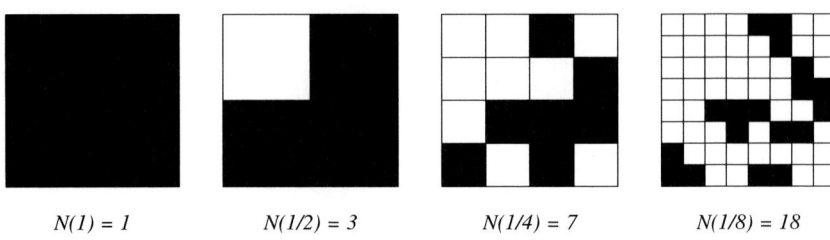

$N(1) = 1$ $N(1/2) = 3$ $N(1/4) = 7$ $N(1/8) = 18$

Fig. 6.4. Box counting method for estimating the fractal dimension.

body knows how to calculate this from data obtained from a simulation, so it (so far) does not have much practical significance.

A related version of dimension is called the *fractal* dimension or the *capacity* dimension of a set A, denoted dim $_F A$. Instead of taking the sum of (diam $U_i)^d$ we impose the condition that diam $U_i \leq \varepsilon$ and replace the sum by $N(\varepsilon)\varepsilon^d$ where $N(\varepsilon)$ is the (minimum) number of sets in the covering. The procedure for defining the 'dimension' apply. However, the resulting 'measure' is not a true measure – it fails to have countable additivity (see Hutchinson's chapter 'Deterministic and random fractals', chapter 5).

A simpler formula can be used than the above:

$$\dim {}_F A = \limsup_{\varepsilon \downarrow 0} \frac{\log N(\varepsilon)}{\log(1/\varepsilon)}.$$

For the meaning of **lim sup** see the Glossary.

Computational methods can be developed for estimating the fractal dimension; the commonest is the *box-counting* method.

The sort of data that is obtained from a simulation is a series of points $x_0, x_1, x_2, \ldots \in \mathbf{R}^n$. These are assumed to belong to some known rectangular region $\{\, x \in \mathbf{R}^n \mid l_i \leq x_i \leq u_i \,\}$ where l_i and u_i are, respectively, lower and upper bounds. The rectangular region is subdivided into 2^n boxes (halving each dimension – see fig. 6.4). The boxes are marked according to whether or not one of the points x_k belongs to the box.

This subdivision can be done recursively, giving a hierarchy of sets of marked boxes. The boxes cover the set $\{x_0, x_1, x_2, \ldots\}$ and as we go to finer and finer subdivisions of the original rectangle, we get better approximations to the actual set.

More formally, if A_k is the union of marked boxes after k subdivisions,

then
$$\overline{\{x_0, x_1, x_2, \ldots\}} = A = \bigcap_{k \geq 0} A_k.$$

If we set N_k to be the number of marked boxes on the level of A_k, then
$$\dim_F A = \limsup_{k \to \infty} \frac{\log N_k}{k \log 2}.$$

While this limit can be estimated directly by looking at the ratio for large k (that is, at fine subdivisions), it is better in practice to fit a line of the form $\log N_k = a + b\, k \log 2$, and to get estimates of a and b. Care should be taken to limit the maximum value of k for such a fit depending on the amount of data. One heuristic that you might use is that at the finest subdivision there should be an average of several points (say > 3) per box. With this maximum value of k, you should pick a minimum value of k over which the fitting should be done. This is a compromise between taking $k_{\min}$ close to $k_{\max}$ to get close to the limit, and choosing $k_{\min}$ small so as to kill some of the noise by fitting over a number of points. You could, for example, take $k_{\min} = k_{\max}/2$.

6.4.3 Other estimates of dimension and their relationships

Before we consider other 'dimensions', we should first note a basic relationship between the Hausdorff dimension and the fractal dimension. In the former, the basic formulae are the same except that the infimum is taken over *all* covers of the set A, while for the latter, the infimum is taken over all covers with diameter ε. Thus, for any A,
$$\dim {}_F A \geq \dim {}_H A.$$

They can and do differ. One of the simpler examples where they differ is the set $B_1 = \{0, 1, 1/2, 1/3, 1/4, \ldots\}$. (Note: this set is closed and can be generated in an obvious way.) In fact, $\dim {}_F B_1 = 1/2$, while $\dim {}_H B_1 = 0$. The latter follows from the fact that the set is countable. The former is an exercise for the reader!

There is a generalised class of such examples $B_\alpha = \{0, 1, 1/2^\alpha, 1/3^\alpha, 1/4^\alpha, \ldots\}$ for $\alpha > 0$. B_α has fractal dimension $1/(1+\alpha)$, but has Hausdorff dimension zero. On the other hand convergent geometric sequences have both fractal and Hausdorff dimension zero.

(These simple observations have implications for estimating the dimensions of attractors. It is common practice to chop off the leading part of the sequence of iterates of a chaotic dynamical system before using them to estimate dimensions. This is almost always appropriate, but if the rate of convergence to the attractor is slower than geometric or exponential, the fractal dimension may be inflated above the true fractal dimension of the attractor.)

As noted above there are a number of definitions of 'dimension', some of which have been invented by people coming to fractals and chaotic systems from different directions:

Correlation dimension   Suppose that we have generated N points of an iteration $x_0, x_1, x_2, x_3, \ldots, x_{N-1}$. For each point x_i count the number of points within a radius of $\varepsilon > 0$ and call this $N_i(\varepsilon)$. The average ratio $N_i(\varepsilon)/N$, where N is large and fixed, should behave like ε^d for some exponent d – this d is the correlation dimension. More formally,

$$\dim {}_C = \limsup_{\varepsilon \downarrow 0} \limsup_{N \to \infty} \frac{\log(\sum_{i=0}^{N-1} N_i(\varepsilon)/N^2)}{\log \varepsilon}.$$

We have not written $\dim_C A$ because it is not really the dimension of a *set* so much as being the dimension of a *probability measure*. For a probability measure μ, the ratio $N_i(\varepsilon)/N$ corresponds to $\mu(x + \varepsilon B_0)$ (B_0 is the unit ball). Thus we get

$$\dim {}_C \mu = \limsup_{\varepsilon \downarrow 0} \frac{\log \int \mu(x + \varepsilon B_0)\, d\mu(x)}{\log \varepsilon}.$$

If the measure is the appropriate Hausdorff measure normalised onto the attractor, then the fractal dimension and the correlation dimensions are identical.

Information dimension   This is more like computing the fractal dimension by the box-counting method, except that this time we assign probabilities to each box. At each level instead of using $\log N_k$ we use $I_k = -\sum_i p_i \log p_i$ where i is an index that ranges over all boxes on level k. Like the correlation dimension this is really a dimension for a probability measure. More formally, the dimension is given by

$$\dim {}_I \mu = \limsup_{k \to \infty} \frac{I_k}{k \log 2}.$$

Equivalent definitions can be given in terms of general (measurable) covers of the attractor.

If μ is a probability measure with bounded support $A \subset \mathbf{R}^n$ we have the inequalities

$$\dim {}_H A \leq \dim {}_F A, \qquad \dim {}_C \mu \leq \dim {}_I \mu \leq \dim {}_F A.$$

Lyapunov dimension   This 'dimension' really belongs in a separate section as it is really a bound on the Hausdorff dimension of an attractor that is computed essentially from the dynamical system and one or more trajectories. More will be said on this in §6.6.

6.4.4 Amount of data

One of the difficulties that must be faced with any method for estimating dimensions is that of limited data. This is particularly important with box-counting and related methods. To get even a rough estimate of the dimension you should get at least hundreds, and usually thousands of points. In practice, for low-dimensional attractors and sets, it is our experience that we can get errors of a few percent by using upwards of $\sim 10^4$ points.

For higher dimensional attractors (dimension > 3, say), the amount of data needed grows somewhat faster.

6.4.5 Time series and embeddings

In recent times, seeing that many deterministic systems are chaotic, and *appear* to be random but are not, there has been a surge of interest in analysing data to see if we can determine whether or not it is (mostly) stochastic or (mostly) deterministic (but possibly chaotic). This leads to the world of 'chaotic time-series'.

If one knew the full state vector then one could quickly determine if a system was stochastic or deterministic – wait until identical (or nearly identical) states re-occurred and look at the difference in the outputs. Deterministic systems will not look random over short time intervals, so randomness can easily be detected or ruled out.

Consider the situation where there is a stream of data coming from some 'real world' process, such as sunspot data, population data, weather (temperatures, rainfall) or the Dow Jones index of the New York Stock Exchange. The problem is to understand what sort of process has generated the data. In a situation like this, not only is the state vector of the process not measurable, we do not even know what dimension it is! Often more information taken at each sampling time will give much more information about the process. For example, other data from the sun might help understand the sunspot cycle, age structure is needed for accurate population prediction and analysis, pressures are useful as well as temperatures in analysing the weather, and a selection of 'key stocks' may give greater insight to the behaviour of stockbrockers. This corresponds to taking a (short) *vector* of measurements at once. As always with data collection, collecting data can be expensive, and there is also the cost of analysing it. A good understanding of the fundamentals of the system should help guide how much and what sort of data is collected. The problem is, given this data, to reconstruct the dynamics of the system, and to obtain further information about, for example, the fractal dimensions of attractors.

A thorough article on this topic is 'Embedology' by Sauer, Yorke and Casdagli (1991). Previous work in this area was done by Takens (1981).

We start with a set $A \subset \mathbf{R}^N$ (A and N are both unknown) which is bounded. We assume that there is a measurement function $h \colon \mathbf{R}^N \to \mathbf{R}^k$. We also assume that A is an invariant set (such as an attractor) for a map $g \colon \mathbf{R}^N \to \mathbf{R}^N$. Sometimes we might have a set (which does not come from a dynamical system) and we just wish to determine its dimension. In this case we just have a measurement function H but no map g.

The following results are taken principally from Sauer *et al.* (1991), whose main results are generalisations of Whitney's embedding theorem (Guilleman & Pollack, 1974, §1.8). To state this we need to define what it means for a property or set to be *generic*. First, consider $\mathbf{R}^m$ for any finite m. On this space we have the **Lebesgue measure** μ_{Leb} which is translation invariant and finite for bounded (measurable) sets. A set E is a *null set* if its Lebesgue measure $\mu_{\text{Leb}}(E)$, or m-dimensional volume, is zero. A property is said to hold *almost everywhere* if the set of exceptions is a null set.

These definitions are usually appropriate for finite dimensional spaces, but for function spaces we cannot define measures in an easy or natural way. So there are some more generally applicable approaches: one is to say that a set is *generic* if it is open and dense (i.e. its closure is the entire space). (Some people define a set to be generic if it is the countable intersection of open dense sets.) In $\mathbf{R}^m$, the two concepts of a property being generic and holding almost everywhere are actually not comparable: neither implies the other. However, if a property is either generic or holds almost everywhere, there is at least something

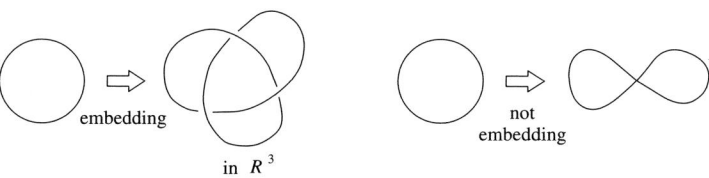

Fig. 6.5. Knotted circle, which can be embedded in three-dimensional space but not in two-dimensional space.

that satisfies the condition – and as close to any given point that you wish.

Sauer, Yorke and Casdagli (1991) define a generalisation of 'almost everywhere' which they call *prevalence*. A set S in a vector space X is called **prevalent** if there is a finite dimensional subspace V where for every $x \in X$ and almost every $v \in V$ the point $x + v$ lies in S.

For sets X and Y, an *embedding* is a smooth map $f: X \to Y$ where f has a smooth inverse on the image $f(X)$. The importance of this idea is that often a set can be embedded in a higher dimensional space in a way that preserves the 'shape' of the set. For example, a circle can be embedded in three dimensional space in a knotted way as shown in fig. 6.5. The main embedding theorem is due to Whitney, and it and related concepts are described in books on differential topology such as Guilleman and Pollack (1974). The Whitney theorem states that if the dimension of a smooth **manifold** X is k, then a *generic* set of smooth maps $f: X \to \mathbf{R}^{2k+1}$ embeds X in $\mathbf{R}^{2k+1}$. This means that for *manifolds*, provided the dimension of the manifold is strictly less than half of the space into which we wish to embed it, almost any smooth function will do.

This result was modified by Sauer and Yorker (1991) in terms of prevalence:

Theorem 2 *If A is a smooth manifold of dimension d, the set of smooth maps $h: A \to \mathbf{R}^m$ which are embeddings is prevalent if $m > 2d$.*

To understand these embedding theorems, the following fact is crucial. If there are two hyperplanes in $\mathbf{R}^m$ where the first has dimension k and the second has dimension l where $k + l < m$, then almost always the hyperplanes will not touch. A simple example to consider is a pair of lines in three-dimensional space. If they do touch, then bumping one of the lines will almost always break the contact. Why is this? For a point to lie on the k dimensional hyperplane is equivalent to that vector satisfying a collection of $m - k$ equations. Similarly for the l dimensional hyperplane. For a point to lie on *both* hyperplanes requires satisfying $(m - k) + (m - l)$ equations with just m unknowns (which are the co-ordinates of the point in $\mathbf{R}^m$). If $k + l < m$, then there are more equations than unknowns, and the system cannot be expected to be solvable in general.

For studying embeddings, the main problem is when the manifold intersects itself in the embedding space. Since the manifold is smooth, it can be approximated by hyperplanes near the point(s) of intersection, and these hyperplanes have the same dimension as the manifold, k. If $2k < m$, then almost any perturbation of the original function will separate the two parts of the same manifold, and the self-intersection will be eliminated – this is the essence of the Whitney embedding theorem.

Sauer *et al.* (1991) went on to prove a related result for the *fractal* dimension of a set. Note that an *immersion* is a smooth map on a manifold whose gradient is one to one on the tangent plane of that manifold.

Theorem 3 *If A is a bounded set with fractal dimension d, then the set of smooth maps $h: \mathbf{R}^N \to \mathbf{R}^m$ satisfying*

(i) *h is one to one on A,*
(ii) *h is an immersion on any smooth manifold in A,*

is prevalent if $m > 2d$.

In practice, we often have just one measurement! (So $h: \mathbf{R}^N \to \mathbf{R}$.) In this case, we have to use a number of measurements to build up an idea of what the state was at the beginning of the series of measurements. However, if we take a long enough sequence we can recreate an diffeomorphic image of the fractal set A:

Theorem 4 *Suppose that g is a diffeomorphism and that A is a compact set in $\mathbf{R}^N$ with fractal dimension d. Assume further that for every positive $p \leq m$, $\dim_F \{\, x \mid g^p(x) = x \,\} < p/2$ and that $\nabla(g^p)(x)$ has distinct eigenvalues whenever $g^p(x) = x$.*

Then for a prevalent set of functions $h: \mathbf{R}^N \to \mathbf{R}$ the map

$$F_m(h, g)(x) = (h(x), h(g(x)), h(g^2(x)), \ldots, h(g^{m-1}(x)))$$

is

(i) *one to one on A, and*

(ii) *an immersion on every smooth manifold in A.*

Once a suitable embedding dimension has been found, parametric models of the local behaviour can be estimated to give a 'model' dynamical system that represents the data. This is a 'hot' topic at the time of writing, and there are quite a few papers coming out which give practical experience with this kind of approach.

So, the general requirement for reconstructing the dynamics of a system is that the embedding dimension must be greater than twice the fractal dimension of the attractor(s) of the system. What should the embedding dimension be if we do not wish to reconstruct the dynamics, but just to estimate the fractal dimension of the attractor? The general conclusion is that it is only necessary to have the embedding dimension $\geq$ the dimension of the attractor(s). However, there are some questions regarding the mathematical validity of this answer.

To preserve the Hausdorff dimension it has been proven that the embedding dimension m just needs to be $\geq$ the Hausdorff dimension of the attractor (Sauer & Yorke, 1991). However, the corresponding result for the fractal dimensions is not true (Sauer & Yorke, 1991). This is not entirely satisfactory, even though for most practical situations the Hausdorff and fractal dimensions are the same.

A standard technique for getting the fractal dimension of an attractor from time series $y_0, y_1, y_2, \ldots$ is to construct short vectors containing m successive values of the ys. The value m is the embedding dimension that is used. These vectors are $x_0 = [y_0, y_1, \ldots, y_{m-1}]$, $x_1 = [y_1, y_2, \ldots, y_m]$, $x_2 = [y_2, y_3, \ldots, y_{m+1}]$, etc. The vectors $x_0, x_1, x_2, \ldots \in \mathbf{R}^m$ are then used as the input data for a fractal-dimension estimating algorithm. The fractal dimension can then be estimated for $m = 1, 2, 3, \ldots$. Since the fractal dimension of $\{x_0, x_1, x_2, \ldots\}$ is (usually) the minimum of m and the fractal dimension of the underlying attractor, if m is large enough, then the fractal dimension of $\{x_0, x_1, x_2, \ldots\}$ is the fractal dimension of the attractor.

As with all dimension estimation techniques for fractals, care must be taken with offsetting the resolution (the ϵ used in the estimates) with the amount of data available, and the embedding dimension m. Asymptotically you need $O(\epsilon^{-d})$ data points where d is the fractal dimension – and more data is better than less. Also, as the embedding dimension increases, the amount of data also needs to increase, although perhaps not as quickly. To get reasonable estimates, 10^4 to 10^5 data points would

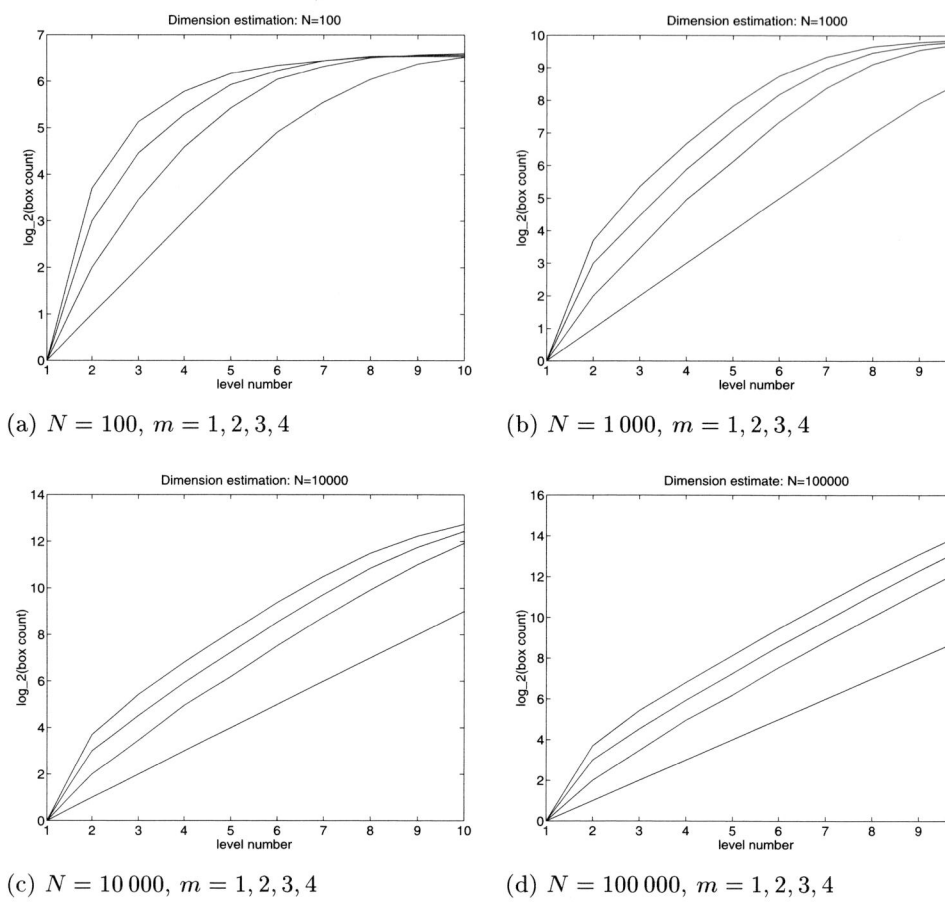

Fig. 6.6. Dimension estimation plots – *note the effect of data size N.*

be needed; but more would not be unwelcome, especially if you wish to have some confidence in the results.

To illustrate these points, fig. 6.6 shows some plots of box counts against $\log_2(1/\epsilon)$ from a box-counting algorithm for data obtained from the Hénon map. Standard parameter values were used: $a = 1.4$ and $b = 0.3$. The fractal dimension of the attractor of the Hénon map is estimated to be ≈ 1.25 from the same box counting approach, using 10^5 points.

For these graphs, the first coordinate was used as the 'measured' quantity from which to estimate fractal dimensions. The main point of these graphs is the amount of data needed to obtain reliable results, and the effect of the embedding dimension. Each graph shows a plot of the logarithm of the box count against the logarithm of the resolution used in the box counting for N data points. The fractal dimension is estimated by the *slope* of the linear part of the curves. The different curves in each graph correspond to different embedding dimensions. The lowest curve is for $m = 1$; the curve above that is the curve for $m = 2$; the curve above that is for $m = 3$, and so on.

For a large amount of data (see the $N = 100\,000$ graph), the linear part is quite clear, and good estimates can be obtained for the fractal dimension. Note that the slope of the curve for $m = 1$ is exactly one; an embedding dimension of one is clearly too small. For the graph with $N = 10\,000$ data points it can be seen that for $m = 3$ and $m = 4$ the slopes are starting to drop off at the finer resolutions. The curve for $m = 2$ would give the best estimate of the fractal dimension. These effects are even more pronounced in the graph for $N = 1000$ data points. With only $N = 100$ data points it has become very difficult to identify a linear region in any of the curves, and is worse for larger values of m.

Clearly, the amount of data needs to be related to both the resolution used, and the embedding dimension. It should also be clear that a great deal of data is needed for obtaining truly reliable results for fractal dimensions.

The problems of limited data are more extensively discussed in several articles (Hunt, 1990, Ramsey & Yuan, 1990, Tong, 1992). Comments on the harder task of reconstructing the dynamics of a system from time series can be found in Casdagli *et al.*(1991) and Kostelich (1992).

6.4.6 Random or chaotic?

One question that arises repeatedly is the question of whether apparently 'noisy' systems are stochastic or are deterministic chaotic systems. Of course, a system could be a combination of the two with truly random noise driving a chaotic system. Trying to answer the question as to whether a system is chaotic from time series data is an extremely challenging one in practice. For a truly random process and an infinite amount of data, the fractal dimension of the data embedded into $\mathbf{R}^m$ is m, as independent random vectors will slowly fill up all of the space. In principle this hypothesis can be tested by embedding and estimating the fractal dimensions. If the estimates of the fractal dimensions for different m increase without bound as m increases, then the system is stochas-

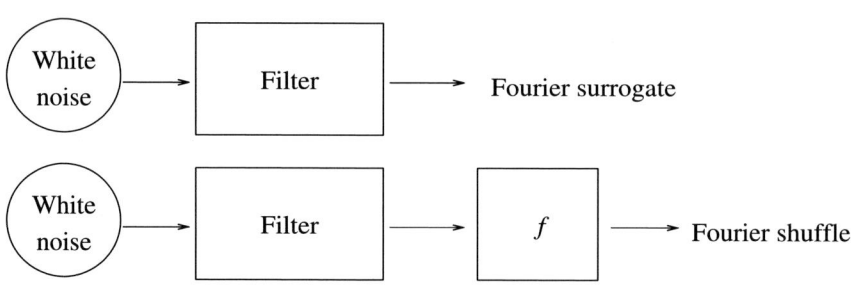

Fig. 6.7. Surrogates for imitating possibly chaotic systems.

tic, since for a chaotic system, the fractal dimensions of embeddings are bounded.

The practical limitation of this approach is that the amount of data needed to do this increases exponentially in m. An alternative approach has been suggested by Sauer (1993) which involves trying to obtain stochastic systems with no chaotic component that mimic the statistical behaviour of the time series data. Such a stochastic system is called a *surrogate* by Sauer (1993) (see fig. 6.7).

One simple example of a surrogate system for a time series is obtained as follows. Take a Fourier transform of the time series data, and save the magnitudes of the Fourier components. The idea is that the frequency components of the original data are preserved if white noise is passed through a filter with a frequency response which is identical to the spectrum (or *periodogram*) of the original data. Such a statistical system is linear, and is clearly not chaotic. To obtain a sample output of such a linear system, for each frequency component independently choose random numbers a and b from a Gaussian distribution with mean zero and standard deviation 1. Multiply the corresponding component of the Fourier transform of the original data by $a + ib$ ($i = \sqrt{-1}$), take the inverse Fourier transform of the result, and ignore the imaginary part of the result. This is the result of filtering white noise in a way that mimics the original data, but in a way that is in no way chaotic. If the test used to distinguish between chaotic and stochastic systems cannot distinguish the surrogate from the original in a statistical sense, then there is no reason to claim that the system is actually chaotic. This 'surrogate test' is stronger than the 'fractal dimensions test' on its own, and can help to identify problems in the latter test due to insufficient data.

6.5 Stable, unstable and centre subspaces and manifolds

Equilibrium points (also called fixed points) are of course very important for understanding a dynamical system, and help to organise its behaviour. Indeed, this is a standard mathematical approach: find the equilibrium points; determine the behaviour near them, and use this information to explore the behaviour further away. This is a fruitful approach whether the system is well-behaved or chaotic. Let us look at an equilibrium point x^*. For concreteness, consider the dynamical system

$$\frac{dx}{dt} = f(x) \quad \text{or} \quad x_{n+1} = g(x_n),$$

with $x \in \mathbf{R}^m$, so that $f(x^*) = 0$ or $g(x^*) = x^*$.

The first way to obtain information about the behaviour of the system near an equilibrium point is to consider the linearisation about x^*: write $x = x^* + y$, where to linear order

$$\dot{y} = \nabla f(x^*) \, y \quad \text{or} \quad y_{n+1} = \nabla g(x^*) \, y_n \, .$$

Determining whether the linear system is stable is a matter of looking at the eigenvalues of the matrix $\nabla f(x^*)$ or of $\nabla g(x^*)$. If λ is an eigenvalue of $\nabla f(x^*)$ with an eigenvector z, then $\nabla f(x^*) \, z = \lambda z$. Then we can look for solutions of $\dot{y} = \nabla f(x^*) \, y$ of the form $y(t) = \psi(t) z$. Substituting this gives

$$\dot{\psi}(t) = \lambda \psi(t) \quad \text{and hence} \quad \psi(t) = e^{\lambda t} \psi(0) \, .$$

If the real part of λ is negative, then $e^{\lambda t}$ goes to zero as $t \to +\infty$; if the real part of λ is positive, then the size of $e^{\lambda t}$ goes to infinity as $t \to +\infty$. A sufficient condition for x^* to be a stable equilibrium for $\dot{x} = f(x)$ is that all eigenvalues of $\nabla f(x^*)$ have negative real part. Furthermore, if any eigenvalue has positive real part, then the original non-linear system is unstable at x^*.

A similar analysis can be made for the discrete time system $x_{n+1} = g(x_n)$ and its linearisation $y_{n+1} = \nabla g(x^*) \, y_n$. Suppose, instead that z is an eigenvector of $\nabla g(x^*)$ with eigenvalue λ:

$$\nabla g(x^*) \, z = \lambda z \, .$$

Then there are solutions of the linearised system with $y_n = \psi_n z$. The equation for the ψ_n is then $\psi_{n+1} = \lambda \psi_n$, which has the solution $\psi_n = \lambda^n \psi_0$. This decays to zero if $|\lambda| < 1$, and grows (to infinity) if $|\lambda| > 1$. Thus the non-linear discrete-time system is stable if all eigenvalues have

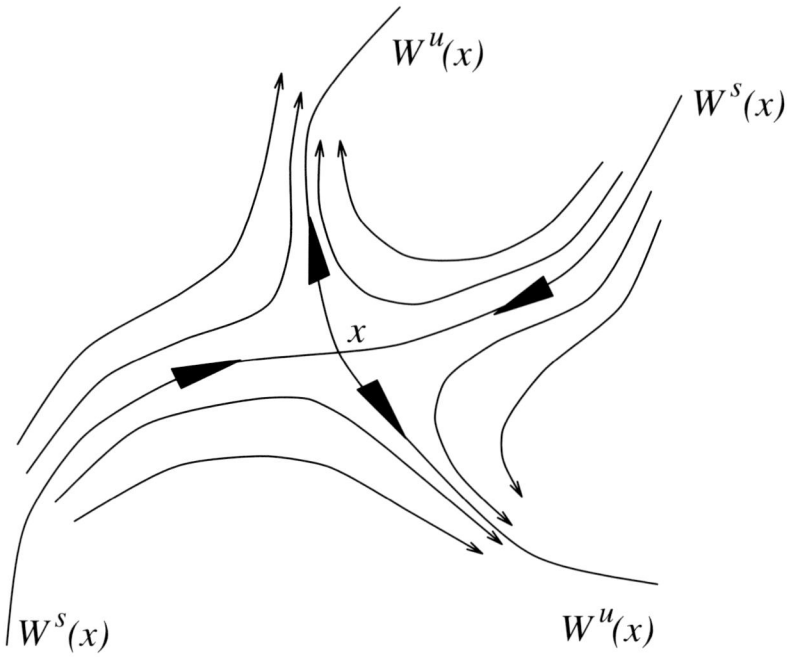

Fig. 6.8. Flow about a hyperbolic point.

magnitude less than one, and unstable if some eigenvalue has magnitude greater than one.

If eigenvalues lie on the stability boundary ($\text{Re}\lambda = 0$ – imaginary axis – for the ODE, or $|\lambda| = 1$ – unit circle – for the discrete-time case) then the analysis of the linearised system is not sufficient to determine the local behaviour of the system. In the alternative case, where no eigenvalue lies on the stability boundary, the local behaviour about x^* *can* be determined from the linearised analysis. Such fixed points are called *hyperbolic*. These systems also exhibit the property of *structural stability* around x^*: that is, small changes to the dynamical system (i.e. perturbing f or g) will not change the (topological) structure of the behaviour near x^*. Equilibrium points which are not hyperbolic are usually associated with *bifurcations* (see §6.7) where the local behaviour can change its structure with arbitrarily small perturbations.

From the nature of the linearised flow, it is clear that there are some directions in which the flow is attracted exponentially toward the equilibrium point and others in which it is repelled exponentially, so that the qualitative nature of the flow is as depicted in fig. 6.8.

Associated with an equilibrium point x^* are its stable and unstable sets which are

$$W^s(x^*) = \{\, x \mid \varphi(x,t) \to x^* \text{ as } t \to +\infty \,\}$$

and

$$W^u(x^*) = \{\, x \mid \varphi(x,t) \to x^* \text{ as } t \to -\infty \,\}$$

respectively. If x^* is a hyperbolic equilibrium point, then these turn out to be smooth **manifolds** – the *stable* and *unstable* manifolds, respectively.

The set of tangent vectors of a smooth manifold form a linear subspace known as the *tangent plane* or *tangent space*. This tangent plane at a point gives the best linear approximation of the manifold nearby to the point. The tangent spaces to $W^s(x^*)$ and $W^u(x^*)$ are denoted $V^s(x^*)$ and $V^u(x^*)$ respectively. (Alternatively, we may just use V^s and V^u if x^* is understood.)

These subspaces are just the eigenspaces associated with the stable and unstable eigenvalues respectively. Clearly, $V^s + V^u = \mathbf{R}^n$ and $V^s \cap V^u = \{0\}$. In order to start computing representations of the stable and unstable manifolds we need to get the stable and unstable subspaces.

6.5.1 Computing the stable and unstable subspaces

One way to get V^s and V^u is to compute the (generalised) eigenvectors for the stable and unstable eigenvalues. To begin with, there may not be as many eigenvectors as the dimension of the space you are working in. In this case the matrix cannot be put into diagonal form. Thus, finding all of the eigenvectors may not be sufficient to compute the right subspaces. In most courses on linear algebra, the **Jordan canonical form** is the standard way to get around this problem. Unfortunately, computing the Jordan canonical form is a numerically unstable process. The most direct indication of this is that the Jordan canonical form is not continuous in the matrix entries.

Consider

$$\begin{bmatrix} 1 & \varepsilon \\ 0 & 1 \end{bmatrix}.$$

This matrix is not diagonalisable for $\varepsilon \neq 0$ and has a single Jordan block;

but if $\varepsilon = 0$ then the matrix is trivially diagonalisable. On the other hand

$$\begin{bmatrix} 1 & 1 \\ 0 & 1+\varepsilon \end{bmatrix}$$

is diagonalisable for all $\varepsilon \neq 0$ but has a trivial Jordan block if $\varepsilon = 0$.

The more stable way of computing eigenspaces is in terms of the Schur decomposition (Golub & Van Loan, 1989, §7.1.2). Given a matrix $A \in \mathbf{C}^{n \times n}$ (where $\mathbf{C}$ is the complex plane) there is a unitary matrix Q and an upper triangular matrix T such that

$$\overline{Q}^T A Q = T,$$

where $(\overline{\cdot})$ denotes complex conjugation and $(\cdot)^T$ denotes the matrix transpose. The eigenvalues of A are the diagonal entries of T. The eigenvalues can be put in any order using unitary similarity transformations by a Schur exchange algorithm (Golub & Van Loan, 1989, §7.1.3).

This sorting technique can be used to separate out the eigenvalues. This gives one way of obtaining orthogonal bases for the subspaces:

$$\begin{bmatrix} T_{11} & T_{12} \\ 0 & T_{22} \end{bmatrix} \begin{bmatrix} \overline{Q}_1^T & 0 \end{bmatrix} = \begin{bmatrix} T_{11}\overline{Q}_1^T & 0 \end{bmatrix}.$$

Alternatively, we can give a linear change of coordinates that is well conditioned but splits the matrix:

$$\begin{bmatrix} I & X \\ & I \end{bmatrix}^{-1} \begin{bmatrix} T_{11} & T_{12} \\ & T_{22} \end{bmatrix} \begin{bmatrix} I & X \\ & I \end{bmatrix} = \begin{bmatrix} T_{11} & T_{12} + T_{11}X - XT_{22} \\ & T_{22} \end{bmatrix}.$$

To obtain the change of coordinates we need to solve the Sylvester equation $T_{12} = XT_{22} - T_{11}X$ so that the upper-right block on the RHS vanishes. Supposing we have already ordered the Schur decomposition so that the stable eigenvalues are in T_{11} and the unstable eigenvalues are in T_{22}, then we can solve this system by the Bartels–Stewart method (Golub & Van Loan, 1989, pp. 386–389): let T_{22} be $q \times q$ and T_{11} be $p \times p$. The qth column of $X = [x_1, \ldots, x_q]$ can be computed by the equation

$$(T_{12})_q = [(T_{22})_{qq}I - T_{11}] x_q.$$

Since $(T_{22})_{qq}$ is not an eigenvalue of T_{11}, this is a well-conditioned system of equations to solve. Once x_q is computed, x_{q-1} can be computed in terms of the $(q-1)$th column of T_{12} and x_q; similarly the remainder of the columns of X can be computed.

If we write $Q = [Q_1, Q_2]$ where Q_1 is $n \times p$ and Q_2 is $n \times q$, then the stable subspace in the original coordinates is span (Q_1), and the unstable subspace is span $(Q_2 + Q_1 X)$.

6.5.2 Equations for the stable manifold

Supposing that we have been able to split the linearisation of the system at the equilibrium point (and so compute bases for the stable and unstable subspaces), we can set up some equations for the stable and unstable manifolds. We will just look at the stable manifold here; the unstable manifold can be defined in similar fashion, but with time reversed. This works provided the dynamics can be reversed. For ODEs this is just a matter of reversing the sign of the derivatives with respect to time; for diffeomorphisms this requires inverting $x_{n+1} = g(x_n)$ to $x_n = g^{-1}(x_{n+1})$. However in general this is not always possible, for example for maps where g^{-1} does not exist as a function, or for dissipative PDEs like the heat equation (§6.2.6) which give rise to semiflows (§6.2). We will look at the ODE case here. The case for diffeomorphisms and maps is mostly taken care of by replacing integrals by sums, and exponentials by powers.

The ODEs can be put in the following form by an affine change of variables:

$$\begin{bmatrix} \dot{y} \\ \dot{z} \end{bmatrix} = \begin{bmatrix} A & \\ & B \end{bmatrix} \begin{bmatrix} y \\ z \end{bmatrix} + \begin{bmatrix} F(y,z) \\ G(y,z) \end{bmatrix}$$

where $F(0,0) = G(0,0) = 0$ and $\nabla F(0,0) = \nabla G(0,0) = 0$. For any $\varepsilon > 0$ we can find a **ball** about the origin of radius $\delta > 0$ for some δ where F and G are **Lipschitz** in this neighbourhood with Lipschitz constant ε. It is also assumed that F and G have a global Lipschitz constant K' to ensure that the ODEs are meaningful.

Since A contains the stable eigenvalues and B the unstable eigenvalues we have bounds of the form $\|e^{At}\| \le Ke^{-\alpha t}$ for $t \ge 0$ and $\|e^{Bt}\| \le Ke^{\alpha t}$ for $t \le 0$ where $\alpha > 0$. The actual value of α must be less than the smallest real part of an eigenvalue of B, and $-\alpha$ greater than the largest real part of an eigenvalue of A.

Then we consider the integral equations for a given y_0

$$y(t; y_0) = e^{At} y_0 + \int_0^t e^{A\tau} F(y(t-\tau), z(t-\tau)) \, d\tau$$

$$z(t; y_0) = -\int_t^\infty e^{B(t-\tau)} G(y(\tau), z(\tau)) \, d\tau.$$

It is easy to check that a solution of this integral equation is also a solution to the above ODEs. Provided $\|y_0\|$ is small enough, then a Picard type iteration will converge. This is because the operator of $y^{(m)}$ and $z^{(m)}$ is a contraction (i.e. the operation reduces distances by a guaranteed ratio) for sufficiently small $y^{(m)}$ and $z^{(m)}$:

$$y^{(m+1)}(t; y_0) = e^{At} y_0 + \int_0^t e^{A\tau} F(y^{(m)}(t-\tau), z^{(m)}(t-\tau))\, d\tau$$

$$z^{(m+1)}(t; y_0) = -\int_t^\infty e^{B(t-\tau)} G(y^{(m)}(\tau), z^{(m)}(\tau))\, d\tau.$$

For details see Perko (1991, §2.7) and Palis & de Melo (1982). The reason why this works is that the integrals are arranged so that the exponentials are both decaying.

The stable manifold is given locally in terms of solutions of this integral equation by the parametrisation $(y_0, z(0; y_0))$. Once the stable manifold has been locally defined, it can be extended to be a global manifold by means of solving the reversed time equations.

6.5.3 Stable and unstable manifolds for maps and diffeomorphisms

Corresponding results can be obtained for stable manifolds for maps and diffeomorphisms. As with the ODE case we need to split the linearisation of the system by a linear change of coordinates:

$$\begin{bmatrix} y_{n+1} \\ z_{n+1} \end{bmatrix} = \begin{bmatrix} A & \\ & B \end{bmatrix} \begin{bmatrix} y_n \\ z_n \end{bmatrix} + \begin{bmatrix} F(y_n, z_n) \\ G(y_n, z_n) \end{bmatrix}$$

where F and G satisfy the same conditions as above. The matrix A contains the stable eigenvalues (with magnitudes less than one) and B the unstable eigenvalues (with magnitude greater than one). The equations for the stable and unstable manifolds are then given by solving the following equations for a given y_0:

$$y_n = A^n y_0 + \sum_{k=0}^{n-1} A^k F(y_{n-k}, z_{n-k})$$

$$z_n = -\sum_{k=n+1}^{\infty} B^{n-k} G(y_k, z_k).$$

We can use a Picard iteration for these equations:

$$y_n^{(m+1)} = A^n y_0 + \sum_{k=0}^{n-1} A^k F(y_{n-k}^{(m)}, z_{n-k}^{(m)})$$

$$z_n^{(m+1)} = -\sum_{k=n+1}^{\infty} B^{n-k} G(y_k^{(m)}, z_k^{(m)}).$$

With zero initial conditions $y_k^{(0)} = 0$, $z_k^{(0)} = 0$ we get convergent sequences $\mathbf{y}^{(m)} \to \mathbf{y}$ and $\mathbf{z}^{(m)} \to \mathbf{z}$ provided y_0 is small enough. The stable manifold is given by $z(y_0) = z_0$.

6.5.4 Stable and unstable manifolds for periodic orbits

Stable and unstable manifolds also exist for periodic orbits, in both the discrete time and continuous time cases, though there are some subtle differences between the two cases. The discrete time case with diffeomorphisms is the easiest.

Consider a periodic orbit: $x_{k+1} = g(x_k)$ for $k = 0, \ldots, p-1$ and $x_0 = g(x_{p-1})$. We can consider this as being equivalent to the equilibrium case with the diffeomorphism $G = g^p$. The linearisation of this map at x_0 is

$$\nabla g(x_{p-1}) \nabla g(x_{p-2}) \ldots \nabla g(x_1) \nabla g(x_0).$$

The first step is to compute the eigenvalues of this product matrix. Care should be taken with long products, as often the larger eigenvalues will be much larger than the smallest eigenvalues – if the ratio between the smallest to the largest is of the same order as machine epsilon, or smaller, then the smallest eigenvalues will be lost in the computational 'noise'. Recall that machine epsilon describes the precision of the floating point computations (§6.3.1).

While the product matrix $\nabla G(x_0) = \nabla g(x_{p-1}) \ldots \nabla g(x_0)$ does depend on which point of the periodic orbit is x_0, the eigenvalues of the product are independent of which point is used as x_0. The reason is that $\nabla G(x_0)$ is similar to

$$\nabla g(x_0) \nabla G(x_0) (\nabla g(x_0))^{-1} = \nabla G(x_1)$$

since $x_0 = g(x_{p-1}) = x_p$.

There are computationally suitable algorithms due to Bojanczyk, Van Dooren and Golub (1992), who compute the (real) Schur decomposition in factored form. Keeping everything in factored form permits a backward error analysis *in terms of the matrix factors*.

Each point in the periodic orbit has its own stable and unstable subspaces and manifolds. The subspaces are related through the Jacobian matrices:

$$V^s(x_{k+1}) = \nabla g(x_k)\, V^s(x_k), \qquad V^u(x_{k+1}) = \nabla g(x_k)\, V^u(x_k).$$

The manifolds are related through the map:

$$W^s(x_{k+1}) = g(W^s(x_k)), \qquad W^u(x_{k+1}) = g(W^u(x_k)).$$

The continuous time case is a bit more difficult, and we have to modify the meaning of 'hyperbolic' for periodic solution of **autonomous** ODEs. In this case we have $\dot{x} = f(x)$ and $x(0) = x(T)$ where $T > 0$ is the period of the orbit. However, the map $x \mapsto \varphi(x,T)$ is *not* hyperbolic at $x(0)$. The reason is that $\nabla_x \varphi(x(0), T)$ has an eigenvalue 1 with the eigenvector $f(x(0))$. The way to prove the existence of stable and unstable manifolds is to set up a local Poincaré section at $x(0)$ perpendicular to (or at least not tangential to) $f(x(0))$, provided only one simple eigenvalue of $\nabla_x \varphi(x(0), T)$ is on the unit circle. Stable and unstable manifolds for the local Poincaré map at $x(0)$ can then be set up, and then these manifolds can be transferred to all the other points in the periodic orbit by the maps $x \mapsto \varphi(x,t)$ for $0 < t < T$.

As with the discrete time case, $\nabla_x \varphi(x(t), T)$ does change with t, but the eigenvalues do not change.

6.5.5 Hyperbolicity for attractors

So far stable and unstable manifolds and the concept of hyperbolicity have been defined for equilibria and periodic orbits. This can be generalised to general (fractal) invariant sets. The notion of an invariant set being hyperbolic is closely related to the idea of *exponential dichotomies* (Coppel, 1978), where a linear system can be split into parts, one of which is exponentially growing, and the other is exponentially decaying. Hyperbolicity of a dynamical system means that the linearisation about any trajectory has an exponential dichotomy whose constants are independent of the trajectory. Where a linear system has exponential dichotomies it turns out that *two point boundary value problems* are well conditioned, even if the corresponding *initial value problems* are not.

Another reference which ties together exponential dichotomies and related topics is Palmer (1988).

Hyperbolicity is important in many studies of dynamics. One example is Smale's *axiom A* (Smale, 1980 (originally 1967)). Another is in shadowing, which we treat later.

The idea of hyperbolicity is that each point has a pair of subspaces of which one contains the exponentially growing modes, and the other contains the exponentially decaying modes. To use this, it is usual to consider the exponentially decaying modes forward in time, and the exponentially growing modes backwards in time. An invariant set A of a discrete time dynamical system $g: \mathbf{R}^n \to \mathbf{R}^n$ is hyperbolic if for each $x \in A$ there is a splitting $\mathbf{R}^n = V^s(x) + V^u(x)$ and constants $K > 0$ and $\alpha < 1$ such that

- $V^s(x)$ and $V^u(x)$ are both continuous in x in the sense that $d_H(V^s(x) \cap B_0, V^s(y) \cap B_0) \to 0$ as $x \to y$ where B_0 is the usual unit ball.
- $\nabla g(x) V^s(x) = V^s(g(x))$ and $\nabla g(x) V^u(x) = V^u(g(x))$.
- if $P^s(x)$ and $P^u(x)$ are the projections onto $V^s(x)$ and $V^u(x)$ respectively, then

$$\|P^s(g^k(x))\nabla(g^k)(x)P^s(x)\| \leq K\alpha^k,$$

$$\|(P^u(g^k(x))\nabla(g^k)(x)P^u(x))^{-1}\| \leq K\alpha^k$$

for all $k \geq 0$.

The projections are completely specified with the additional condition that $P^s(x) + P^u(x) = I$ for all x. An analogous definition applies to autonomous continuous time systems, although it must be modified to take account of the (simple) eigenvalue of $\nabla_x \varphi(x, t)$ of 1. This simple eigenvalue, at least for periodic trajectories, corresponds to the direction of motion for the ODE. The behaviour of most concern regarding continuous–time systems is the behaviour orthogonal to the direction of travel.

To illustrate the use of this idea, consider axiom A diffeomorphisms:

Definition 2 *A diffeomorphism* $g: \mathbf{R}^n \to \mathbf{R}^n$ *satisfies* axiom A *if the non-wandering set*

$$\Omega = \{\, x \mid \text{for every open } U \text{ containing } x,\ g^k(U) \cap U \neq \emptyset \text{ for some } k \neq 0 \,\}$$

is hyperbolic and is the closure of a set of periodic orbits.

(See Smale 1980 (originally 1967), §I.6.) If an attractor A is hyperbolic, is the closure of a set of periodic orbits, and has a dense orbit, then the flow on A is structurally stable. It is usually thought that the attractors that appear in numerical simulations for, say, Lorenz' equations, are structurally stable. Theorems such as this are sometimes used to explain why attractors obtained by simulations so often appear to be quite robust, not only to numerical errors, but also to quite large changes in parameters. Unfortunately, the theoretical situation is rather unsatisfactory. On the one hand the theoretical tools assume the existence of structures such as hyperbolicity. On the other hand, the existence of these structures has been verified only for specialised cases. For 'real' problems the situation remains unclear in spite of considerable numerical analysis and computational work.

6.5.6 Non-hyperbolic fixed points – centre manifolds

Although the case where the fixed point x^* is hyperbolic is generic, the non-hyperbolic case is important because it occurs at bifurcations (see §6.7). Thus, consider the case in which one or more of the real parts of the eigenvalues of $\nabla f(x^*)$ is zero, or one or more of the eigenvalues of $\nabla g(x^*)$ is on the unit circle. Associated with these eigenvalues there is then a *centre subspace* $V^c(x^*)$, in addition to the stable and unstable subspaces $V^s(x^*)$ and $V^u(x^*)$. $V^c(x^*)$ is tangent to an invariant manifold, $W^c(x^*)$ (Carr, 1981, Guckenheimer & Holmes, 1983, Shub, 1987), at least locally, but this *centre manifold* is not unique – there is typically an infinity of centre manifolds of any given degree of differentiability which become exponentially close in the neighbourhood of x^*.

For example, consider the two-dimensional dynamical system

$$(\dot{x}, \dot{y}) = (x, y^2) \; .$$

The vector field (x, y^2) is plotted in fig. 6.9, as well as the trajectories $(x, y) = \varphi_t(x_0, y_0)$ for initial values $(x_0, y_0) = (\pm 0.5, -1)$, with $t \in (0, 0.7)$, and initial values $(\pm 0.5, 1)$, $(\pm 1, 0.5)$, $t \in (-10, 0)$. All the trajectories in the upper half-plane are tangential to the y-axis at the fixed point (the origin). These, taken together with the origin and the negative y-axis, all form centre manifolds – cf. Arrowsmith and Place (1990, p. 96).

When there are both stable and unstable manifolds, one also defines two more invariant manifolds (Shub, 1987) the centre stable and centre unstable manifolds, tangent respectively to $V^s(x^*) \oplus V^c(x^*)$ and $V^c(x^*) \oplus V^u(x^*)$. The centre manifold is then the intersection of these.

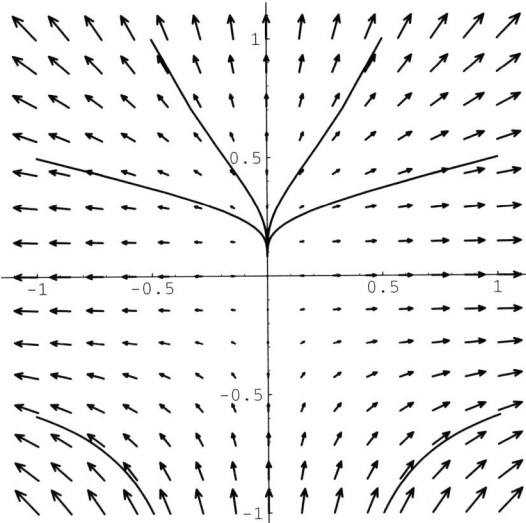

Fig. 6.9. Plot of the vector field (x, y^2) in the (x, y) plane, showing the fixed point at the origin and the unstable manifold on the x-axis. Also shown are some typical trajectories described in the text.

In applications one is primarily concerned however with stable fixed points.

The question of stability of a non-hyperbolic fixed point is much subtler than in the hyperbolic case since it is no longer sufficient for stability that $V^u(x^*) = \{0\}$. However (Carr, 1981, theorem 2(a)) in this case stability can be completely determined by restricting consideration to dynamics on one of the centre manifolds. This can be a very important simplification since the centre manifold may be of much lower dimension than the original dynamical system (which may even be infinite-dimensional).

In these delicate cases it is necessary to have a definition of stability that is not based on exponential divergence or contraction. Instead we use *stability in the sense of Lyapunov*:

Definition 3 *A fixed point, x^*, of a flow φ_t is said to be* stable *if, for every neighbourhood N of x^*, there is a neighbourhood $N' \subseteq N$ of x^* such that, if $x \in N'$, then $\varphi_t(x) \in N$ for all $t > 0$.*

Typically a stable fixed point x^* has a stable manifold on which exponential attraction toward x^* occurs, but Lyapunov stability does not imply that *most* orbits are attracted toward x^*, merely that they are not repelled from x^*. However any solution of a stable dynamical system (started sufficiently close to x^*) *is* attracted toward the centre manifold, i.e. it approaches some solution on the chosen centre manifold with an error decaying exponentially as time increases (Carr, 1981). Thus, apart from losing initial transients, we can understand the general dynamics completely by restricting attention to motion on the centre manifold. Note that the motion on the centre manifold can in principle be arbitrarily complicated. The centre manifold is a particular case of an inertial manifold (see §6.8), appropriate to parameter ranges close to bifurcation points.

Applications of centre-manifold theory to spatio-temporal chaos can be found in Eckmann and Procaccia (1991) and Roberts (1992). Both papers discuss model equations for convection in which several convective rolls with close wave vector can be simultaneously linearly unstable, and interact non-linearly, so that interference can produce 'defects' which move chaotically. The former assumes a small aspect ratio (convection in a channel of finite width) while the latter assumes an aspect ratio of order unity. In this case the dimension of the centre manifold tends to infinity as the size of the box tends to infinity; Roberts (1992) argues that a simplified PDE can be derived whose centre manifold is the same as that of the original system and which approximates the dynamics more accurately than previous attempts. This is supported by numerical calculation in a box of finite size using a spectral method (projection of the PDE onto a finite Fourier basis).

6.6 Lyapunov exponents

Lyapunov exponents (Temam, 1988) describe the long term exponential rates at which the dynamical system stretches and squeezes sets. They provide considerable information about the behaviour of the system and can be used to estimate, amongst other things, the Hausdorff dimension of the attractor, the entropy rate, and estimates of the prediction horizon. In fact, one definition of chaos is simply that there is a positive Lyapunov exponent (and bounded trajectories).

The paper by Eckmann and Ruelle (1985) gives considerable information about Lyapunov exponents, including some information about how to estimate them. See also Ruelle (1989, Ch. 9).

To define the Lyapunov exponent for a trajectory with initial state x_0 consider the Jacobian matrix $\Phi(t) = \nabla_x \varphi(x_0, t)$. The ith Lyapunov exponent is given by

$$\lambda_i = \limsup_{t\to\infty} \frac{\log \sigma_i(\Phi(t))}{t}$$

where $\sigma_i(A)$ is the ith largest singular value of the matrix A (for the definition of **lim sup** see the glossary). The singular values are just the square roots of the eigenvalues of $A^T A$. (The matrix $A^T A$ is positive semidefinite so the square roots of its eigenvalues are real and non-negative.)

Dynamical systems where the above 'lim sup' can be replaced with just 'lim' are called *normal*. It is believed that most dynamical systems are normal on the basis of numerical simulations.

If we consider a small sphere about x_0 then its image at time t is (very nearly) a small ellipse. The singular values are the ratios of the semimajor axes of this ellipse to the radius of the initial sphere. Thus the Lyapunov exponents measure the (exponential) rates of stretching and squeezing.

A bound on the Hausdorff dimension of an invariant set A generated by a trajectory can be found in terms of the Lyapunov exponents $\lambda_1 \geq \lambda_2 \geq \ldots \geq \lambda_n$. Let j be the largest index where $\lambda_1 + \lambda_2 + \ldots + \lambda_j > 0$. (We take $\lambda_{n+1} = -\infty$ to ensure that there is such a j.) Then the Hausdorff dimension of A is bounded by

$$\dim{}_H A \leq j + \frac{\lambda_1 + \lambda_2 + \ldots + \lambda_j}{|\lambda_{j+1}|}.$$

The quantity on the right is sometimes referred to as the *Lyapunov dimension* of A.

Lyapunov exponents can also be used to estimate the *entropy rate*, or *Kolmogorov–Sinai entropy* (Lichtenberg & Lieberman, 1992, p. 304) of a dynamical system. This rate is the rate at which information about the initial conditions is lost. If we have a dynamical system with bounded trajectories we can imagine carving up the state space into cubes (say) of size $\varepsilon > 0$. Any trajectory of a map can be followed to within a tolerance of ε by noting the sequence of cubes that the trajectory touches. The set of all such sequences of boxes of given length N is clearly finite, and a particular sequence can be described by a string of bits of length $b(\varepsilon, N)$. The limit $\limsup_{N\to\infty} b(\varepsilon, N)/N$ is the average number of bits needed to specify a particular trajectory within a tolerance of ε. The entropy rate of the map is

$$\rho = \limsup_{\varepsilon \downarrow 0} \limsup_{N \to \infty} \frac{b(\varepsilon, N)}{N \log_2 \varepsilon}.$$

Lyapunov exponents give a way of estimating the entropy. An upper bound for the entropy is given by $(\lambda_1 + \ldots + \lambda_j)/\log 2$ where j is the largest index of a positive Lyapunov exponent.

There are some instances where Lyapunov exponents can be computed very quickly. For example, with constant coefficient systems $\dot{x} = Ax$, the Lyapunov exponents are just the real parts of the eigenvalues of A. Trajectories of **autonomous** ODEs that are bounded but do not approach equilibrium points must have a zero Lyapunov exponent. This Lyapunov exponent is associated with the direction of the flow.

6.6.1 Computing Lyapunov exponents

A survey of methods for computing Lyapunov exponents is Geist, Parlitz and Lautersborn (1990).

In principle the above definition of Lyapunov exponents gives a numerical procedure for computing Lyapunov exponents, at least about a particular trajectory. However, because of the exponential growth and decay of the singular values, the smaller singular values can quickly be lost in the numerical 'noise'. One way to get around this is to perform the computations in extended precision arithmetic – with about $\log_{10}(\sigma_1(\Phi(t))/\sigma_n(\Phi(t)))$ digits! For a product of several hundred matrices this would require several hundreds to thousands of digits! Needless to say, this is a rather expensive way of doing these computations.

There are much better methods which don't require extended precision arithmetic but still give quite accurate results. The best known is described by Eckmann and Ruelle (1985); this method is better known in the numerical analysis literature as *treppeniteration*. This method does not compute the singular values of the product, but rather what is known as the *QR factorisation*: the QR factorisation of a matrix A is a factorisation $A = QR$ where Q is orthogonal and R is upper triangular. This can be computed accurately in a suitable backward error sense. (For details see (Golub & Van Loan, 1989, ch. 5).) The QR factorisation is well known in numerical linear algebra and is widely used for least-squares problems, finding orthonormal bases for null-spaces and as a component of algorithms for computing eigenvalues (Golub & Van Loan, 1989).

Treppeniteration gives a QR factorisation of the product

$$\nabla_x \varphi(x, k) = \nabla g(x_{k-1}) \nabla g(x_{k-2}) \ldots \nabla g(x_0)$$

as follows. Set $C_0 = \nabla g(x_0)$ and factorise $C_0 = Q_0 R_0$ by the standard QR factorisation. Then form $C_1 = \nabla g(x_1) Q_0$ and then factorise $C_1 = Q_1 R_1$. We repeat the process, forming $C_2 = \nabla g(x_2) Q_2$ and factorising $C_2 = Q_2 R_2$. Repeating until we reach $C_k = \nabla g(x_k) Q_{k-1}$ and factorising $C_k = Q_k R_k$ we have

$$\nabla_x \varphi(x, k) = Q_k R_k R_{k-1} \ldots R_2 R_1 = Q^{(k)} R^{(k)}.$$

Since the product of upper triangular matrices is upper triangular, this is a QR factorisation of the product.

If we start with a 'randomisation' step $C_0 = \nabla g(x_0) Q_{-1}$ with Q_{-1} a random orthogonal matrix, then we have the theorem:

Theorem 5 *For almost all Q_{-1} the Lyapunov exponents are given by*

$$\lambda_i = \limsup_{k \to \infty} \frac{\log |r_{ii}^{(k)}|}{k}.$$

(The original version of treppeniteration sets $Q_{-1} = I$.) The reason for the initial randomisation step is that the columns of Q_{-1} may not have non-zero components in the associated Lyapunov directions. Usually this randomisation is ignored as (1) failure happens only with probability zero in exact arithmetic, and (2) numerical errors usually generate vectors with non-zero components in the right directions.

Note that the diagonal entries of $R^{(k)}$ are just the products of the diagonal entries of the factors, so $r_{ii}^{(k)} = \prod_{j=0}^{k}(R_j)_{ii}$. Thus

$$\lambda_i = \limsup_{k \to \infty} \frac{\sum_{j=0}^{k} \log |(R_j)_{ii}|}{k}.$$

Treppeniteration has a backward error analysis which can be developed from the backward error analysis of the ordinary QR factorisation. The computed factorisation $\widehat{Q}^{(k)} \widehat{R}_k \widehat{R}_{k-1} \ldots \widehat{R}_0$ satisfies the property that there is an exactly orthogonal matrix $\widetilde{Q}^{(k)}$ near $\widehat{Q}^{(k)}$ where

$$\widetilde{Q}^{(k)} \widehat{R}_k \widehat{R}_{k-1} \ldots \widehat{R}_0$$

is the exact QR factorisation of

$$(\nabla g(x_k) + E_k)(\nabla g(x_{k-1}) + E_{k-1}) \ldots (\nabla g(x_0) + E_0)$$

where $\|E_i\|_2 = O(\mathbf{u}) \|\nabla g(x_i)\|_2$ where $\mathbf{u}$ is 'machine epsilon' or 'unit roundoff' (§6.3.1). This corresponds to a time-dependent perturbation of the map of size $O(\mathbf{u})$ relative to the size of $\nabla g(x)$.

Alternative algorithms for estimating Lyapunov exponents exist, with some tuned for the ODE case. One is based on the *stabilised march* which is used for solving boundary value problems. If $\dot{x} = f(x(t))$ then the variational equation is

$$\Phi'(t) = \nabla f(x(t))\,\Phi(t) = A(t)\,\Phi(t).$$

The idea is to set up ODEs for the Q and R factors of $\Phi(t)$.

Write $\Phi(t) = Q(t)R(t)$. Differentiating gives the system

$$Q'R + QR' = AQR.$$

Since $Q(t)$ is orthogonal, $Q(t)^T Q(t) = I$ for all t. Differentiating this gives

$$(Q')^T Q + Q^T Q' = 0.$$

Thus $Q(t)^T Q'(t)$ is an antisymmetric matrix $S(t)$, and $Q'(t) = Q(t)S(t)$. Hence $QSR + QR' = AQR$. Pre-multiplying by Q^T and post-multiplying by R^{-1} gives

$$S + R'R^{-1} = Q^T A Q.$$

Taking the strict lower triangular part of this matrix ODE gives

$$s_{ij}(t) = q_i(t)^T A(t) q_j(t) \qquad \text{for } i > j.$$

This can be extended to a unique antisymmetric matrix. Thus we have a well-defined ODE for $R(t)$. As we are really only interested in the diagonal entries of $R(t)$ we take the diagonal part of the ODE for R. This gives

$$\frac{r'_{ii}(t)}{r_{ii}(t)} = \frac{d}{dt} \log r_{ii}(t) = q_i(t)^T A(t) q_i(t).$$

This is clearly just an integration once $Q(t)$ is computed via $Q' = QS$.

Care must be taken here to ensure that Q remains orthogonal, by suitable integration schemes (Deici *et al.*, 1994) or by periodically forcing $Q(t)$ to be orthogonal. If this is not done, then as the set of orthogonal matrices may be unstable for the ODE for Q, the method breaks down.

There are also algorithms for computing the *singular value decomposition* (SVD) of a product of matrices. This is the factorisation

$$A = U\Sigma V^T$$

where U and V are orthogonal matrices and Σ is a diagonal matrix diag $(\sigma_1, \ldots, \sigma_n)$ with $\sigma_1 \geq \sigma_2 \geq \ldots \geq \sigma_n$. (See Golub and Van Loan (1989, §2.5,§8.3).) The value σ_i is the ith singular value of A. Bojanczyk, Ewerbring, Luk and Van Dooren (1991) describe a method based on Jacobi rotations. Stewart (1994a, 1994b) and Abarbanel et al. (1992) give methods based on treppeniteration.

Computing SVDs by setting up differential equations for the singular values, analogously to the above method for the QR factorisation, suffers from breakdown when singular values cross over. This breakdown does not occur for the continuous QR factorisation algorithm.

6.7 Continuation and bifurcation

Continuation is a technique for solving non-linear equations. This is done by embedding the equations in a one-parameter family of equations $H(x, \lambda) = 0$. Starting at one parameter value, a solution can be followed along the curves in (x, λ) space, hopefully until a solution of the original problem is found. This technique can be used to follow branches of equilibrium points, periodic solutions and particular trajectories. Where continuation breaks down there are usually bifurcations.

Bifurcations occur in both non-linear equations and in dynamical systems, though what can happen in dynamical systems is usually much richer and more complex. Bifurcations are essentially local phenomena; they are about what happens as a parameter varies in the neighbourhood of a bifurcation point.

The theory of bifurcations is a very rich subject and goes well beyond the scope of this chapter. A good introduction is the review by Crawford (1991). All we can hope to do here is to have a look at some of the simpler and more important bifurcations.

Regarding numerical techniques, two sources are Keller (1987) and Crouzeix and Rappaz (1990).

6.7.1 Bifurcation jargon

If we picked a random matrix we would find that it is non-singular with probability one. This is called the *generic* case. We might say that matrices are *generically* non-singular. In dynamical systems theory an equilibrium point x_0 of a smooth ODE or map is generically hyperbolic (i.e. with no eigenvalues on the stability boundary – see §6.5). What is of interest is what happens when one or more eigenvalue(s) cross(es) the stability boundary. Where this happens determines the sort of bifurcation.

Often bifurcations are described as being *codimension n* bifurcations. This means that they (usually) occur on submanifolds of the parameter space with dimension n less than the full dimension. This idea can be made precise for spaces of functions by a similar technique to that used to define prevalence. Another way of understanding the codimension is as the number of parameters that are needed to make the bifurcation happen *for some set of parameter values.*

The simplest bifurcations are called *codimension* 1 bifurcations. The next simplest are *codimension* 2 bifurcations.

If we considered square matrices, almost all matrices are non-singular. But amongst the singular matrices, the matrices with a one-dimensional null space are generic. The matrices with a one-dimensional null space are said to have codimension 1 in the space of all matrices.

The way the solutions change as the parameters vary is called the *unfolding* of the bifurcation. If all unfoldings are equivalent to a given unfolding, or some subset of it, that unfolding is called *universal*.

Usually the dimensionality of the bifurcation can be reduced to a few 'essential' dimensions by means of a variant of the implicit function theorem. This is called *Lyapunov–Schmidt reduction.*

The connections between the branches of solutions and their bifurcations is often displayed on a bifurcation diagram, which plots some relevant quantity against a parameter. Stable branches are usually indicated by solid lines and unstable branches by dotted or dashed lines. An example can be seen in fig. 6.10 which represents a fold bifurcation. The top branch is unstable while the bottom branch is stable.

6.7.2 Some basic bifurcations

These bifurcations are all codimension 1 bifurcations.

Fold bifurcation In dynamical systems terms, this happens when two equilibria collapse and disappear. A universal unfolding for this comes from the equations (for an equilibrium point):

$$x^2 - \lambda = 0.$$

Here x is the state variable and λ is the parameter that is being varied. For $\lambda > 0$ there are two equilibria at $+\sqrt{\lambda}$ and at $-\sqrt{\lambda}$. If we consider

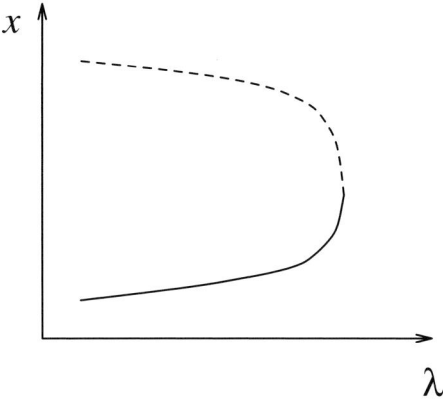

Fig. 6.10. Bifurcation diagram for a fold.

the ODE

$$\dot{x} = x^2 - \lambda$$

the equilibrium at $-\sqrt{\lambda}$ is stable while the equilibrium at $+\sqrt{\lambda}$ is unstable. A bifurcation diagram for this is shown in fig. 6.10.

Other bifurcations for ODEs and maps come from a single equilibrium when its stability changes. These bifurcations are described in terms of where eigenvalues cross the stability boundary. Generically, only one real eigenvalue or a complex conjugate pair cross the stability boundary at a time. For a system of ODEs the stability boundary is the imaginary axis so either a real eigenvalue crosses the axis at zero, or a complex conjugate pair crosses the imaginary axis. The former case is called a *pitchfork* bifurcation, the latter a *Hopf bifurcation*. For maps, there are both pitchfork (one real eigenvalue crossing at one) and Hopf bifurcations (a complex conjugate pair cross the unit circle), but as well there is the possibility of an eigenvalue leaving the unit circle at minus one. This is called a *flip* bifurcation.

Pitchfork bifurcation This results from a real eigenvalue going through zero (for an ODE) or through one (for a map). This results in the creation of a pair of equilibrium points with opposite stability. A standard example for this is

$$\dot{x} = -x^3 + \lambda x.$$

For $\lambda < 0$ the equilibrium $x = 0$ is stable and there are no other equi-

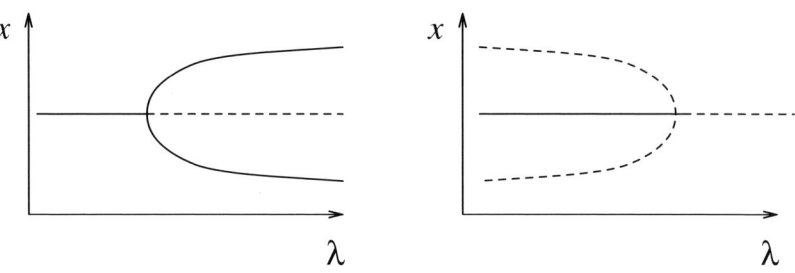

Fig. 6.11. Pitchfork bifurcation diagrams.

libria. As λ becomes positive $x = 0$ becomes unstable and two stable equilibria arise at $x = \pm\sqrt{\lambda}$. Alternatively, an unstable equilibrium can become stable while throwing off two new unstable equilibria. An example is the ODE $\dot{x} = +x^3 + \lambda x$ with λ going from being positive to negative. Bifurcation diagrams for this are shown in fig. 6.11. The direction in which the 'fork' goes (to the right or the left) depends on higher order (cubic) terms.

Hopf bifurcation This results from a complex conjugate pair going through the imaginary axis (for an ODE) or through the unit circle (for a map). For **autonomous** ODEs this results in a new **limit cycle** with the same stability as the original equilibrium point. The equilibrium point usually changes stability. The period of the new orbit is $2\pi/\omega$ where the eigenvalues cross at $\pm i\omega$. For maps the situation is more complex; there is an invariant circle, and the behaviour on this circle is like a circle map. And with circle maps, whether or not there can be a periodic orbit depends on whether the rotation number is rational or not. Of course the angle of the critical eigenvalues to the real axis is changing continuously with the parameter, so the situation is considerably more subtle here.

An example of a Hopf bifurcation for ODEs is given by the following system:

$$\begin{aligned}\dot{x} &= -y + x(\lambda - (x^2 + y^2)) \\ \dot{y} &= +x + y(\lambda - (x^2 + y^2)).\end{aligned}$$

For $\lambda \leq 0$ the origin is a stable equilibrium and there are no periodic solutions. For $\lambda > 0$ the origin becomes an unstable equilibrium and there is a stable limit cycle which is formed by a circle of radius $\sqrt{\lambda}$. For $\lambda > 0$ the period of the limit cycle of this example is 2π. Of course, in general, the period of the limit cycle will only *approach* $2\pi/\omega$ as the parameter approaches the bifurcation point.

The bifurcation diagrams associated with Hopf bifurcations are essentially like the pitchfork bifurcation diagrams; often Hopf bifurcation diagrams are drawn in like one-sided pitchfork diagrams.

Flip bifurcation   This bifurcation can only occur for maps. It occurs when an eigenvalue goes through the unit circle at -1. The equilibrium point involved changes stability as the eigenvalue moves in or out of the unit circle, but once the bifurcation point is crossed there is a new orbit of period 2. This is the bifurcation view of why period doubling is so common. (This is especially true in one dimension as then Hopf bifurcation of maps is not possible.) An example of such a map is

$$x_{n+1} = -(1+\lambda)x_n + x_n^2$$

and at the equilibrium point $x = 0$ as λ crosses through zero. The periodic orbit for $\lambda > 0$ consists of two points at distance $\approx \sqrt{\lambda}$ from the equilibrium point.

Flip bifurcation diagrams, like Hopf bifurcation diagrams, are often drawn as one-sided pitchfork bifurcations.

Transcritical bifurcation   Transcritical bifurcations occur when two equilibrium or periodic points points cross each other, one stable and the other unstable. As a result, the two equilibria exchange stability. An example of such a bifurcation is

$$\dot{x} = x(x - \lambda).$$

There are two equilibria for $\lambda \neq 0$, which are $x = 0$ and $x = \lambda$. For $\lambda < 0$, the equilibrium point $x = 0$ is stable and the equilibrium point $x = \lambda$ is unstable. But for $\lambda > 0$, $x = 0$ is now unstable and $x = \lambda$ is stable. This is illustrated in fig. 6.12. Note that a combination of a transcritical bifurcation and a fold bifurcation can approximate a pitchfork bifurcation.

6.7.3 Continuation

A good reference for this section is the recent book *Numerical Continuation Methods: an Introduction* (Allgower & Georg, 1990). Other references include Garcia and Zangwill (1981).

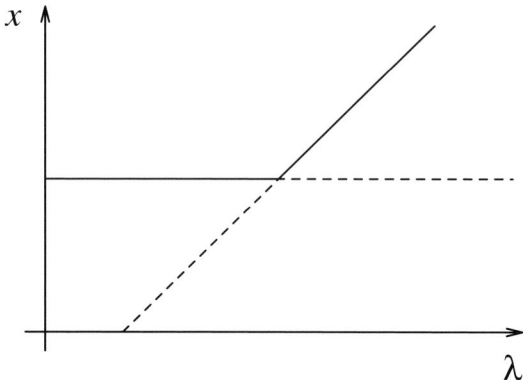

Fig. 6.12. Transcritical bifurcation.

Continuation is a technique that is used for solving non-linear equations. Modern continuation techniques provide globally convergent methods for solving difficult and highly non-linear equations. They are backed up with considerable theory, and there are a number of software packages available which perform continuation. The package HOMPACK (Watson et al., 1987) is one example of a suite of continuation routines which is publicly available. Continuation software has been incorporated into a number of packages for analysing dynamical systems; these use continuation to get to bifurcation points; the bifurcation is analysed and solution branches are identified; finally, continuation is used to track the behaviour of the different solution branches. This is done, for example, in the AUTO package (Doedel, 1986) and in BIFPACK (Seidel, 1989) and other packages.

The general set up for continuation is that we wish to solve a system of equations $F(x) = 0 \in \mathbf{R}^n$, $x \in \mathbf{R}^n$, starting from a computationally trivial system $G(x) = 0$. The idea is to set up a *homotopy* between the two problems $H(x, \lambda)$ where $\lambda \in [0, 1]$ with $H(x, 0) = G(x)$ and $H(x, 1) = F(x)$. Then we can try to follow the *path* $H(x, \lambda) = 0$ in (x, λ) space, from the trivial solution of $G(x) = 0$ ($\lambda = 0$) to the solution we want, $F(x) = 0$ ($\lambda = 1$). Old style continuation assumed that one could write $x = x(\lambda)$ and monotonically increase λ from zero to one. However, this often fails due to the curve $H(x, \lambda) = 0$ 'folding back'. If the curve does fold back then you might try to change the homotopy and/or start from a different G. This can work sometimes, but it fails more often than it succeeds.

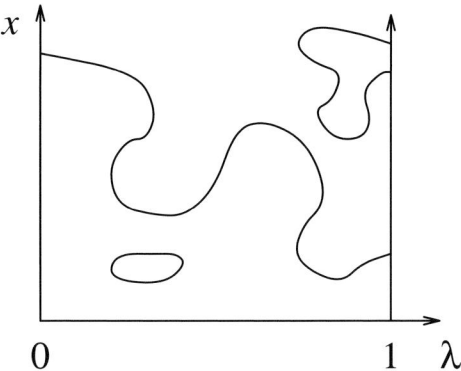

Fig. 6.13. Possible zero curves of $H(x, \lambda)$.

If you try doing continuation in (x, λ) space and simply try to follow the curve, whether or not it folds back, then you can do much better than using old style continuation. As long as the $(n+1) \times n$ Jacobian matrix $\nabla_{(x,\lambda)} H(x, \lambda)$ has **full rank** then the curve, locally, really is a smooth curve and not something more complex (like a pitchfork bifurcation or something nastier). If the homotopy $H(x, \lambda)$ is given, such as when you want to follow a branch of solutions, then bifurcations, or other problems where $\nabla_{(x,\lambda)} H(x, \lambda)$ fails to have full rank, may be unavoidable. This sort of continuation is called *natural continuation*.

However, if you are willing to follow a path merely to get to the other end, then we can incorporate some randomness which ensures that there is a smooth path almost every time the algorithm is run. The trick is to incorporate an extra vector of parameters into H so that the matrix $\nabla_{(x,\lambda,a)} H(x, \lambda; a)$ always has full rank, except possibly at $\lambda = 1$. For example, we could take the linear homotopy

$$H(x, \lambda; a) = (1 - \lambda)(x - a) + \lambda F(x).$$

(At $\lambda = 0$, $x = a$ is the obvious, trivial, solution.) By a generalisation of the Morse–Sard lemma, for almost all a, the matrix $\nabla_{(x,\lambda)} H(x, \lambda; a)$ has full rank whenever $H(x, \lambda; a) = 0$. The introduction of this extra parameter makes this kind of continuation *artificial parameter continuation*.

6.7.4 Applications of continuation

Tracking equilibria Equilibria for ODEs and maps are simply solutions of (non-linear) functions given in the ODE or map. For the ODE case, they are given by $f(x) = 0$; in the case of maps, we should look for fixed points: $x - g(x) = 0$.

Most codimension 1 bifurcations in the dynamics are either not a problem for modern continuation methods (fold bifurcations) or are ignored completely (Hopf and flip bifurcations). However, at pitchfork bifurcations either $\nabla f(x)$ or $I - \nabla g(x)$ is singular, for the ODE and map cases respectively.

Tracking periodic solutions Periodic solutions of maps, or periodically forced ODEs, can be accomplished easily within the continuation framework. In this case the period T is predetermined, and it is just a matter of solving the equations $x - \varphi(x,T) = 0$ where $\varphi(\cdot,T)$ is the map or flow generated by the ODE or map at time T. Tracking periodic solutions of autonomous ODEs is more complex. This is because there is an extra parameter (the period T) which needs to be computed, and there is some redundancy as there is no need to distinguish different points on the same periodic orbit. Local continuation can be done, provided we restrict perturbations to be (for example) perpendicular to the direction of the vector field. However, global properties cannot be guaranteed, at least in 'probability one' form.

Tracking homoclinic and heteroclinic orbits Homoclinic orbits and heteroclinic orbits can be tracked for varying parameters *if the stable and unstable manifolds intersect transversally* by setting up two point boundary value problems. If the stable and unstable manifolds don't intersect transversally then the situation is not structurally stable, and we can't expect numerical methods to work.

Suppose we want to compute a connecting orbit between two hyperbolic equilibria, from x to y. (We could have $x = y$ for the case of a homoclinic orbit.) Suppose also that the stable and unstable manifolds intersect transversally so that the intersection of $W^u(x)$ and $W^s(y)$ occurs at an angle and that $\dim W^u(x) + \dim W^s(y) \geq n$. Small perturbations *of the manifolds* will not destroy transversal intersection. Close to x we can approximate $W^u(x)$ by the linear manifold $x + V^u(x)$, and similarly we can approximate $W^s(y)$ near y by $y + V^s(y)$. (Taylor series approaches can improve on this, although the arithmetic is terrible!)

Non-linear dynamics 227

For a suitably large $T > 0$ we consider the boundary value problem

$$\dot{u} = f(u), \qquad u(0) \in x + V^u(x), \qquad u(T) \in y + V^s(y)$$

or an analogous system for maps. While this is potentially an extremely ill-conditioned problem if treated as an initial value problem, it can still be very well behaved if multiple shooting type methods are used. Pick an integer $N > 0$ and we split the interval $[0,T]$ into pieces $[kT/N, (k+1)T/N]$, $k = 0, \ldots, N-1$, so that the matrix functions $\nabla_x \varphi(\cdot, T/N)$ are always well conditioned. This can be done if the differential equation has a Lipschitz constant K and we pick N to make $N = O(KT)$. In practice, it would be desirable to have $N \leq \alpha KT$ where α lies in the range zero to five. Then we have to solve a $(N-1)n$ system of equations

$$u_{n+1} - \varphi(u_n, T/N) = 0$$

together with the boundary conditions $u_0 \in x + V^u(x)$ and $u_N \in y + V^s(y)$. We also need to make sure that u_0 is close enough to x, and that u_N is close enough to y; if it isn't we need to make T larger. The value of T does not have to be enormous, as the trajectories will converge to x and y at an exponential rate. This can be fast enough that the linear approximation to $W^u(x)$ and $W^s(y)$ is all that is needed rather than taking higher order approximations: the distances to the equilibria are $\exp(-O(T-T_0))$ where T_0 is the time to travel between reference points near to x and y.

Provided x and y are hyperbolic equilibria, and the intersection is transversal, the Newton equations to solve are well conditioned. If we have dim $W^u(x)$ + dim $W^s(y) > n$ then we have an underdetermined system. This can be solved using QR factorisations which can be used to give the solution with minimal 2-norm. For hyperbolic homoclinic orbits we have dim $W^u(x)$ + dim $W^s(y) = n$ so the Newton system is a square system of Nn equations in Nn unknowns.

6.7.5 Continuation techniques

Given a homotopy function $H(x, \lambda)$ whose zero curve we wish to follow, to actually perform continuation we need the following:

- a way to step along the path;
- a way to see that we are keeping the right direction and haven't accidentally reversed;
- a way to correct for drifting from the path.

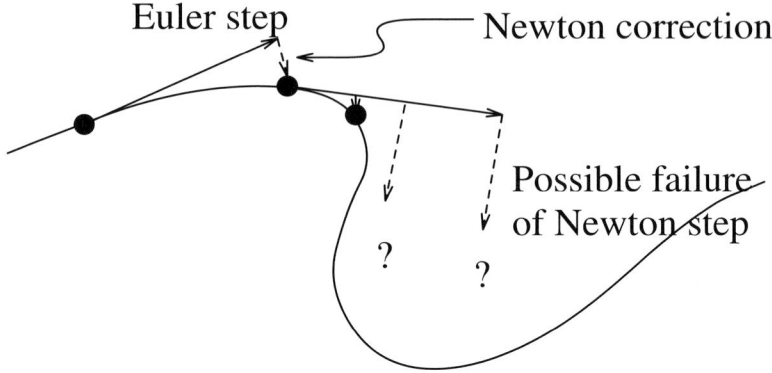

Fig. 6.14. Prediction–correction for path following.

For the remainder of this section we use $z = (x, \lambda)$ and write $H(z)$ for $H(x, \lambda)$ and assume the gradient is with respect to z unless otherwise indicated.

Stepping schemes   The simplest way to obtain a step for a homotopy method is to step along the tangent plane to the curve. We will need to work out the correct direction, but that will be dealt with next. It isn't hard to compute the tangent plane for $H(z) = 0$. It is just the kernel of the $n \times (n+1)$ matrix $\nabla H(z)$. This can be determined by computing the QR factorisation of $\nabla H(z)^T$; the last column of the resulting Q matrix is a unit vector in the kernel.

Once the direction has been corrected, we can do the simplest thing and step a fixed amount in that direction. Applied to ODEs this is just Euler's method. This is a rather dangerous strategy; it is better to check on, say, $H(z + \alpha \Delta z)$, and reduce the step if this is too large (fig. 6.14). Alternatively, this step control strategy could be built into the correction step which takes us back close to the true path as shown in fig. 6.14.

If we think of $z = (x, \lambda) = (x(s), \lambda(s)) = z(s)$ where s is the arc length of the curve in (x, λ) space, then dz/ds is just the unit direction vector computed for that point. This can be thought of as defining an ODE for the curve. Techniques from ODEs have been applied to get better stepping schemes, such as using Runge–Kutta or multistep methods. Also interpolation methods have been applied. One of the best seems to be extrapolation based on Hermite interpolants as they use not only position information but also derivatives at the endpoints.

Keeping the right direction   The unit vector computed from the QR factorisation could be in the right direction or the opposite. There are two ways of determining if we are heading in a consistent direction. The simplest is just to take the inner product with the previous computed direction. If the inner product is negative then something is quite wrong. In fact, this inner product is used in some codes to detect sharp bends; there is some tolerance *tol* such that if $(\dot{z}_{old})^T \dot{z}_{new} < 1 - tol$, the step size would be reduced.

To start this method off, we need to know if the first direction is in the right direction. This is easy. We just want $\dot{\lambda} > 0$. If the computed direction has $\dot{\lambda} < 0$, then it must be reversed.

There is a more sophisticated way of ensuring that the direction stays consistent. That is to compute the sign of $\det[\nabla H(z)^T \dot{z}]$. This sign should be constant. The reference sign is computed as the start of the path following, and thereafter used to check directions. The sign of the determinant can in fact be computed quite quickly in terms of the Q and R factors.

Keeping to the narrow path   Provided we are not too far from the path, then Newton's method can be used to correct and bring us closer to the true path. This is a slightly unusual Newton method as it is an under-determined system of equations:

$$\nabla H(z)\Delta z = -H(z) \in \mathbf{R}^n, \qquad z \in \mathbf{R}^{n+1}$$

where $z = (x, \lambda)$. As noted above, this can be dealt with by means of QR factorisations (yet, again!).

Since Newton's method is only guaranteed to converge if it is started close enough to the true solution, care must be taken to prevent problems caused by taking steps that are 'too large'. Careful step size control strategies should be built into the Newton method to make it robust and reliable. A standard technique is known as the Armijo line search: Compute the step Δz from Newton's method. While $\|H(z + \Delta z)\| > \|H(z)\|$, set $\Delta z \leftarrow \Delta z/2$. Now set $z \leftarrow z | \Delta z$. This ensures that $\|H(z)\|$ does not increase. Scaling strategies can also be important. In addition the strategies for ensuring that the computed points don't stray too far should be incorporated into the stepping algorithm, as discussed above.

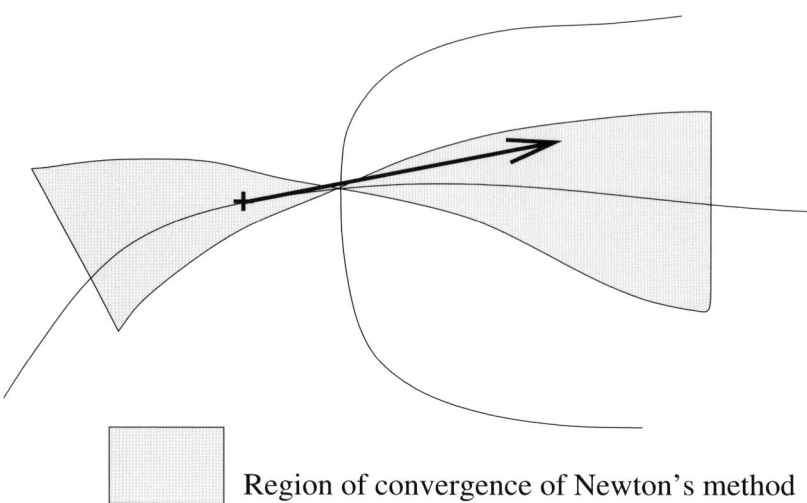

Fig. 6.15. Jumping over a pitchfork bifurcation point.

6.7.6 When a path meets a bifurcation ...

Provided simple bifurcations occur, and the type of bifurcation can be identified, then it is not too hard to jump over the bifurcation point and continue on the other side. For example, in fig. 6.15 is shown a pitchfork bifurcation, which has the property that the far branch is tangential to the near branch at the bifurcation. At the bifurcation point the equations we wish to solve become degenerate and the Jacobian matrix becomes singular. The main trick is to identify (roughly) where the bifurcation point is, and to jump far enough (but not too far!) so that the predicted point lies within the domain of convergence of the Newton method. This is illustrated in fig. 6.15.

6.8 Inertial manifold theory

The world abounds with dynamical systems that are not only chaotic, but have very large dimension. Perhaps one of the best known is the weather. It is not only now known to be chaotic, but it is described by *partial differential equations*. That is, the equations describing the weather (temperatures, pressures, humidities, etc.) not only describe how these quantities vary over time, but also how they vary over space. To describe a single 'snapshot' of the earth's weather fields requires a vast amount of information. The number of simple scalar quantities that are needed to describe such a snapshot is indeed very large. This means that weather is a *large-scale* or *high-dimensional* dynamical system. Indeed, in most mathematical formulations of the way weather behaves, it is an *infinite-dimensional* system.

In spite of the fact that a detailed description of the state of the weather system requires vast numbers of simple scalar quantities, it is commonly observed that a few crucial (selected or averaged) quantities can be remarkably good at predicting the behaviour of the overall system. Perhaps the most successful example of this is the 'southern oscillation' index which is used to quantify the 'El-Niño' effect. Of course, this is just one of a number of quantities that could be used to give a more-or-less complete description of the weather state. This suggests that, despite the infinite number of degrees of freedom of the system, it is behaving *as if it were a low-dimensional system.*

The idea can be put into a mathematical form in terms of *inertial manifolds*. Inertial manifolds are manifolds or 'surfaces' in the state space of a dynamical system which

- are finite dimensional;
- are smooth, or at least Lipschitz;
- attract all trajectories at an exponential rate.

If such a manifold exists, the long-term behaviour of the large-scale system can be reduced to asking about the long-term behaviour of the system on its inertial manifold. This reduces an infinite-dimensional problem (or very large-scale problem) to a finite-dimensional, and therefore, in principle, tractable problem.

Even if existence can be proved, there are of course a number of practical issues which need to be considered, such as the difficulty of finding an inertial manifold for a large-scale system, and the dimension of the inertial manifold so constructed.

The minimum dimension of any inertial manifold a system might have is the lowest integer greater than or equal to the fractal or Hausdorff dimension of its attractor. Since any trajectory is attracted exponentially fast to an inertial manifold, the attractor(s) must lie within the inertial manifold. The dimension of the attractor must therefore be no more than that of an inertial manifold.

Proving the existence of inertial manifolds is not a trivial exercise, and involves a great deal of technical machinery. However, the basic idea is not so difficult to describe. In some form or other, there needs to be some form of *dissipation* in the dynamical system. Some systems do not have dissipation and do not have inertial manifolds. A simple example is the wave equation

$$\frac{\partial^2 u}{\partial t^2} = c^2 \frac{\partial^2 u}{\partial x^2}.$$

(This is a wave in one space dimension. Here c is the speed of the wave.) These equations describe the vibrations of a guitar string and waves of electromagnetic radiation. Wave equations have the property of being *reversible*. This means that if you take the film of a solution of the wave equation, and run it backwards, the resulting picture would also be a solution of the wave equation. Since the equation is reversible, there is no loss of information, and no attraction to some part of the state space. This means that there is no reduction in the number of quantities needed to accurately describe the state of the system, and since the starting conditions cannot be completely described with a finite number of quantities, there cannot be an inertial manifold.

However, if some sort of dissipation is introduced (air-resistance or friction for the guitar string, or electrical resistance for the electromagnetic wave), then the situation changes radically. Such dissipation quickly removes the rapidly oscillating parts of the solution leaving the lower frequency components, which decay much more slowly.

If the system were left to itself, even the slowly decaying modes would eventually damp out, and the system would relax to a state of maximum uniformity. For example, in the case of weather, if it were not for the energy input to the atmosphere from the sun and the perturbing effect of the departure of the solid earth from a spherical ball, the system would relax to one of spherical symmetry.

However, if low frequency energy is injected into the system the lower frequency components may not decay away completely, and in fact may be generated spontaneously from bifurcations breaking the symmetry of the relaxed, background state. For example, in the case of weather the heating of the land during the day may drive large scale convective instabilities. The degrees of freedom corresponding to rapidly decaying motions (the 'slave modes') will still be excited as an adiabatic response to slow driving by the large scale modes (the 'master modes'), but their response will be smaller the farther the driving is from their natural frequency.

The effect of damping is to pull the solution towards some finite dimensional manifold within the infinite dimensional state space of the fluid equations, on which the coefficients of the master modes form a coordinate system. This is the inertial manifold. The effect of the response of the slave modes is to distort this manifold away from its tangent space at the origin, the tangent space being the linear vector space spanned by the eigenmodes with very small growth or damping (the master modes) found by linearising about the stationary background state.

A full treatment of weather should also include the turbulent boundary layer near the surface of the earth. The treatment of turbulence requires the recognition of an intermediate scale, between the molecular scale and the macroscopic scale of the weather map, on which instabilities produce small-scale eddies which modify the transport equations describing the dynamics on the macroscopic scale. This is an example of a hierarchical complex system in which the macroscopic elements on one scale are microscopic on another.

6.8.1 Technical ideas

The first mathematical presentation of inertial manifold theory appears to be Mañé (1977). The technical basis of the existence theory of inertial manifolds can be found in Constantin et al. (1988, 1989).

The basic idea is to split the equation describing the dynamics of the system into a dominant linear part and a weaker non-linear part. The linear part has to have a strongly dissipative character; in addition, *spectral gap* or *spectral barrier* conditions are assumed so that there is a separation of time scales between the slow dynamics and the slaved response.

To illustrate how the theory works, consider

$$\frac{du}{dt} = -Au + f(u)$$

where A is a linear, positive definite, symmetric operator, and f is possibly non-linear. An example of such systems are the *reaction-diffusion* equations

$$\frac{\partial u_i}{\partial t} = \nu \nabla^2 u_i + f_i(u_1, \ldots, u_m), \qquad i = 1, \ldots, m.$$

Here

$$\nabla^2 v = \frac{\partial^2 v}{\partial x_1^2} + \cdots + \frac{\partial^2 v}{\partial x_n^2}$$

in n-dimensional space. The term $\nu \nabla^2 u_i$ is due to diffusion, and $f_i(u_1, \ldots, u_m)$ is an ordinary function which gives the reaction rates for the concentration u_i. The quantity ν is the diffusion coefficient for the system. Reaction-diffusion equations incorporate spatial dependence with the equations for a reaction. This should be compared with continuous stirred-tank reactors, where the components are assumed to be so thoroughly mixed that there is no spatial dependence. In this case the equations reduce to ordinary differential equations:

$$\frac{du_i}{dt} = f_i(u_1, \ldots, u_m), \qquad i = 1, \ldots, m.$$

With spatial variation new phenomena can be observed such as travelling waves, moving spirals and so on.

The theory developed by Constantin, Foias, Nicolaenko and Témam starts by considering the eigenvalues $\lambda_1 \leq \lambda_2 \leq \ldots$ of A. Provided a gap $\lambda_{k+1} - \lambda_k$ can be found in the spectrum which is larger than the Lipschitz constant of f, an inertial manifold exists with dimension k. This idea is refined in Constantin et al.(1988, 1989) and is extended to a number of different circumstances such as the Navier–Stokes equations for fluid flow in two dimensions and the Kuramoto–Sivashinsky and other related equations. However, there is still a need in this theory to have gaps in the spectrum that are in some sense 'large enough' to control the nonlinear parts of the equations. In a sense, this theory is a global version of the theory of hyperbolic equilibrium and periodic points. For hyperbolic equilibrium points there has to be a separation of the eigenvalues away from the stability boundary to control the local non-linearities; here there has to be a gap in the eigenvalues to control the global behaviour of the non-linearities.

The need for spectral gaps or barriers of some kind has limited the theory to a low number of spatial dimensions, and there have been a number of counter-examples which show that the theory cannot be extended in its current form to even reaction–diffusion equations in four or more dimensions (Mallet-Paret & Sell, 1986).

A different approach, discussed in the next section, has also been developed which does not require spectral gaps, but gives the existence only of an *approximate inertial manifold*.

6.8.2 Approximate inertial manifolds

Approximate inertial manifolds for a dynamical system are manifolds that

- are finite dimensional;

- are smooth, or at least Lipschitz;
- attract all trajectories to within a small distance at an exponential rate;
- have trajectories starting on the manifold that make a small angle to the manifold.

Of crucial importance are the answers to the questions 'How close will trajectories be attracted?'. and 'How small an angle will the trajectories make to the manifold?' However, this weakening means that sensible results can be obtained even where there are no large spectral gaps. All that is required is that the eigenvalues of A grow without bound. With this condition the answer to the question 'How close will trajectories be attracted?' is 'As close as you please'. The cost, however, is the increasing dimension of the approximate inertial manifold as the 'tolerance' for closeness of trajectories is reduced.

Computational methods for finding such approximate inertial manifolds have been developed by Debussche and Marion (1992), Goubet (1992) and Marion (1989).

6.9 Finding shadows

The final numerical analysis section of this chapter is on finding *shadows* of numerically computed orbits. Recall that numerically we do not compute true trajectories but rather δ-pseudo-trajectories: $\|x_{k+1} - g(x_k)\| \leq \delta$ for all k. As was noted in §6.1.5, the exponential divergence of trajectories means that errors, even those due to the precision of the floating point arithmetic, quickly grow and overwhelm the solution. Nevertheless, the overall behaviour of the system seems to be remarkably stable.

One way to resolve this paradox is through shadowing: even though we have computed something which diverges from the true trajectory, we may be close to some trajectory with slightly different initial conditions. An ε-shadow of x_k is a true trajectory y_k where $\|x_k - y_k\| < \varepsilon$ for all k (see fig. 6.16).

When do shadows exist? When and how can we compute them? When can we compute a value of ε for which there is a shadow? There are both theoretical and numerical approaches to this problem. The theoretical approach is due to Anosov and Bowen (1967, 1975) and is based on hyperbolicity. The numerical approach avoids assuming hyperbolicity, but instead explicitly computes the conditioning of the matrices that hyperbolicity implies are stable.

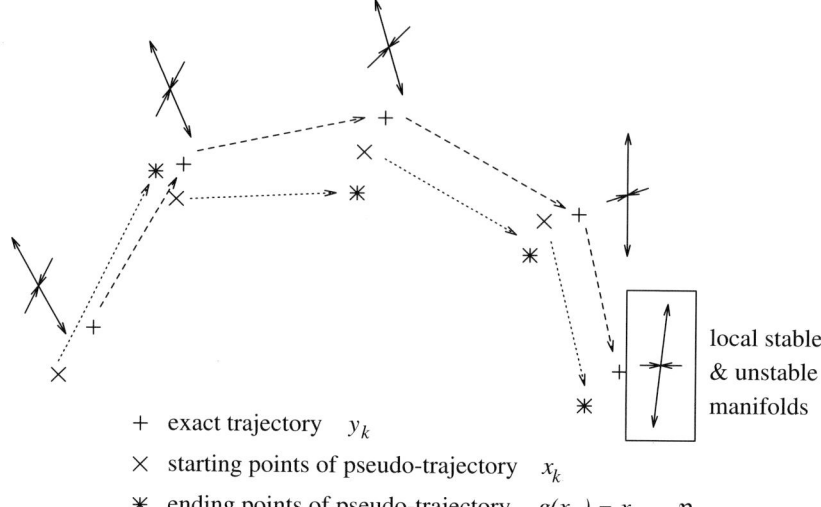

Fig. 6.16. Exact trajectory y_k shadowing a pseudo-trajectory x_k resulting from the combined effect of the map g and the addition of an error vector η_k at each step: $x_{k+1} = g(x_k) + \eta_k$.

The basic shadowing lemma is due to Anosov and Bowen.

Theorem 6 *If a map g is hyperbolic then for every $\varepsilon > 0$ there is a $\delta > 0$ such that every δ-pseudo-orbit has an ε-shadow.*

Of course, g might be hyperbolic on an attractor, even if not globally, and provided the numerically computed orbit stays on the attractor, then a shadowing lemma applies here as well.

6.9.1 Why shadowing can be done

The basic idea behind shadowing is to consider a trajectory, not as the result of an iteration, but 'all at once' as a single entity satisfying a system of equations:

$$g(y_n) - y_{n+1} = 0 \in \mathbf{R}^n.$$

Let us write $\mathbf{y}$ for the sequence (y_k). Then we have $\mathbf{y} \in l^\infty(\mathbf{Z}, \mathbf{R}^n)$ where the space $l^\infty(\mathbf{Z}, X)$ is the set of all bounded sequences $\mathbf{z} = (z_k \mid k \in \mathbf{Z})$ such that $z_k \in X$ and with the norm

$$\|\mathbf{z}\|_\infty = \sup_{k \in \mathbf{Z}} \|z_k\|_X < +\infty,$$

Non-linear dynamics

where $\|\cdot\|_X$ is the norm on X. We assume $X = \mathbf{R}^n$ and use the usual 2-norm in $\mathbf{R}^n$ as this norm. As is standard, $\mathbf{Z}$ denotes the set of all integers, positive negative and zero.

Consider the map $\mathcal{G}: l^\infty(\mathbf{Z}, \mathbf{R}^n) \to l^\infty(\mathbf{Z}, \mathbf{R}^n)$ defined by

$$\mathcal{G}(\mathbf{z})_k = z_{k+1} - g(z_k).$$

That is, $\mathcal{G}$ generates the sequence of errors given an arbitrary sequence $\mathbf{z} \in l^\infty(\mathbf{Z}, \mathbf{R}^n)$. For the shadowing true orbit, $\mathbf{y}$, the errors are zero, so we need to solve $\mathcal{G}(\mathbf{y}) = 0$. Our strategy will be to solve this iteratively, taking as initial guess $\mathbf{z} = \mathbf{x}$ and ending up, we hope, with $\mathbf{z} = \mathbf{y}$. We know already that Newton's method is a powerful method for solving non-linear equations, even when simple iterative methods diverge. Clearly we have a non-linear equation, but can we use Newton's method on it?

Well, Newton's method is given by

$$\nabla \mathcal{G}(\mathbf{z})\, \delta\mathbf{z} + \mathcal{G}(\mathbf{z}) = 0$$

where

$$(\nabla \mathcal{G}(\mathbf{z})\, \delta\mathbf{z})_k = \delta z_{k+1} - \nabla g(z_k)\, \delta z_k.$$

The next step is to show that this linear operator $\nabla \mathcal{G}(\mathbf{z})$ has a bounded inverse. This requires hyperbolicity.

Recall that hyperbolicity says that there are continuous families of projections $P^s(x)$ and $P^u(x)$ where $P^s(x) + P^u(x) = I$, which satisfy consistency conditions

$$\nabla g(x)\, \text{range } P^s(x) = \text{range } P^s(g(x))$$

and similarly for P^u. The other condition is that there are constants $K > 0$ and $0 < \alpha < 1$ where

$$\|P^s(g^k(x))\nabla(g^k)(x)P^s(x)w\| \leq K\alpha^k \|P^s(x)w\|,$$
$$\|P^u(g^k(x))\nabla(g^k)(x)P^u(x)w\| \geq (1/(K\alpha^k))\|P^u(x)w\|$$

for all $k > 0$. Split $\delta\mathbf{z} = \mathbf{u} + \mathbf{v}$ where $u_k \in \text{range } P^s(z_k)$ and $v_k \in \text{range } P^u(z_k)$.

Put $\eta_k = P^s(z_{k+1})\mathcal{G}(\mathbf{z})_k$ and $\xi_k = P^u(z_{k+1})\mathcal{G}(\mathbf{z})_k$. Then the equations for Newton's method become

$$u_{k+1} - \nabla g(z_k) u_k = \eta_k$$
$$v_{k+1} - \nabla g(z_k) v_k = \xi_k.$$

Now we solve for u_k by going forward in time, and for v_k by going backward in time. If we write $\nabla_s g(z_k)$ for $P^s(z_{k+1})\nabla g(z_k)P^s(z_k)$ and similarly for $\nabla_u g(z_k)$ we can write

$$u_k = \sum_{l<k} \nabla_s g(z_k)\ldots\nabla_s g(z_l)\,\eta_l$$

$$v_k = \sum_{k<l} (\nabla_u g(z_k))^{-1}\ldots(\nabla_u g(z_l))^{-1}\,\xi_l$$

where the inverses $(\nabla_u g(z_k))^{-1}$ are understood as linear maps:

$$\text{range } P^u(z_{k+1}) \to \text{range } P^u(z_k).$$

Thus we get

$$\|u_k\| \leq \sum_{l<k} K\alpha^{l-k}\|\eta_l\| \leq \frac{K\alpha}{1-\alpha}\|\boldsymbol{\eta}\|_\infty$$

$$\|v_k\| \leq \sum_{k<l} (K\alpha^{l-k})^{-1}\|\xi_l\| \leq \frac{1}{K(\alpha-1)}\|\boldsymbol{\xi}\|_\infty.$$

Thus there is a constant C such that $\|\delta\mathbf{z}\|_\infty \leq C\|\mathcal{G}(\mathbf{z})\|_\infty \leq C\delta$ for a δ-pseudo-orbit.

To complete the proof we need to use the Newton–Kantorovich lemma which gives a lower bound on the radius of convergence of Newton's method in terms of $\|(\nabla\mathcal{G}(\mathbf{z}))^{-1}\|_\infty$ and the Lipschitz constant of $\nabla\mathcal{G}(\mathbf{z})$. To get the second quantity we need to look at

$$((\nabla\mathcal{G}(\mathbf{z}) - \nabla\mathcal{G}(\mathbf{y}))\mathbf{u})_k = (\nabla g(y_k) - \nabla g(z_k))u_k.$$

Suppose that L is a Lipschitz constant for ∇g, then

$$\|\nabla\mathcal{G}(\mathbf{z}) - \nabla\mathcal{G}(\mathbf{y})\|_\infty \leq L\|\mathbf{z}-\mathbf{y}\|_\infty.$$

(Actually, we just need $L \geq \sup_k L_k$ where L_k is the sum of local Lipschitz constants for ∇g at z_k and y_k. Thus, provided we have z_k and y_k bounded and g is C^1 we can get a suitable local constant for $\nabla\mathcal{G}$.)

Given this Lipschitz constant L, the Newton method will converge provided $\delta < 1/(C^2 L)$. The value of ϵ is of the order of $C\delta$.

6.9.2 Numerical methods for shadows

Hammel *et al.* (1987) is an early paper on numerical shadowing in the case of the logistic map $x_{n+1} = g(x_n;a) = ax_n(1-x_n)$. In this one-dimensional case they can use a version of *interval arithmetic*.

This works by computing a finite sequence (x_k) for $k = 0, 1, 2, \ldots, N$ which is a δ-pseudo-orbit. In practice the δ is a modest multiple of machine epsilon. Then a sequence of intervals is constructed, starting with $I_N = [x_N, x_N]$. Given I_n we compute I_{n-1} to be an interval containing x_{n-1} such that $I_n \subseteq g(I_{n-1})$. This nesting condition allows the computation of a sequence of intervals whose width generally doesn't grow as n decreases. This is because the logistic map is generally expansive, and their are no stable subspaces.

Hammel, Yorke and Grebogi claim that with the initial pseudo-orbit of 10^7 iterates calculated with $a = 3.8$ and $\delta \approx 3 \times 10^{-14}$, there was a shadowing radius $\varepsilon = 10^{-8}$. To get this sort of accuracy with a straightforward direct iteration would require computations to be done with about 10^{10} digits.

Multidimensional systems require a more delicate approach which separates out the stable and the unstable subspaces. A Newton method can be set up using a truncated version of $\mathcal{G}$ and $\nabla \mathcal{G}$, and in fact works well, except that the Newton equations are underdetermined with nN unknowns and $n(N-1)$ variables. A standard QR approach is not optimal here, as we want to minimise the $\|\cdot\|_\infty$ norm of the correction, not the 2-norm as is usual. For getting a Newton correction, a stable factorisation procedure such as LU with partial pivoting or QR factorisation should be used.

6.9.3 Shadowing autonomous ODEs

Shadowing **autonomous** ODEs has some subtleties that are not present in the problem of finding shadows for maps. The main problem is that the maps generated by flows are not actually hyperbolic; there is one Lyapunov exponent which is zero corresponding to the direction of the flow. This is not just an artefact, but a real effect. To see this, suppose we have a **limit cycle** for, say, the van der Pol equation (see §6.2.5) or some other equation. If we perturb one of the parameters in the van der Pol equation we would get a new limit cycle. Can we shadow an orbit in one limit cycle by an orbit in the other limit cycle? If the two limit cycles have different periods, the answer is no. No matter how small the difference is, if you wait long enough then the two cycles will go out and then back in phase. Shadowing autonomous ODEs has to also involve stretching or compressing the time variable as well.

Shadowing algorithms for autonomous ODEs have been developed, and these work by 'factoring out' the time aspect. The simplest and most direct way to do this is in terms of what are effectively Poincaré cross-sections. About every point in a numerically computed flow $\widehat{x}(t_k)$, a local Poincaré cross-section can be set up which is the plane passing through $\widehat{x}(t_k)$ normal to $f(\widehat{x}(t_k))$. The shadow trajectory that is computed (or whose existence is verified) crosses this cross-section at a different time $t_k + \tau_k$ where $|\tau_k - \tau_{k-1}|$ is always small. The lack of hyperbolicity shows up in the fact that we can have $\tau_k \to \infty$ as $k \to \infty$, at least at a linear rate.

6.10 Concluding remarks

Inertial manifold theory shows how dissipative infinite-dimensional dynamical systems, such as the Navier–Stokes equation, can give rise to low-dimensional dynamical systems that can be studied using the tools developed in this article. In the Introduction we pointed out that non-equilibrium statistical mechanics takes this idea a step further back, by seeking to explain the origin of the dissipation in terms of irreversibility produced by coarse-graining of the fundamental, microscopic dynamics. This is a metaphor, and perhaps an instructive paradigm, for attacking other complex systems in which there is an overwhelmingly large number of microscopic elements.

Acknowledgements

We acknowledge with thanks the comments of Professor D. J. Evans on our discussion of the relation between nonequilibrium statistical mechanics and chaos, and of Professor A. J. Roberts on the subject of inertial manifolds and related methods.

Glossary

This glossary defines, rather heuristically, basic terms assumed known in the text. For a deeper understanding of their meaning the reader is referred to appropriate textbooks. Words defined in this glossary are flagged by **bold** type when first used in a section. For concepts defined in the text, please consult the index of this book. We assume that the basic set theory symbols ∈ (element of), ⊂ (subset of), ⊃ (superset of), ∪ (union) and ∩ (intersection) are understood.

autonomous: An autonomous dynamical system is one in which there is no explicit time dependence in the equations of motion. Physically this corresponds to a closed system – one not driven by external forces, but evolving purely due to the mutual interactions of its component parts.

ball: The interior of the sphere of a given radius (in a space with a metric, so that length is defined). See **unit ball**.

diffeomorphism: A C^n-diffeomorphism is a **map** that is n-times **differentiable** and whose inverse function exists and is also n-times differentiable. For instance, if we take a rubber sheet and deform it by stretching it in an arbitrarily non-uniform but smooth way, without tearing it or folding it, then we are applying a diffeomorphism to the points on the sheet.

differentiable: A function $f(x)$ is n-times differentiable (f belongs to C^n) if its nth derivative is finite and continuous.

full rank: An $m \times n$ matrix is *full rank* for $m \geq n$ if all of its columns are linearly independent, and full rank for $n \geq m$ if all of its rows are linearly independent. If a matrix is full rank then the number of linearly independent rows or columns is the maximum possible.

inf: The infimum, or greatest lower bound, $a = \inf a_n$ of a sequence $\{a_n\}$ is the greatest number such that $a \leq a_n$ for all n.

irreversibility: is the phenomenon whereby closed macroscopic systems are observed always to evolve, as time increases, toward a more disordered state (a state of higher entropy). This means that dynamical systems describing macroscopic motions are typically not invariant under time-reversal (due to terms describing transport effects like diffusion or viscous dissipation).

Jordan canonical form: Any square matrix A can be put into Jordan canonical form by a similarity transformation $J = XAX^{-1}$ for some non-singular (i.e., invertible) matrix X. The eigenvalues and related properties of J are the same as A, but the eigenvectors usually are not. The Jordan canonical form J consists of Jordan blocks on the diagonal of J; each Jordan block has the form

$$J_i = \begin{bmatrix} \lambda_i & 1 & & & \\ & \lambda_i & 1 & & \\ & & \ddots & \ddots & \\ & & & \lambda_i & 1 \\ & & & & \lambda_i \end{bmatrix}.$$

Lebesgue measure: This is the ordinary "volume" measure. On the real line (**R**) the Lebesgue measure of an interval is its length. In the plane ($\mathbf{R}^2$) the Lebesgue measure of a set is its area; in three-dimensional space ($\mathbf{R}^3$) the Lebesgue measure of a set is its volume.

Lipschitz: A function $f\colon \mathbf{R}^n \to \mathbf{R}^m$ is *Lipschitz continuous* if there is a constant L (the Lipschitz constant for f) where $\|f(x) - f(y)\| \leq L\|x - y\|$ for every x and y. Note that $\|u - v\|$ denotes the distance between points u and v.

lim inf a_n: Also denoted $\underline{\lim}\, a_n$. This is the least limit point of an infinite sequence $\{a_n\}$, which, though bounded below, may not have a unique limit (e.g. a_n may wander chaotically as $n \to \infty$). This is distinct from plain inf a_n because lim inf a_n must be approached arbitrarily closely *infinitely often* as $n \to \infty$, whereas inf a_n must take into account points visited once or a finite number of times (see **inf**).

lim sup a_n: Also denoted $\overline{\lim}\, a_n$. This is the greatest limit point of an infinite sequence $\{a_n\}$. See the discussion of the analogous lower limit point **lim inf** above.

limit cycle: A simple periodic attractor (a closed loop in the phase plane) that occurs in two-dimensional dissipative continous-time dynamical systems (which cannot exhibit chaos because the Poincaré–Bendixson theorem shows that fixed points and limit cycles exhaust the possibilities for attractors of two-dimensional flows).

manifold: A mathematical space that can locally be described by a Cartesian coordinate system. In general the local coordinates cannot be joined smoothly to form a single Cartesian coordinate system. (We shall assume that the manifold itself is everywhere smooth – it is a differentiable manifold.) The problem is a topological one, e.g. Mercator's projection is an attempt to describe a sphere by a Cartesian coordinate system, but it becomes singular at the poles. A manifold can always be regarded as a surface embedded in a higher dimensional space, though it is not always natural to do this, e.g. what space is the curved space-time of general relativity embedded in?

map: A map (also called a mapping or function) f from a space A to a space B (abbreviated $f : A \to B$) is a rule for associating one or more elements of B with any element of A. If the map is one-to-one and onto (i.e. each element of B is associated uniquely with an element in A) then the inverse function $f^{-1} : B \to A$ exists and f is called a *bijection*. If, further, f and f^{-1} are continuous, then f is called a *homeomorphism*, while if they are smooth in some sense then it is called a **diffeomorphism**.

$\mathbb{R}$: The set of real numbers or the real line. $\mathbf{R}^n$ is an n-dimensional space describable globally by a Cartesian coordinate system.

sup: The supremum, or least upper bound, $a = \sup a_n$ of a sequence $\{a_n\}$ is the least number such that $a \geq a_n$ for all n.

torus: An n-torus is an n-dimensional **manifold** such that n *topologically distinct* (i.e. not deformable into one another) closed curves can be drawn on it. A 1-torus is a (topological) circle, and a 2-torus has the topology of the familiar donut or anchor ring shape which gives rise to the generic name. A coordinate system on an n-torus consists of n angle-like variables.

unit ball B_0: A **ball** of unit radius. It is used in conjunction with the notion of addition of sets in a vector space, $A + rB_0$, to imply the 'padding out' of the set A by taking the union of all balls of radius r centred on the points comprising A.

Bibliography

Abarbanel, H. D. I., Brown, R., & Kennel, M. B. (1992). Local Lyapunov exponents computed from observed data. *Chaos: An Int. J. of Nonlinear Science*, **2**, 343–365.

Allgower, E., & Georg, K. (1990). *Numerical Continuation Methods: an introduction.* Springer-Verlag.

Anderson, P. W. (1972). More is different. *Science*, **177**, 393–396.

Anosov, D. V. (1967). Geodesic flows and closed Riemannian manifolds with negative curvature. *Proc. Steklov Instit. Math.*, **90**.

Arnol'd, V. I. (1963). Small denominators. II Proof of A.N. Kolmogorov's theorem on the conservation of conditionally periodic motions with a small variation in the Hamiltonian. *Usp. Mat. Nauk. (English translation:* Russian Math. Surv*)*, **18:5**, 13–39 (English 9–36).

Arnol'd, V. I. (1980). *Mathematical Methods of Classical Mechanics.* Graduate Texts in Mathematics, no. 60., Translated by K. Vogtmann and A. Weinstein. Springer-Verlag, New York.

Arrowsmith, D. K., & Place, C. M. (1990). *An Introduction to Dynamical Systems.* Cambridge University Press.

Arrowsmith, D. K., & Place, C. M. (1992). *Dynamical Systems: Differential equations, maps and chaotic behaviour.* Chapman and Hall, London.

Artuso, R., Cvitanovic, P., & Casati, G. (eds). (1991). *Chaos, Order and Patterns.* NATO ASI Series B: Physics, No. 280. Page 25.

Block, L. S., & Coppel, W. A. (1992). *Dynamics in One Dimension.* Lecture Notes in Mathematics, no. 1513. Springer-Verlag, Berlin.

Bojanczyk, A., Golub, G. H., & Van Dooren, P. (1992). *The Periodic Schur Decomposition: Algorithms and Applications.* Tech. rept. NA-92-07. Stanford University.

Bojanczyk, A. W., Ewerbring, M., Luk, F. T., & Dooren, P. Van. (1991). An accurate product SVD algorithm. *Pages 113–131 of:* Vaccaro, R. J. (ed), *SVD and Signal Processing II.* Elsevier Science, Amsterdam, New York.

Bowen, R. (1975). ω-limit sets for Axiom-A diffeomorphisms. *J. Differential Equations*, **18**, 333–339.

Carr, J. (1981). *Applications of Centre Manifold Theory*. Applied Mathematical Sciences, no. 35. Springer-Verlag, New York.

Casdagli, M., Eubank, S., Farmer, J. D., & Gibson, J. (1991). State space reconstruction in the presence of noise. *Physica D*, **51**, 52–98.

Chapman, S., & Cowling, T. G. (1939). *The Mathematical Theory of Non-Uniform Gases*. 2nd edn. Cambridge University Press.

Cohen, E. G. D. (1993). Fifty years of kinetic theory. *Physica A*, **194**, 229–257.

Constantin, P., Foias, C., Nicolaenko, B., & Témam, R. (1988). *Integral Manifolds and Inertial Manifolds for Dissipative Partial Differential Equations*. Appl. Math. Sciences Ser, no. 70. Springer-Verlag, New York.

Constantin, P., Foias, C., Nicolaenko, B., & Témam, R. (1989). Spectral barriers and inertial manifolds for dissipative partial differential equations. *J. Dynamics and Differential Equations*, **1**, 45–73.

Coppel, W. A. (1978). *Dichotomies in Stability Theory*. Lecture Notes in Mathematics, no. 629. Springer-Verlag, Berlin.

Crawford, J. D. (1991). Introduction to bifurcation theory. *Rev. Mod. Phys*, **63**, 991–1037.

Crouzeix, M., & Rappaz, J. (1990). *On Numerical Approximation in Bifurcation Theory*. Recherches en mathématiques appliquées, no. 13. Springer-Verlag, Berlin.

Crutchfield, J. P., & Kaneko, K. (1987). Phenomenology of Spatio-Temporal Chaos. *Pages 272–353 of:* Hao, Bai-lin (ed), *Directions in Chaos*. Directions in Condensed Matter Physics, vol. 1, no. 3. World Scientific, Singapore.

Debussche, A., & Marion, M. (1992). On the construction of families of approximate inertial manifolds. *J. Differential Equations*, **100**, 173–201.

Deici, L., Russell, R. D., & Van Vleck, E. S. (1994). Unitary integrators and applications to continuous orthonormalisation techniques. *SIAM J. Numer. Anal.*, **31**, 261–281.

Delbourgo, R. (1992). Relations Between Universal Scaling Constants in Dissipative Maps. *Pages 231–256 of: Proc. 4th Physics Summer School, 1991 Australian National University, Canberra, Australia*. World Scientific, Singapore.

Devaney, R. L. (1989). *An Introduction to Chaotic Dynamical Systems*, 2nd edn. Addison-Wesley, Redwood City.

Dewar, R. L., & Henry, B. I. (1992). Nonlinear Dynamics and Chaos. *Pages 7–25 of:* Dewar, R. L., & Henry, B. I. (eds), *Proc. 4th Physics Summer School, January, 1991, Australian National University, Canberra, Australia*. World Scientific, Singapore.

Dewar, R. L., & Meiss, J. D. (1992). Flux-minimizing curves for reversible area-preserving maps. *Physica D*, **57**, 476–506.

Doedel, E. J. (1986). *AUTO: Software for Continuation and Bifurcation Problems in Ordinary Differential Equations*. Tech. rept. California Institute of Technology. Software available from E. Doedel on request; email `keqfe82@vax2.concordia.ca`.

Eckmann, J.-P., & Procaccia, I. (1991). Spatio-Temporal Chaos. *Pages 135–172 of:* Proc. NATO Advanced Study Institute, 1990, Lake Como, Italy.

Eckmann, J.-P., & Ruelle, D. (1985). Ergodic theory of chaos and strange attractors. *Rev. Modern Physics*, **57**, 617–656.

Evans, D., Sarman, S., Baranyai, A., Cohen, E. G. D., & Morriss, G. P. (1992). Lyapunov Exponents and Bulk Transport Coefficients. *Pages 285–99 of:* Proc. NATO Advanced Study Institute, Alghero, Sardinia.

Evans, D. J., & Morriss, G. P. (1990). *Statistical Mechanics of Nonequilibrium Liquids*. Academic Press, London.

Evans, D. J., Cohen, E. G. D., & Morriss, G. P. (1990). Viscosity of a simple fluid from its maximal Lyapunov exponents. *Phys. Rev. A*, **42**, 5990–5997.

Feigenbaum, M. J. (1978). Quantitative universality for a class of nonlinear transformations. *J. Stat. Phys*, **19**, 25–52.

Fermi, E., Pasta, J. R., & Ulam, S. (1965). Studies of Nonlinear Problems. *Collected Works of Enrico Fermi*, vol. 2. University of Chicago Press, Chicago.

Garcia, C. B., & Zangwill, W. I. (1981). *Pathways to Solutions, Fixed Points and Equilibria*. Computational Mathematics. Prentice Hall, Englewood Cliffs, N. J.

Geist, K., Parlitz, U., & Lautersborn, W. (1990). Comparison of different methods for computing Lyapunov exponents. *Progress Theoret. Physics*, **83**, 875–93.

Goldstein, H. (1980). *Classical Mechanics*. 2nd edn. Addison-Wesley, Reading, Mass.

Golub, G., & Van Loan, C. (1989). *Matrix Computations*. 2nd edn. Johns Hopkins Press, Baltimore.

Goubet, O. (1992). Construction of approximate inertial manifolds using wavelets. *SIAM J. Math. Anal.*, **23**, 1455–81.

Grauer, R. (1989). Nonlinear interactions of tearing modes in the vicinity of a bifurcation point of codimension two. *Physica D*, **35**, 107–126.

Greene, J. M. (1979). A method for determining a stochastic transition. *J. Math. Phys.*, **20**, 1183–201.

Guckenheimer, J., & Holmes, P. (1983). *Nonlinear Oscillations, Dynamical Systems, and Bifurcations of Vector Fields*. Springer-Verlag, New York.

Guckenheimer, J., Myers, M., Worfolk, F. J., & Wicklinand, P. A. (1992). dstool: *A Dynamical System Toolkit with an Interactive Graphical Interface*. Tech. rept. Center For Applied Mathematics, Cornell University.

Guilleman, V., & Pollack, A. (1974). *Differential Topology*. Prentice Hall, Englewood Cliffs, N. J.

Haken, H. (1978). *Synergetics: An Introduction* 2nd edn. Springer Series in Synergetics 1: Nonequilibrium Phase Transitions and Self-Organization in Physics, Chemistry and Biology. Springer-Verlag, Berlin.

Haken, H. (1983). *Advanced Synergetics*. Springer Series in Synergetics 20: Instability Hierarchies of Self-Organizing Systems and Devices. Springer-Verlag, Berlin.

Hale, J. K., Lin, X.-B., & Raugel, G. (1988). Upper semicontinuity of attractors for approximation of semigroups and partial differential equations. *Math. Comp.*, **50**, 89–123.

Hammel, S. M., Yorke, J. A., & Grebogi, C. (1987). Do numerical orbits of chaotic dynamical processes represent true orbits? *J. Complexity*, **3**, 136–45.

Hao, Bai-lin (ed). (1987). *Directions in Chaos*. Directions in Condensed Matter Physics, vol. 1, no. 3. World Scientific, Singapore.

Hao, Bai-Lin (ed). (1990). *Chaos II*. World Scientific, Singapore.

Hénon, M. (1976). A two-dimensional mapping with a strange attractor. *Commun. Math. Phys.*, **50**, 69–77.

Hoover, W. G. (1993). Nonequilibrium molecular dynamics: the first 25 years. *Physica A*, **194**, 450–61.

Hunt, F. (1990). Error analysis and convergence of capacity dimension algorithms. *SIAM J. Appl. Math.*, **50**, 307–21.

Keller, H. B. (1987). *Lectures on Numerical Methods in Bifurcation Problems*. Springer-Verlag, Berlin.

Kloeden, P. E., & Lorenz, J. (1986). Stable attracting sets sets in dynamical systems and their one-step discretizations. *SIAM J. Numerical Analysis*, **23**, 986–95.

Kolmogorov, A. N. (1954). Preservation of conditionally periodic movements with small change in the hamilton function. *Akad. Nauk. SSSR Doklady*, **98**, 527–530.

Kostelich, E. J. (1992). Problems in estimating dynamics from data. *Physica D*, **58**, 138–52.

Landau, L. D., & Lifshitz, E. M. (1971). *Mechanics*. Pergamon, Oxford.

Lanford, O. E. (1981). The hard sphere gas in the Boltzmann-Grad limit. *Physica A*, **106**, 70–6.

Lebowitz, J. L. (1993a). Boltzmann's entropy and time's arrow. *Physics Today*, **46**, 22–29.

Lebowitz, J. L. (1993b). Macroscopic laws, microscopic dynamics, time's arrow and Boltzmann's entropy. *Physica A*, **194**, 1–27.

Lichtenberg, A. J., & Lieberman, M. A. (1992). *Regular and Chaotic Dynamics*. e2. Applied Mathematical Sciences, no. 38. Springer-Verlag, New York.

Mañé, R. (1977). *Reduction of Semilinear Parabolic Equations to Finite Dimensional C^1 Flows*. Lecture Notes in Mathematics, no. 597. Springer-Verlag, Berlin. Pages 361–78.

MacKay, R. S., & Meiss, J. D. (eds). (1987). *Hamiltonian Dynamical Systems*. Adam Hilger, Bristol.

MacKay, R. S., Meiss, J. D., & Percival, I. C. (1985). Transport in Hamiltonian systems. *Physica D*, **13**, 55–81.

Mallet-Paret, J., & Sell, G. R. (1986). The principle of spatial averaging and inertial manifolds for reaction-diffusion equations. *Nonlinear Semigroups, Partial Differential Equations and Attractors*. Lecture Notes in Mathematics, no. 1248. Springer-Verlag, Berlin.

Manneville, P. (1990). *Dissipative Structures and Weak Turbulence*. Perspectives in Physics. Academic Press, Boston.

Mareschal, M., & Holian, B. (eds). (1992). *Microscopic Simulations of Complex Hydrodynamic Phenomena*. Plenum, New York.

Marion, M. (1989). Approximate inertial manifolds for

reaction-diffusion equations in higher space dimension. *J. Dynamics and Differential Equations*, **1**, 245–67.

Meiss, J. D. (1992). Symplectic maps, variational principles, and transport. *Rev. Mod. Phys.*, **64**, 795–848.

Moler, C., Little, J., & Bangert, S. (1990). *Pro-MATLAB User's Guide.* The MathWorks Inc. and South Natick, Mass.

Montgomery, D. C., & Tidman, D. A. (1964). *Plasma Kinetic Theory.* McGraw-Hill, New York.

Moon, F. C. (1992). *Chaotic and Fractal Dynamics: An Introduction for Applied Scientists and Engineers.* John Wiley, New York.

Moser, J. (1962). On invariant curves of area-preserving mappings of an annulus. *Nachr. Akad. Wiss. Göttingen Math.-Phys*, **Kl. II**, 1–20.

Muncaster, R. G. (1983). Invariant manifolds in mechanics I: the general construction of coarse theories from fine theories. *Arch. Rat. Mech. and Anal*, **83**, 353–73.

Nusse, H. E., & Yorke, J. A. (1993). *Dynamics: Numerical Explorations.* Springer-Verlag, New York.

Oppenheim, I., Shuler, K. E., & Weiss, G. H (eds). (1977). *Stochastic Processes in Chemical Physics: The Master Equation.* MIT Press, Cambridge, Mass.

Ott, E. (1993). *Chaos in Dynamical Systems.* Cambridge University Press.

Palis, J., & de Melo, W. (1982). *Geometric Theory of Dynamical Systems: An Introduction.* Springer-Verlag, Berlin.

Palmer, K. (1988). Exponential dichotomies, the shadowing lemma and transversal momoclinic points. *Pages 265–306 of: Dynamics Reported,* Vol. 1. John Wiley, New York.

Parker, R. D., Dewar, R. L., & Johnson, J. L. (1990). Symmetry breaking bifurcations of a current sheet. *Phys. Fluids B*, **2**, 508–15.

Parker, T. S., & Chua, L. O. (1989). *Practical Numerical Algorithms for Chaotic Systems.* Springer-Verlag, Berlin.

Perko, L. (1991). *Differential Equations and Dynamical Systems.* Texts in Appl. Maths, no. 17. Springer-Verlag, Berlin.

Poincaré, H. (1892). *Les Méthodes Nouvelles de la Méchanique Céleste.* Gauthiers-Villars, Paris.

Ramsey, J. B., & Yuan, H.-J. (1990). The statistical properties of dimension calculations using small data sets. *Nonlinearity*, **3**, 155–76.

Raugel, G., & Hale, J. K. (1989). Continuity of attractors: Attractors, inertial manifolds and their approximation. *RAIRO Modél. Math. Anal. Numér*, **23**, 519–33.

Roberts, A. J. (1990). Low-dimensionality – Approximations In Mechanics. *Pages 715–722 of:* Noye, J., & Hogarth, W (eds), *Computational Techniques and Applications: CTAC–89.* Hemisphere.

Roberts, A. J. (1992). Planform evolution in convection – an embedded centre manifold. *J. Austral. Math. Soc. B*, **34**, 174–198.

Roberts, J. A. G., & Quispel, G. R. W. (1992). Chaos and time-reversal symmetry. order and chaos in reversible dynamical suystems. *Phys. Rep.*, **216**, 63–177.

Ruelle, D. (1989). *Chaotic Evolution and Strange Attractors.* Cambridge University Press.

Sanz-Serna, J. M. (1992). Symplectic integrators for Hamiltonian problems: an overview. *Pages 273–279 of:* Iserles, A. (ed), *Acta Numerica*. Cambridge University Press.

Sauer, T. (1993). Private Communication.

Sauer, T., & Yorke, J. A. (1991). *Statistically Self Similar Sets*. Notes in preparation.

Sauer, T., Yorke, J. A., & Casdagli, M. (1991). Embedology. *J. Statistical Physics*, **65**, 579–616.

Scheck, F. (1990). *Mechanics: From Newton's Laws to Deterministic Chaos*. Springer-Verlag, Berlin.

Schweber, S S. (1993). Physics, community and the crisis in physical theory. *Physics Today*, **46**, 34–40.

Seidel, R. (1989). *BIFPACK, a Program for Continuation, Bifurcation and Stability Analysis*. Tech. rept. University of Würzburg.

Shub, M. (1987). *Global Stability of Dynamical Systems*. Springer-Verlag, New York.

Smale, S. (1980 (originally 1967)). Differentiable dynamical systems. *Pages 1–84 of: The Mathematics of Time*. Springer-Verlag, Berlin.

Stewart, D. E. (1994a). A new algorithm for the SVD of a long product of matrices. *Submitted to SIAM J. Matrix Analysis and Applications*.

Stewart, D. E. (1994b). *Product Algorithms for Eigensystems*. Tech. rept. ACTR-31-03-94. Australian National University, Centre for Mathematics and its Applications.

Sudarshan, E. C. G., & Mukunda, N. (1974). *Classical Dynamics: A Modern Perspective*. Wiley-Interscience, New York. Chap. 16, pages 275–281.

Swinney, H. L., & Gollub, J. P. (1981). *Hydrodynamic Instabilities and the Transition to Turbulence*. Topics in Applied Physics, no. 45. Springer-Verlag, Berlin.

Takens, F. (1981). *Detecting Strange Attractors in Turbulence*. Lecture notes in Mathematics, no. 898. Springer-Verlag, Berlin.

Temam, R. (1988). *Infinite-Dimensional Dynamical Systems in Mechanics and Physics*. Applied Mathematical Sciences, no. 68. Springer-Verlag, Berlin.

Tong, H. (1992). Some comments on a bridge between nonlinear dynamicists and statisticians. *Physica D*, **58**, 299–303.

van der Pol, B. (1927). Forced oscillations in a circuit with nonlinear resistance. *London, Edinburgh and Dublin Philosoph. Mag.*, **3**, 65–80.

Watson, L. T., Billups, S. C., & Morgan, A. P. (1987). Algorithm 652: HOMPACK: A suite of codes for globally convergent homotopy algorithms. *ACM Trans. on Math. Software*, **13**, 281–310.

Wessen, K. P. (1993). *Application of the Invariant Manifold Reduction to Plasma Instabilities*. Ph.D. thesis, The Australian National University, Canberra ACT 0200, Australia.

Wiggins, S. (1992). *Chaotic Transport in Dynamical Systems*. Interdisciplinary Applied Mathematics, no. 2. Springer-Verlag, New York.

Wolfram, S. (1991). Mathematica: *A System for Doing Mathematics by a Computer*. 2nd edn. Addison-Wesley, Redwood, CA.

Zwanzig, R. (1964). On the identity of three generalized master equations. *Physica*, **30**, 1109–1123.

7
Non-linear control systems

MATTHEW R. JAMES

Department of Systems Engineering,
Research School of Information Sciences and Engineering,
Australian National University,
Canberra, ACT 0200, AUSTRALIA

7.1 Introduction

Control systems are prevalent in nature and in man-made systems. Natural regulation occurs in biological and chemical processes, and may serve to maintain the various constituents at their appropriate levels, for example. In the early days of the industrial revolution, governors were devised to regulate the speed of steam engines, while in modern times, computerised control systems have become commonplace in industrial plants, robot manipulators, aircraft and spacecraft, etc. Indeed, the highly maneuverable X-29 aircraft using forward swept wings is possible only because of its control systems, and moreover, control theory has been crucial in NASA's Apollo and Space Shuttle programmes. Control systems such as in these examples use in an essential way the idea of *feedback*, the central theme of this chapter.

Control theory is the branch of engineering/science concerned with the design and analysis of control systems. Linear control theory treats systems for which an underlying linear model is assumed, and is a relatively mature subject, complete with firm theoretical foundations and a wide range of powerful and applicable design methodologies (Anderson & Moore, 1990), (Kailath, 1980). In contrast, non-linear control theory deals with systems for which linear models are not adequate, and is relatively immature, especially in relation to applications. In fact, linear systems techniques are frequently employed in spite of the presence of non-linearities. Nonetheless, non-linear control theory is exciting and vitally important, and is the subject of a huge and varied range of research worldwide.

The aim of this chapter is to convey to readers of *Complex Systems* something of the flavour of the subject, the techniques, the computational issues, and some of the applications. To place this chapter in perspective, in relation to the other chapters in this book, it is worthwhile citing Brockett's remark that control theory is a *prescriptive science*, whereas physics, biology, etc, are *descriptive sciences* (Brockett, 1976). Computer science shares some of the prescriptive qualities of control theory, in the sense that some objective is prescribed, and means are sought to fulfill it. It is this *design* aspect that is most important here. Indeed, control systems are designed to *influence* the behaviour of the system being controlled in order to achieve a desired level of performance. Brockett categorised control theory briefly:

(i) To express models in input–output form, thereby identifying those variables which can be manipulated and those which can be observed.

(ii) To develop methods for regulating the response of systems by modifying the dynamical nature of the system—e.g. stabilisation.

(iii) To optimise the performance of the system relative to some performance index.

In addition, feedback design endevours to compensate for disturbances and uncertainty. This chapter attempts to highlight the fundamental role played by feedback in control theory. Additional themes are stability, robustness, optimisation, information and computational complexity.

It is impossible in a few pages or even in a whole book to do justice to our aims, and the material presented certainly omits many important aspects. In addition, there are new ideas and technologies emerging continually, many of which have close links to computer science, mathematics and physics. Examples include: discrete event systems, neural networks, fuzzy logic control, massively parallel computation, and chaos. We briefly discuss some of these topics below.

In §7.2, some of the basic ideas of feedback are introduced and illustrated by example. The basic definition of the models commonly used is given in §7.3, and this framework is used throughout this chapter. Differential geometry provides useful tools for non-linear control, as will be seen in §7.4 on the important system theoretic concepts of controllability and observability. The possibility of feedback linearisation of non-linear systems is discussed in §7.5, and the problem of stabilisation is the topic of §7.6. Optimisation-based methods are very important in control theory, and are discussed in the remaining sections (§§7.7–7.13). The basic ideas of deterministic and stochastic optimal control are reviewed, and issues such as robustness and stabilisation are discussed. Some results concerning output feedback problems are briefly touched upon, and computational methods and issues of computational complexity are covered.

This chapter is written in a tutorial style, with emphasis on intuition. Some parts of this chapter are more technically demanding than other parts, and it is hoped that the references provided will be of use to interested readers. A number of standard beginning graduate level texts are available, including: Athans and Falb (1965), Bertsekas (1987), Craig (1989), Khalil (1992), Kumar and Variaya (1986), Nijmeijer and van der Schaft (1990), Sontag (1990) and Vidyasagar (1993). The books by Isidori (1985), Fleming and Rishel (1975), and Fleming and Soner (1993) are excellent references. Finally, the report (1988) contains

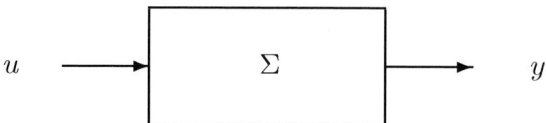

Fig. 7.1. Open loop system.

a lot of interesting discussion regarding the state of affairs in control theory.

7.2 The power of feedback

7.2.1 What is feedback?

The conceptual framework of system theory depends in a large part on the distinction between different types of variables in a model, in particular, the inputs and outputs. It may happen that certain variables have natural interpretations as inputs since they are readily manipulated, while other measured variables serve as outputs. For instance, in a robot arm, the torque applied by the joint actuators is an input, whereas the measured joint angles are outputs. However, the distinction is not always clear and is a matter of choice in the modelling process. In econometrics, is the prime interest rate an input, or an output, i.e. is it an independent variable, or one which responds to other influences? For us, a *system* Σ is a causal map which assigns an output $y(\cdot)$ to each input $u(\cdot)$, drawn in fig. 7.1. Causality means that past values of the output do not depend on future values of the input.

Feedback occurs when the *information* in the output is made available to the input, usually after some processing through another system. A general feedback system may involve additional inputs and outputs with specific meanings (fig. 7.2).

In the configuration of fig. 7.2, the system Σ is often called the *plant* (the system being controlled), while the system C is called the *controller*. The inputs w may include reference signals to be followed, as well as unknown disturbances (such as modelling errors and noise), and the

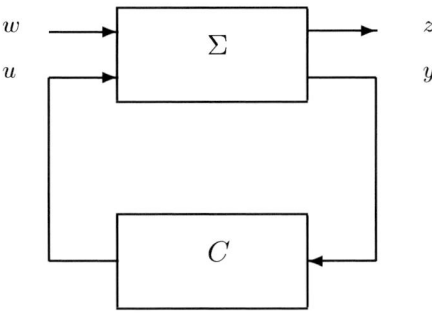

Fig. 7.2. General closed loop feedback system.

output z is a performance variable, say an error quantity with ideal value zero. The feedback system shown in fig. 7.2 is called a *closed loop* system, for obvious reasons, whereas, the system Σ alone as in fig. 7.1 is called an *open loop* system. Thus a feedback system is characterised by the information flow that occurs or is permitted among the various inputs and outputs. The modelling process consists of building a model of this form, and once this is done, the task of designing the controller to meet the given objectives can be tackled.

7.2.2 Feedback design

To illustrate some of the benefits of feedback, we consider the problem of trying to keep an inverted pendulum (or one-link robot manipulator) in its vertical *state of rest* or *equilibrium*. As the pendulum can easily fall, this equilibrium configuration is not *stable* (just try to balance a pencil on its tip!). We want to *stabilise* this equilibrium position, using a *feedback* control system. This type of problem is very common. Equation (7.1) is a simple non-linear model of the inverted pendulum:

$$m\ddot{\theta} - mg\sin\theta = \tau. \tag{7.1}$$

The angle θ is measured from the vertical ($\theta = 0$ corresponds to the vertical equilibrium, i.e. the state of rest or absence of motion), and τ is the torque applied by a motor to the revolute joint attaching the pendulum to a frame (fig. 7.3). The motor is the active component, and the pendulum is controlled by adjusting the motor torque appropriately. Thus the input is $u = \tau$. If the joint angle is measured, then the output

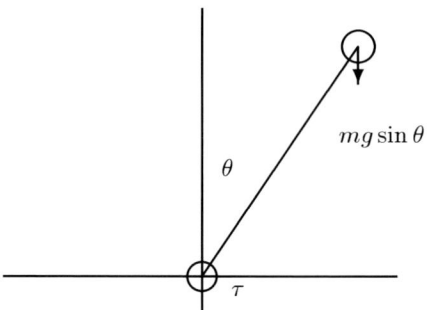

Fig. 7.3. Inverted pendulum or 1-link robot arm.

is $y = \theta$. If the angular velocity is also measured, then $y = (\theta, \dot\theta)$. The pendulum has length 1 m, mass m kg, and g is the acceleration due to gravity.

The model neglects effects such as friction, motor dynamics, etc.

We begin by analysing the stability of the equilibrium $\theta = 0$ of the homogeneous system

$$m\ddot\theta - mg\sin\theta = 0 \qquad (7.2)$$

corresponding to zero motor torque (no control action). Note that if $\theta(0) = 0$ and $\dot\theta(0) = 0$, then ideally $\theta(t) = 0$ for all t. The linearisation of (7.2) at $\theta = 0$ is

$$m\ddot\theta - mg\theta = 0, \qquad (7.3)$$

since $\sin\theta \approx \theta$ for small θ. This equation has general solution $\theta(t) = \alpha e^{\sqrt{g}t} + \beta e^{-\sqrt{g}t}$ (where α and β depend on the initial conditions $\theta(0)$ and $\dot\theta(0)$), and, because of the first term (exponential growth), $\theta = 0$ is not a stable equilibrium for (7.2). This means that if the pendulum is initially set-up in the vertical position, then a small disturbance can cause the pendulum to fall. We would like to design a control system to prevent this from happening, i.e. to restore the pendulum to its vertical position in case a disturbance causes it to fall.

One could design a stabilising feedback controller for the linearised system

$$m\ddot\theta - mg\theta = \tau \qquad (7.4)$$

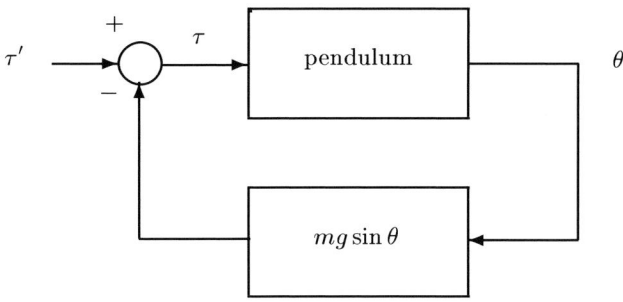

Fig. 7.4. Feedback linearisation of the pendulum.

and use it to control the non-linear system (7.1). This will result in a locally stable system, which may be satisfactory for small deviations θ from 0.

Instead, we adopt an approach which is called the computed torque method in robotics, or feedback linearisation (see §7.5). This method has the potential to yield a globally stabilising controller. The non-linearity in (7.1) is not *neglected* as in the linearisation method just mentioned, but rather it is *cancelled*. This is achieved by applying the *feedback control law*

$$\tau = -mg\sin\theta + \tau' \tag{7.5}$$

to (7.1), where τ' is a new torque input, and results in the *closed loop system* relating θ and τ':

$$m\ddot{\theta} = \tau'. \tag{7.6}$$

Equation (7.6) describes a linear system with input τ', shown in fig. 7.4.

Linear systems methods can now be used to choose τ' in such a way that the pendulum is stabilised. Indeed, let's set

$$\tau' = -k_1\dot{\theta} - k_2\theta.$$

Then (7.6) becomes

$$m\ddot{\theta} + k_1\dot{\theta} + k_2\theta = 0. \tag{7.7}$$

If we select the *feedback gains*

$$k_1 = 3m, \quad k_2 = 2m,$$

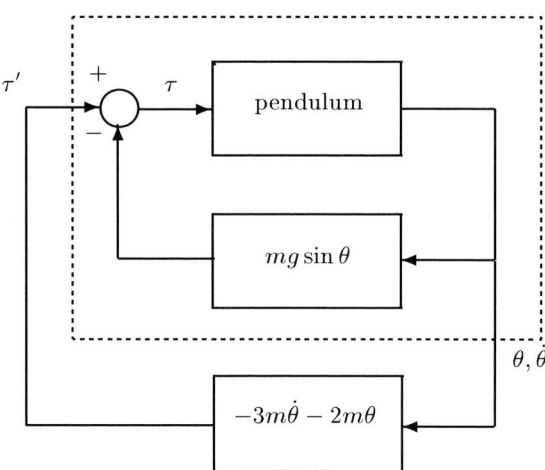

Fig. 7.5. Feedback stabilisation of the pendulum using feedback linearisation.

then the system (7.7) has general solution $\theta(t) = \alpha e^{-t} + \beta e^{-2t}$, and so $\theta = 0$ is now globally stable. Thus the effect of any disturbance will decay exponentially, and the pendulum will be restored to its vertical position.

The feedback controller just designed and applied to (7.1) is

$$\tau = -mg\sin\theta - 3m\dot\theta - 2m\theta. \tag{7.8}$$

The torque τ is an explicit function of the angle θ and its derivative, the angular velocity $\dot\theta$. The controller is a *feedback* controller because it takes measurements of θ and $\dot\theta$ and uses this *information* to adjust the motor torque in a way which is stabilising, see fig. 7.5. For instance, if θ is non-zero, then the last term (proportional term) in (7.8) has the effect of forcing the motor to act in a direction opposite to the natural tendency to fall. The second term (derivative term) in (7.8) responds to the speed of the pendulum. An additional integral term may also be added to remove any steady state errors. The proportional-integral-derivative (PID) combination is widespread in applications.

It is worth noting that the feedback controller (7.8) has fundamentally altered the dynamics of the pendulum. With the controller in place, it is no longer an unstable non-linear system. Indeed, in a different context, it was reported in the article (Hunt & Johnson, 1993) that it is possible to remove the effects of *chaos* with a suitable controller, even using a simple proportional control method.

Note that this design procedure requires explicit knowledge of the model parameters (length, mass, etc.), and if they are not known exactly, performance may be degraded. Similarly, unmodelled influences (such as friction, motor dynamics) also impose significant practical limitations. Ideally, one would like a design which is *robust* and tolerates these negative effects.

The control problem is substantially more complicated if measurements of the joint angle and/or angular velocity are not available. The problem is then one of partial state information, and an output feedback design is required. These issues are discussed in §7.4.2 and §7.12.

It is also interesting to note that the natural state space (corresponding to $(\theta, \dot{\theta})$) for this example is the cylinder $M = S^1 \times \mathbf{R}$ – a manifold and not a vector space. Thus it is not surprising to find that differential geometric methods provide important tools in non-linear control theory. What this means is that in certain applications (e.g. robotics, satellite attitude control) the natural geometry can be very important and usefully exploited. Differential geometry deals with spaces which can be curved, and not simply flat like $\mathbf{R}^n$.

7.3 State space models

The systems discussed in §7.2.1 were represented in terms of block diagrams showing the information flows among key variables, the various inputs and outputs. No internal details were given for the blocks. In §7.2.2 internal details were given by the physical model for the pendulum, expressed in terms of a differential equation for the joint angle θ. In general, it is often convenient to use a differential equation model to describe the input–output behaviour of a system, and we will adopt this type of description throughout this chapter. Systems will be described by first-order differential equations in *state space* form, viz:

$$\begin{cases} \dot{x} &= f(x, u) \\ y &= h(x), \end{cases} \qquad (7.9)$$

where the state x takes values in $\mathbf{R}^n$ (or in some manifold M in which case we interpret (7.9) in local coordinates), the input $u \in U \subset \mathbf{R}^m$, the output $y \in \mathbf{R}^p$. Intuitively, by adjusting u we can steer the system by changing the direction in which the vector $f(x,u)$ points. When the system is in state x, $h(x)$ represents the measurements taken. In the case of the pendulum, the choice for a state mentioned above was $x = (\theta, \dot{\theta})$. The controlled vector field f is assumed sufficiently smooth, as is the observation function h. (In this chapter we will not concern ourselves with precise regularity assumptions.)

The state x need not necessarily have a physical interpretation, and such representations of systems are far from unique. The state space model may be derived from fundamental principles, or via an *identification* procedure. In either case, we assume such a model is given. If a control input $u(t)$, $t \geq 0$, is applied to the system, then the state trajectory $x(t)$, $t \geq 0$, is determined by solving the differential equation (7.9), and the output is given by $y(t) = h(x(t))$, $t \geq 0$.

The state summarises the internal status of the system, and its value $x(t)$ at time t is sufficient to determine future values $x(s)$, $s > t$ given the control values $u(s)$, $s > t$. The state is thus a very important quantity, and the ability of the controls to manipulate it as well as the information about it available in the observations are crucial factors in determining the quality of performance that can be achieved.

A lot can be done at the input–output level, without regard to the internal details of the system blocks. For instance, the *small gain theorem* concerns stability of a closed loop system in an input–output sense, see (Vidyasagar, 1993). However, in this chapter, we will restrict our attention to state space models and methods based on them.

7.4 Controllability and observability

In this section we look at two fundamental issues concerning the nature of non-linear systems. The model we use for a non-linear system Σ is

$$\begin{cases} \dot{x} &= f(x) + g(x)u \\ y &= h(x), \end{cases} \tag{7.10}$$

where f, g and h are smooth. In equation (7.10), $g(x)u$ is short for $\sum_{i=1}^{m} g_i(x)u_i$, where g_i, $i = 1, \ldots, m$, are smooth vector fields. Note that the control u appears affinely, making the analysis a bit easier without loosing much generality. It is of interest to know how well the controls u influence the states x (controllability and reachability), and how much information concerning the states is available in the measured outputs y (observability and reconstructability).

Controllability and observability have been the subject of a large amount of research over the last few decades; we refer the reader to the paper by Hermann and Krener (1977) and the books by Isidori (1985), Nijmeijer and van der Schaft (1990) and Vidyasagar (1993), and the references contained in them. The differential geometric theory for controllability is 'dual' to the corresponding observability theory, in that the former uses vector fields (column vector functions) and distributions (column vector subspaces) while the latter uses 1-forms (row vector functions) and codistributions (row vector subspaces). In both cases, the Frobenius theorem is a basic tool, see Boothby (1975). The Frobenius theorem concerns the existence of (integral) submanifolds (e.g. surfaces) which are tangent at each point to a given field of subspaces (of the tangent spaces), and is a generalisation of the idea that solutions to differential equations define one dimensional submanifolds (i.e. curves) tangent to the one dimensional subspaces spanned by the vector field.

7.4.1 Controllability

Here is a basic question. Given two states x^0 and x^1, is it possible to find a control $t \mapsto u(t)$ which steers Σ from x^0 to x^1? If so, we say that x^1 is *reachable* from x^0, or that x^0 is *controllable* to x^1. This question can be answered (at least locally) using differential geometric methods.

Consider the following special situation with two inputs and no drift f:

$$\dot{x} = u_1 g_1(x) + u_2 g_2(x), \tag{7.11}$$

where g_1 and g_2 are smooth vector fields in $\mathbf{R}^3$ ($n = 3$). The problem is to determine, locally, the states which can be reached from a given point x^0.

Suppose that in a neighbourhood of x^0 the vector fields g_1, g_2 are linearly independent, so that (7.11) can move in two independent directions: setting $u_2 = 0$ and $u_1 = \pm 1$ causes Σ to move along solutions of $\dot{x} = \pm g_1$, and similarly movement along g_2 is effected by setting $u_1 = 0$ and $u_2 = \pm 1$. Clearly Σ can move in the direction of any linear combination of g_1 and g_2. But can Σ move in a direction independent of g_1 and g_2? The answer is yes provided the following algebraic condition holds:

$$g_1(x^0),\ g_2(x^0),\ [g_1, g_2](x^0)\ \text{ linearly independent.} \tag{7.12}$$

Here, the vector field $[g_1, g_2]$ denotes the Lie bracket of the two vector

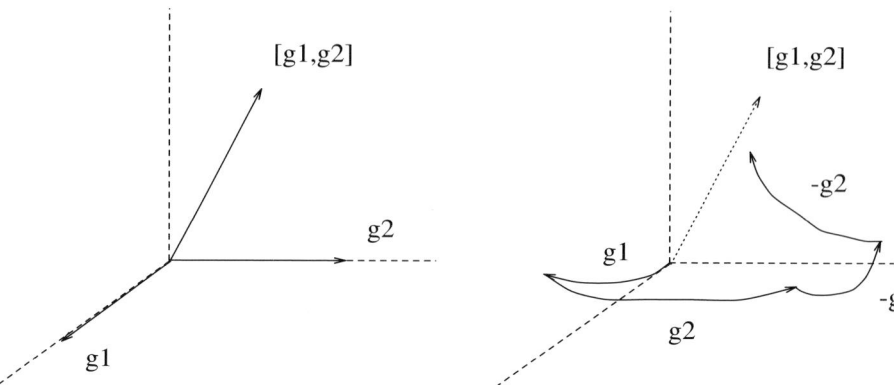

Fig. 7.6. Lie bracket approximated by switching.

fields g_1 and g_2, defined by

$$[g_2, g_2] = \frac{\partial g_2}{\partial x} g_1 - \frac{\partial g_1}{\partial x} g_2, \qquad (7.13)$$

see Boothby (1975). To move in the $[g_1, g_2]$ direction, the control must be switched between g_1 and g_2 appropriately. To define the desired switching, i.e. sequentially changing directions, let

$$u(t) = (u_1(t), u_2(t)) = \begin{cases} (1, 0) & \text{if } 0 \leq t \leq \tau \\ (0, 1) & \text{if } \tau \leq t \leq 2\tau \\ (-1, 0) & \text{if } 2\tau \leq t \leq 3\tau \\ (0, -1) & \text{if } 3\tau \leq t \leq 4\tau. \end{cases}$$

Then the solution of (7.11) satisfies

$$x(4\tau) = x^0 + \tfrac{1}{2}\tau^2 [g_1, g_2](x^0) + O(\tau^3)$$

for small τ, readily verified using Taylor's formula. This is illustrated in fig. 7.6. Note that $u(t)$ is discontinuous, but the resulting trajectory $x(t)$ is continuous. (By more complicated switching, it is possible to move along higher-order Lie brackets.) Using the inverse function theorem, it is possible to show that condition (7.12) implies that the set of points reachable from x^0 contains a non-empty open set.

The general theory involves a generalisation of the rank condition (7.12), and includes the drift f: if $\mathcal{R}$ is the vector space (Lie algebra) obtained by calculating all possible Lie brackets, the *reachability rank condition* says

$$\dim \mathcal{R}(x^0) = n. \tag{7.14}$$

If the reachability rank condition fails, motion may be restricted to lower-dimensional submanifolds.

Example 7.1 Consider the system in $\mathbf{R}^2$: $\dot{x}_1 = u$, $\dot{x}_2 = x_1^2$. Here $f(x_1, x_2) = (0, x_1^2)$ and $g(x_1, x_2) = (1, 0)$. Then $[f, g] = -(0, 2x_1)$ and $[[f, g], g] = (0, 2)$, and consequently $\dim \mathcal{R}(x_1, x_2) = 2$ for all $(x_1, x_2) \in \mathbf{R}^2$. Therefore the reachability rank condition is satisfied everywhere. For any open neighbourhood V of $x^0 = (0, 0)$, the set of states reachable from $(0, 0)$ in time up to T and staying inside V equals $V \cap \{x_2 > 0\}$. Note that points in the upper half plane are reachable from $(0, 0)$, but is not possible to control such points to $(0, 0)$. Thus systems with drift are not generally reversible. □

7.4.2 Observability

For many reasons it is useful to have knowledge of the state of the system being controlled. However, it is not always possible to measure all components of the state, and so knowledge of them must be inferred from the outputs or measurements that are available. In control system design, filters or observers are used to compute an estimate $\hat{x}(t)$ of the state $x(t)$ by processing the outputs $y(s)$, $0 \leq s \leq t$. See §7.12.1. We now ask what information about the states can be obtained from such an output record. This can be answered (locally at least) using a rank condition, not unlike the reachability rank condition.

Two states x^0 and x^1 are *indistinguishable* if no matter what control is applied, the corresponding state trajectories always produce the same output record – there is no way of telling the two states apart by watching the system.

For simplicity, we begin with an uncontrolled single-output system in $\mathbf{R}^2$,

$$\begin{cases} \dot{x} &= f(x) \\ y &= h(x), \end{cases} \tag{7.15}$$

and consider the problem of distinguishing a state x^0 from its neighbours.

This is possible if the following algebraic condition holds:

$$dh(x^0), \ dL_f h(x^0) \text{ linearly independent.} \tag{7.16}$$

Here, for a real valued function φ, $d\varphi = (\frac{\partial \varphi}{\partial x_1}, \ldots, \frac{\partial \varphi}{\partial x_n})$ (a 1-form or row vector), and $L_f \varphi$ denotes the Lie derivative (directional derivative along f)

$$L_f \varphi = d\varphi \cdot f = \sum_{i=1}^n \frac{\partial \varphi}{\partial x_i} f_i.$$

Let $x^i(t, x^i)$ and $y^i(t, x^i)$, $i = 1, 2$, denote the state and output trajectories corresponding to the initial states x^0, x^1. If these two states are indistinguishable, then

$$y^0(t, x^0) = y^1(t, x^1), \text{ for all } t \geq 0. \tag{7.17}$$

Evaluating (7.17) at $t = 0$ gives

$$h(x^0) = h(x^1), \tag{7.18}$$

and differentiating (7.17) with respect to t and setting $t = 0$ gives

$$L_f h(x^0) = L_f h(x^1). \tag{7.19}$$

Define $\Phi : \mathbf{R}^2 \to \mathbf{R}^2$ by $\Phi(x) = (h(x) - h(x^0), L_f h(x) - L_f h(x^0))$. Then $\frac{\partial \Phi}{\partial x}(x^0) = (dh(x^0), dL_f h(x^0))$, and so the condition (7.16) implies, by the inverse function theorem, the existence of an open set V of x^0 such that states in V are distinguishable from x^0. This follows from (7.18) and (7.19) because if x^1 is indistinguishable from x^0, then $\Phi(x^1) = \Phi(x^0) = (0, 0)$. The inverse function theorem implies that Φ is a local diffeomorphism, forcing $x^1 = x^0$ for any indistinguishable x^1 near x^0.

In the general situation, the *observability rank condition* says that

$$\dim d\mathcal{O}(x^0) = n, \tag{7.20}$$

where $d\mathcal{O}$ is the row vector space of differentials $d\varphi$ of the space $\mathcal{O}$ of functions φ which contains $h_1, \ldots, h_p$ and is closed with respect to Lie differentiation by $f, g_1, \ldots, g_m$. In cases where the observability rank condition fails, indistinguishable states may be contained in integral submanifolds of the observability codistribution $d\mathcal{O}$.

Example 7.2 Consider the 1-dimensional system: $\dot{x} = u$, $y = \sin x$. Here $f(x) = 0$, $g(x) = 1$, and $h(x) = \sin x$. The observation space $\mathcal{O}$ is the span (over $\mathbf{R}$) of $\{\sin x, \cos x\}$, and $d\mathcal{O}(x) = \text{span}\{\cos x, \sin x\} = T_x^* \mathbf{R} \equiv \mathbf{R}$ for all $x \in \mathbf{R}$. Thus the observability rank condition is everywhere satisfied. Thus the system is locally observable in $(-\pi, \pi)$, but not globally, since the set of points indistinguishable from 0 equals $\{2\pi k, k \text{ any integer}\}$. □

7.5 Feedback linearisation

In §7.2 we saw that it was possible to linearise the inverted pendulum (or one-link manipulator) using non-linear state feedback. The most general results involve state space transformations as well (Hunt & Meyer, 1983, Nijmeijer & van der Schaft, 1990, Isidori, 1985, Vidyasagar, 1993, Khalil, 1992). Feedback linearisation methods have been successfully applied in the design of a helicopter control system. The advantage of this approach is that well-established linear design methods can be applied to the feedback-linearised system.

For clarity, we consider the following single-input system Σ ($m = 1$):

$$\dot{x} = f(x) + g(x)u, \tag{7.21}$$

where f and g are smooth vector fields with $f(0) = 0$. The (local) feedback linearisation problem is to find, if possible, a (local) diffeomorphism S on $\mathbf{R}^n$ with $S(0) = 0$, and a state feedback law

$$u = \alpha(x) + \beta(x)v \tag{7.22}$$

with $\alpha(0) = 0$ and $\beta(x)$ invertible for all x such that the resulting system

$$\dot{x} = f(x) + g(x)\alpha(x) + g(x)\beta(x)v \tag{7.23}$$

transforms under $z = S(x)$ to a controllable linear system

$$\dot{z} = Az + Bv. \tag{7.24}$$

Thus the closed loop system relating v and z is linear, see fig. 7.7.

The main results concerning this single-input case assert that the non-linear system (7.21) is feedback linearisable to the controllable linear system (7.24) under certain technical conditions involving Lie derivatives. Fortunately, they are easily satisfied in example 7.3 below.

Once a system has been feedback linearised, further control design can take place using linear systems methods to obtain a suitable control law $v = Kz = KS(x)$ achieving a desired goal (e.g. stabilisation as in §7.2).

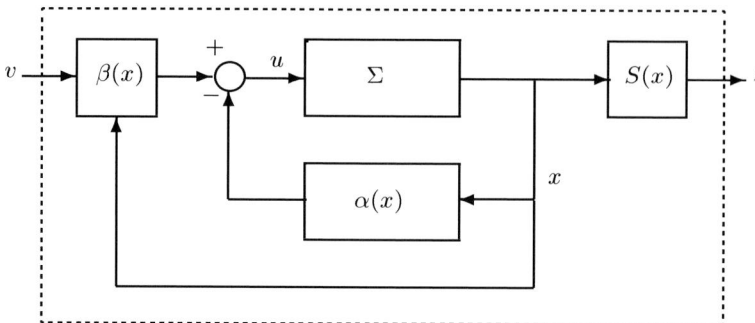

Fig. 7.7. Feedback linearisation.

Example 7.3 In robotics, this procedure is readily implemented, and is known as the *computed torque method*, see (Craig, 1989). The dynamic equations for an n-link manipulator are

$$M(\theta)\ddot{\theta} + V(\theta, \dot{\theta}) + G(\theta) = \tau, \qquad (7.25)$$

where θ is an n-vector of joint angles, $M(\theta)$ is the (invertible) mass matrix, $V(\theta, \dot{\theta})$ consists of centrifugal and coriolis terms, $G(\theta)$ accounts for gravity, and τ is the n-vector of joint torques applied at each joint. Feedback linearisation is achieved by the choice

$$\tau = M(\theta)\tau' + V(\theta, \dot{\theta}) + G(\theta). \qquad (7.26)$$

The resulting linear system is

$$\ddot{\theta} = \tau'. \qquad (7.27)$$

The state $x = (\theta, \dot{\theta})$ is $2n$-dimensional, and the reader can easily express the dynamics and feedback law in state space form. □

7.6 Feedback stabilisation

Stabilisation is one of the most important problems in control theory. After all, one cannot do much with a system that is wildly out of control! In this section we take a quick look at some of the issues that arise.

The (local) feedback stabilisation problem for the system

$$\dot{x} = f(x) + g(x)u, \qquad (7.28)$$

where f and $g = g_1, \ldots, g_m$ are smooth vector fields with $f(0) = 0$, is to find, if possible, a function $\alpha : \mathbf{R}^n \to \mathbf{R}^m$ such that the state feedback law

$$u = \alpha(x) \qquad (7.29)$$

results in a (locally) asymptotically stable closed loop system:

$$\dot{x} = f(x) + g(x)\alpha(x). \qquad (7.30)$$

That is, for any initial condition x^0 (in a neighbourhood of 0), the resulting closed-loop trajectory of (7.30) should satisfy $x(t) \to 0$ as $t \to \infty$. As is apparent from our discussion so far, systems which are feedback linearisable can be stabilised by smooth feedback simply by using a stabilising linear feedback for the linearised system in conjunction with the linearising transformation and feedback.

According to classical ordinary differential equation (ODE) theory, in order that solutions to (7.30) exist and are unique, the function α must satisfy some smoothness requirements, say Lipschitz continuity. Thus it is not unreasonable to restrict one's search to include only sufficiently smooth α. Also, considering the definition of controllability, one may guess that some form of this property is needed. However, a necessary condition was given in (Brockett, 1983) which implies that controllability is not enough for there to exist a continuous stabilising feedback of the form (7.29). Indeed, there does not exist a continuous stabilising feedback for the drift-free system

$$\dot{x} = g(x)u, \qquad (7.31)$$

when $m < n$, even if controllable. Such systems cannot even be stabilised by continuous dynamic feedback. Thus one is forced to relax the smoothness requirement, or look for different types of feedback functions.

Recently, some remarkable results have been obtained which assert that satisfaction of the reachability rank condition is enough to imply the existence of a stabilising *time-varying periodic* feedback

$$u = \alpha(x,t) \qquad (7.32)$$

for drift-free systems. It is very interesting that time-varying feedback laws can succeed when time-invariant ones fail. Generalisations of these results to the case of systems with drift have been obtained.

Example 7.4

Consider the following drift-free system in $\mathbf{R}^3$ (Pomet, 1992):

$$\begin{pmatrix} \dot{x}_1 \\ \dot{x}_2 \\ \dot{x}_3 \end{pmatrix} = \begin{pmatrix} 1 \\ 0 \\ 0 \end{pmatrix} u_1 + \begin{pmatrix} 0 \\ x_1 \\ 1 \end{pmatrix} u_2. \qquad (7.33)$$

Since $m = 2 < n = 3$, this system cannot be stabilised by a continuous feedback. This system satisfies the reachability rank condition, since, as the reader can verify, the vector fields g_1, g_2 and $[g_1, g_2]$ are everywhere linearly independent. It is stabilised by the periodic feedback

$$u_1 = x_2 \sin t - (x_1 + x_2 \cos t)$$
$$u_2 = -(x_1 + x_2 \cos t) x_1 \cos t - (x_1 x_2 + x_3). \qquad (7.34)$$

The reader can check that this feedback is stabilising by showing that the function

$$V(t, x) = \tfrac{1}{2}(x_1 + x_2 \cos t)^2 + \tfrac{1}{2}x_2^2 + \tfrac{1}{2}x_3^2 \qquad (7.35)$$

is a *Lyapunov function* for the time-varying closed-loop system:

$$\begin{aligned}\frac{d}{dt}V(t, x(t)) = & \\ & -(x_1(t) + x_2(t) \cos t)^2 \\ & -(x_1(t) x_2(t) + x_3(t) + x_1(t)(x_1(t) + x_2(t) \cos t) \cos t)^2 \\ & < 0. \end{aligned} \qquad (7.36)$$

This implies that the 'energy' $V(t, x(t))$ is (strictly) decreasing, a fact which can be used to show that $x(t) \to 0$ as $t \to \infty$. Note that the equilibrium state 0 has zero energy: $V(t, 0) = 0$. (For a detailed discussion of Lyapunov methods, see, e.g. (Vidyasagar, 1993, Khalil, 1992).) □

A number of other interesting approaches to the stabilisation problem are available in the literature. To give one more example, *artificial neural networks* have been proposed as a means for constructing stabilising feedback functions α (Sontag, 1992, Sontag, n.d.). Because, in general, continuous stabilisers do not exist, one has to go beyond single-hidden-layer nets with continuous activation. In fact (asymptotic) controllability is enough to ensure the existence of stabilising controllers using two-hidden-layer nets (Sontag, 1992).

In §7.9 we shall see that stabilising controllers can sometimes be obtained using optimal control methods.

7.7 Optimisation-based design

Dynamic optimisation techniques constitute a very powerful and general design methodology. Optimisation techniques are frequently used in engineering and other application areas. The basic idea is to formulate the design problem in terms of an optimisation problem whose solution is regarded as the 'best' design. The engineering component is the formulation of the optimisation problem, whereas the procedure for solving the optimisation problem relies on mathematics. The practical utility of dynamic optimisation methods for non-linear systems is limited by the computational complexity involved, see §7.13.

Optimal control and game theory are subjects with long and rich histories, and there are many excellent books discussing various aspects (Athans & Falb, 1965, Basar & Olsder, 1982, Bellman, 1957, Bertsekas, 1987, Fleming & Rishel, 1975, Fleming & Soner, 1993, Pontryagin, *et al.*, 1962) and the references contained therein. The remainder of the chapter is concerned with optimisation-based design.

7.8 Deterministic optimal control

In deterministic optimal control theory, there is no uncertainty, and the evolution of the system is determined precisely by the initial state and the control applied. We associate with the non-linear system

$$\dot{x} = f(x, u) \qquad (7.37)$$

a *performance index* or *cost function*

$$J = \int_{t_0}^{t_1} L(x(s), u(s))\, ds + \psi(x(t_1)), \qquad (7.38)$$

where $x(\cdot)$ is the trajectory resulting from the initial condition $x(t_0) = x^0$ and control $u(\cdot)$. The optimal control problem is to find a control $u(\cdot)$ (open loop) which minimises the cost J.

The function $L(x, u)$ assigns a cost (or energy) along the state and control trajectories, while ψ is used to provide a penalty for the final state $x(t_1)$. For example, if one wishes to find a controller which steers the system from x^0 to near the origin with low control energy, then simply choose

$$L(x, u) = \tfrac{1}{2}|x|^2 + \tfrac{1}{2}|u|^2 \qquad (7.39)$$

and $\psi(x) = \tfrac{1}{2}|x|^2$. Some weighting factors can (and often are) included. The cost function penalises trajectories which are away from 0 and use large values of u. In general, these functions are chosen to match the problem at hand.

There are two fundamental methods for solving and analysing such optimal control problems. One is concerned with necessary conditions, the *Pontryagin minimum principle* (PMP) (Pontryagin et al., 1962), while the other, *dynamic programming* (DP) (Bellman, 1957), is concerned with sufficient conditions and verification. As we shall see, *feedback arises naturally in dynamic optimisation*.

7.8.1 Pontryagin minimum principle

Let's formally derive the necessary conditions to be satisfied by an optimal control $u^*(\cdot)$. Introduce Lagrange multipliers $\lambda(t)$, $t_0 \leq t \leq t_1$, to include the dynamics (7.37) (a constraint) in the cost function J:

$$\bar{J} = \int_{t_0}^{t_1} [L(x(s), u(s)) + \lambda(s)'(f(x(s), u(s)) - \dot{x}(s))]\, ds + \psi(x(t_1)).$$

Integrating by parts, we get

$$\bar{J} = \int_{t_0}^{t_1} [H(x(s), u(s), \lambda(s)) + \dot{\lambda}(s)' x(s)]\, ds \tag{7.40}$$
$$\quad -\lambda(t_1)' x(t_1) + \lambda(t_0)' x(t_0) + \psi(x(t_1)), \tag{7.41}$$

where $H(x, u, \lambda) = L(x, u) + \lambda' f(x, u)$. Consider a small control variation δu, and apply the control $u + \delta u$. This leads to variations δx and $\delta \bar{J}$ in the state trajectory and augmented cost function. Indeed, we have

$$\delta \bar{J} = [\psi_x - \lambda']\delta x|_{t=t_1} + \lambda' \delta x|_{t=t_0} + \int_{t_0}^{t_1} [(H_x + \dot{\lambda}')\delta x + H_u \delta u]\, ds.$$

If u^* is optimal, then it is necessary that $\delta \bar{J} = 0$. These considerations lead to the following equations which must be satisfied by an optimal control u^* and resulting trajectory x^*:

$$\dot{x}^* = f(x^*, u^*), \quad t_0 \leq t \leq t_1, \tag{7.42}$$

$$\dot{\lambda}^{*\prime} = -H_x(x^*, u^*, \lambda^*), \quad t_0 \leq t \leq t_1, \tag{7.43}$$

$$H(x^*, u^*, \lambda^*) = \min_u H(x^*, u, \lambda^*), \quad t_0 \leq t \leq t_1, \tag{7.44}$$

$$\lambda^*(t_1) = \psi_x(x^*(t_1))', \quad x^*(t_0) = x^0. \tag{7.45}$$

The PMP gets its name from (7.44), which asserts that the Hamiltonian H must be minimised along the optimal trajectory. The optimal control is open loop (it is not a feedback control).

To solve an optimal control problem using the PMP, one first solves the two-point boundary value problem (7.42)–(7.45) for a candidate optimal control, state and adjoint trajectories. One must then check the optimality of the candidate. Note that there is no guarantee that an optimal control will exist in general, and this should be checked. In practice, solving (7.42)–(7.45) must be done numerically, and there are many methods available for doing this (Jennings et al., 1991, Teo et al., 1991, Hindmarsh, 1983). An interesting application of this method is to the Moon Landing Problem; see the example discussed in the book (Fleming & Rishel, 1975).

7.8.2 Dynamic programming

Dynamic programming is very common and appears in many guises in many applications. For instance, dynamic programming is used by computer chess programs to determine the best move to make at the current time by analysing future scenarios. Key to this is the definition of a value function which represents the best cost at the current time and in the current state. If such a function is known, the best move to make can be determined readily, as the dynamic optimisation has been reduced to a static optimisation. For an introduction to dynamic programming ideas in discrete-time (Bertsekas, 1987, Kumar & Varaiya, 1986).

The DP approach considers a family of optimisation problems parameterised by the initial time t and state $x(t) = x$:

$$J(x,t;u) = \int_t^{t_1} L(x(s), u(s))\, ds + \psi(x(t_1)). \qquad (7.46)$$

The *value function* describes the optimal value of this cost:

$$V(x,t) = \min_{u(\cdot)} J(x,t;u), \qquad (7.47)$$

where the minimum is over all (admissible) open loop controls. The fundamental principle of dynamic programming asserts that for any $r \in [t, t_1]$,

$$V(x,t) = \min_{u(\cdot)}[\int_t^r L(x(s), u(s))\, ds + V(x(r), r)]. \qquad (7.48)$$

Thus the optimal cost starting at time t equals the optimal value of the accumulated cost on the interval $[t, r]$, plus the optimal cost obtained if we started again at time r, but from the state $x(r)$. This principle leads to a partial differential equation (PDE) to be satisfied by V, called the dynamic programming equation (DPE) or Hamilton–Jacobi–Bellman (HJB) equation:

$$\begin{cases} \frac{\partial V}{\partial t} + \min_u [\nabla_x V \cdot f(x,u) + L(x,u)] = 0 \text{ in } \mathbf{R}^n \times (t_0, t_1), \\ V(x, t_1) = \psi(x) \text{ in } \mathbf{R}^n, \end{cases} \quad (7.49)$$

where ∇_x denotes the gradient in the x variable. The DPE here is a non-linear first-order PDE, the significance of which is the *verification theorem* (Fleming & Rishel, 1975, Fleming & Soner, 1993). If there exists a continuously differentiable solution V of the DPE (7.49) (satisfying some additional technical conditions) and if $\mathbf{u}^*(x,t)$ is the control value attaining the minimum in (7.49), then

$$u^*(t) = \mathbf{u}^*(x^*(t), t), \quad t_0 \leq t \leq t_1, \quad (7.50)$$

is optimal, where $x^*(\cdot)$ is the resulting optimal state trajectory. If V is known, finding $\mathbf{u}^*(x)$ is a static optimisation problem (for each x). For instance, if $f(x,u) = f(x) + g(x)u$ and L is as in (7.39), then a direct calculation gives

$$\mathbf{u}^*(x, t) = -g(x)' \nabla_x V(x, t).$$

It is very important to note that *the optimal control is a state feedback controller!* This turns out to be a general principle in DP, and so, as was said earlier, *feedback arises naturally in dynamic optimisation.*

To see why (7.50) is optimal, let $u(\cdot)$ be any control. If V is continuously differentiable, then

$$V(x(t_1), t_1) = V(x, t) + \int_t^{t_1} [\frac{\partial V}{\partial t}(x(s), s) + DV(x(s), s) \cdot f(x(s), u(s))] \, ds,$$

which together with (7.49) implies

$$\begin{aligned} V(x,t) &= -\int_t^{t_1} [\frac{\partial V}{\partial t}(x(s), s) + DV(x(s), s) \cdot f(x(s), u(s))] \, ds + \psi(x(t_1)) \\ &\leq \int_t^{t_1} L(x(s), u(s)) \, ds + \psi(x(t_1)) \end{aligned}$$

with equality if $u = u^*$. Setting $t = t_0$ and $x(t_0) = x^0$ we get

$$J = V(x^0, 0)$$
$$= \int_{t_0}^{t_1} L(x^*(s), u^*(s))\, ds + \psi(x^*(t_1))$$
$$\leq \int_{t_0}^{t_1} L(x(s), u(s))\, ds + \psi(x(t_1))$$

for any other control u.

In general, the DPE (7.49) does not have smooth solutions, and so the verification theorem no longer applies in this form. In fact, the DPE must be interpreted in a generalised sense, viz. the viscosity sense (Crandall & Lions, 1984, Fleming & Soner, 1993). However, more sophisticated verification theorems are available using the methods of non-smooth analysis (Clarke, 1983). The controllers obtained are not, in general, continuous.

Example 7.5 To see that optimal controllers can easily be discontinuous, consider the classical minimum time problem. The dynamics are $\dot{x}_1 = x_2$, $\dot{x}_2 = u$, and the controls take values in the compact set $U = [-1, 1]$. The problem is to find the controller which steers the system from an initial state $x = (x_1, x_2)$ to the origin in minimum time. The value function is simply the minimum time possible:

$$T(x) = \inf_{u(\cdot)} \left\{ \int_0^{t_f} 1\, dt \; : \; x(0) = x, \; x(t_f) = 0 \right\}. \tag{7.51}$$

For a number of reasons, it is convenient to use a transformed minimum time function

$$S(x) = 1 - e^{-T(x)},$$

with DPE

$$\begin{cases} S(x) = \min_{u \in [-1,1]} \{DS(x) \cdot f(x, u) + 1\} \\ S(0) = 0. \end{cases} \tag{7.52}$$

The value function $S(x)$ is not differentiable at $x = 0$, rather, it is only Holder continuous there (with exponent $\frac{1}{2}$), see fig. 7.8. The optimal state feedback $\mathbf{u}^*(x)$ equals either -1 or $+1$, and has a single discontinuity across the so-called switching curve, see fig. 7.9.

□

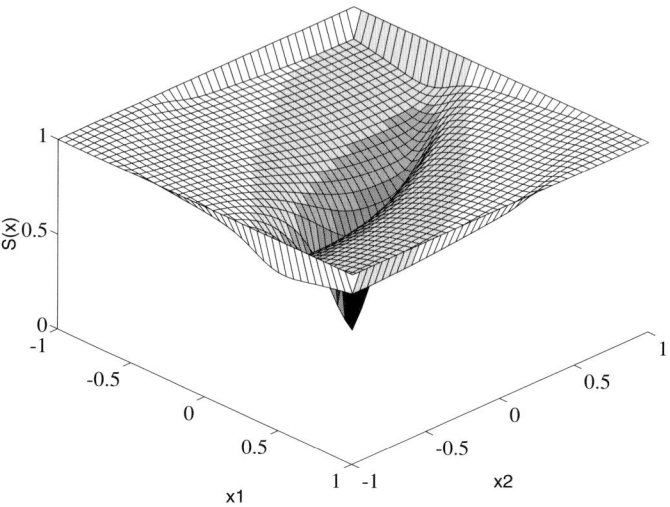

Fig. 7.8. A non-smooth value function.

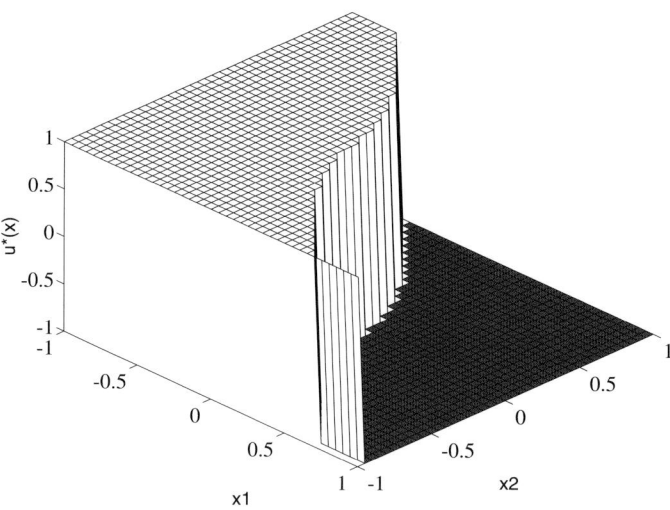

Fig. 7.9. A discontinuous optimal feedback controller.

Non-linear control systems 273

7.9 Optimal stabilisation

The basic optimal control problem described in §7.8 is a *finite horizon* problem, because it is defined on a finite time interval. *Infinite horizon* problems are defined on $[0, \infty)$, and so the asymptotic behaviour of the trajectories and controls is an issue. Indeed, infinite horizon problems can be used to obtain stabilising state feedback controllers. (The minimum time problem is also an infinite horizon problem, but of a different type.)

To this end, we consider the following infinite horizon value function:

$$V(x) = \min_{u(\cdot)} \int_0^\infty L(x(s), u(s)) \, ds. \tag{7.53}$$

The DPE is now of the following stationary type

$$\min_u [\nabla_x V \cdot f(x, u) + L(x, u)] = 0 \text{ in } \mathbf{R}^n, \tag{7.54}$$

and, assuming the validity of the verification theorem, the optimal controller $\mathbf{u}^*(x)$ is the control value achieving the minimum in (7.54):

$$u^*(t) = \mathbf{u}^*(x^*(t)), \quad t \geq 0. \tag{7.55}$$

Under suitable hypotheses, this controller is a stabilising state feedback controller. For instance, a sufficiently strong form of controllability would be needed, the cost term $L(x, u)$ must adequately reflect the behaviour of the states (observability), and $\mathbf{u}^*(x)$ must be sufficiently regular. However, it can easily happen that no controller is capable of making the cost (7.53) finite. Often, a discount factor is used, especially in the stochastic case.

Example 7.6 Consider the inverted pendulum discussed in §7.2. A state space form of the dynamic equation (7.1) is

$$\begin{pmatrix} \dot{x}_1 \\ \dot{x}_2 \end{pmatrix} = \begin{pmatrix} x_2 \\ g \sin x_1 \end{pmatrix} + \begin{pmatrix} 0 \\ g/m \end{pmatrix} \tau. \tag{7.56}$$

Choose the cost term L as in (7.39), in which case the value function (7.53) takes the form

$$V(x) = \min_{\tau(\cdot)} \int_0^\infty \tfrac{1}{2} \left(|x_1(s)|^2 + |x_2(s)|^2 + |\tau(s)|^2 \right) ds. \tag{7.57}$$

Since the inverted pendulum is (exponentially) feedback stabilisable (with smooth feedback), the value function will be finite, but not necessarily smooth. The optimal state feedback controller, if it exists, need

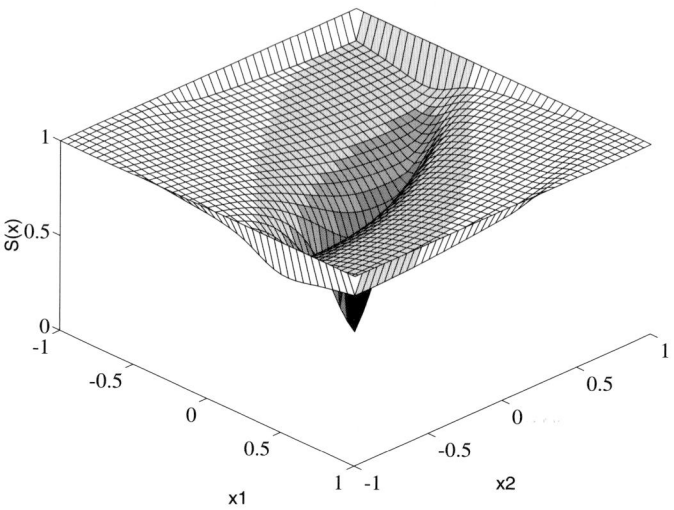

Fig. 7.10. Value function for the pendulum.

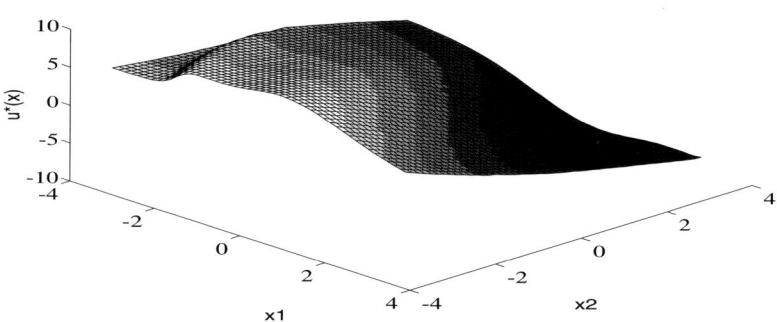

Fig. 7.11. Optimal stabilising feedback for the pendulum.

not necessarily be smooth. However, it is possible to solve this problem numerically by solving the DPE (7.54) using the methods described in §7.13. Figs. 7.10–7.12 show the value function, optimal state feedback control, and a simulation showing that it is stabilising.

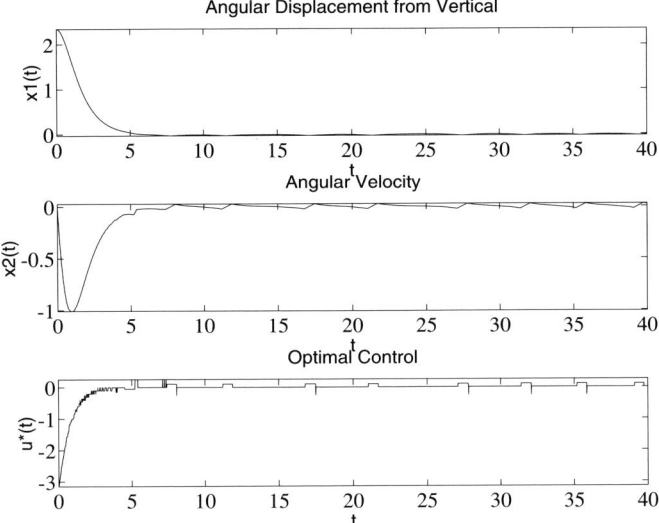

Fig. 7.12. Optimal trajectories for the stabilised pendulum.

From fig. 7.11, it is evident that the control is linear near the origin, and so is close to the optimal controller obtained by using the linearisation $\sin x_1 \approx x_1$ there. Away from the origin, the control is definitely non-linear. The control appears to be continuous in the region shown in fig. 7.11. Note that the initial angle in the simulation, fig. 7.12, is well away from the linear region. □

7.10 Stochastic optimal control

The simple deterministic optimal control problems discussed above do not take into account disturbances which may affect performance. In stochastic control theory disturbances are modelled as random or stochastic processes, and the performance index averages over them.

The non-linear system (7.37) is replaced by the *stochastic differential equation* model

$$dx = f(x, u)\, dt + \varepsilon\, dw, \qquad (7.58)$$

where w is a standard Brownian motion. Formally, $\frac{dw}{dt}$ is white noise, and models disturbances as random fluctuations. The solutions to (7.58) have a general trend due to the drift f, plus fluctuations due to the noise. If the noise variance ε^2 is small, then the solution of (7.58) is close to that of (7.37) (c.f. semiclassical limits in physics). The cost function is

$$J = \mathbf{E}\left[\int_{t_0}^{t_1} L(x(s), u(s))\, ds + \psi(x(t_1))\right], \qquad (7.59)$$

where the expectation denotes integration with respect to the probability distribution associated with w.

It turns out that DP methods are very convenient for stochastic problems, although there are PMP-type results available. The value function is defined by

$$V(x,t) = \min_{u(\cdot)} \mathbf{E}_{x,t}\left[\int_t^{t_1} L(x(s), u(s))\, ds + \psi(x(t_1))\right], \qquad (7.60)$$

and the DPE is

$$\begin{cases} \frac{\partial V}{\partial t} + \frac{\varepsilon^2}{2}\Delta V + \min_u[\nabla_x V \cdot f(x,u) + L(x,u)] = 0 \text{ in } \mathbf{R}^n \times (t_0, t_1), \\ V(x, t_1) = \psi(x) \text{ in } \mathbf{R}^n, \end{cases}$$
(7.61)

where $\Delta = \sum_{i=1}^n \partial^2/\partial x_i^2$ is the Laplacian. This PDE is a non-linear second-order parabolic equation, and in this particular case smooth solutions often exist. Again, the optimal control is a state feedback controller $\mathbf{u}^{\varepsilon*}(x,t)$ obtained by minimising the appropriate expression in the DPE (7.61):

$$u^{\varepsilon*}(t) = \mathbf{u}^{\varepsilon*}(x^*(t), t), \quad t_0 \leq t \leq t_1. \qquad (7.62)$$

Under certain conditions, the controller $\mathbf{u}^{\varepsilon*}(x,t)$ is related to the deterministic controller $\mathbf{u}^*(x,t)$ obtained in §7.8.2 by

$$\mathbf{u}^{\varepsilon*}(x,t) = \mathbf{u}^*(x,t) + \varepsilon^2 \mathbf{u}_{stoch}(x,t) + \ldots, \qquad (7.63)$$

for small $\varepsilon > 0$ and for some term $\mathbf{u}_{stoch}$ which, to first-order, compensates for the disturbances described by the stochastic model.

7.11 Robust control

An alternative approach to stochastic control is to model the disturbance as an unknown but deterministic signal. This is known as robust H_∞ control, and is discussed, e.g. in the book (Basar & Bernhard, 1991). What we discuss here is a simple and very special situation closely related to game theory.

The system model is

$$\dot{x} = f(x, u) + w, \quad (7.64)$$

where w models disturbances and uncertainties such as plant model errors and noise; w is not random, but rather an unknown deterministic signal. Associated with (7.64) is a performance variable

$$z = \ell(x, u). \quad (7.65)$$

For instance, one could take $|z|^2$ to be the cost term $L(x, u)$.

The problem is to find, if possible, a stabilising controller which ensures that the effect of the disturbance w on the performance z is bounded in the following sense:

$$\int_{t_0}^{t_1} |z(s)|^2 \, ds \leq \gamma^2 \int_{t_0}^{t_1} |w(s)|^2 \, ds + \beta_0(x(t_0)) \quad (7.66)$$

for all w, and all $t_1 \geq t_0$, for some finite β_0. Here, $\gamma > 0$ is given, and is interpreted as essentially a bound on the size of the input–output map $w \mapsto z$. When $w = 0$, the ideal situation (as in deterministic control), the size of z is limited by the initial state (via β_0), but if $w \neq 0$, the resulting effect on z is limited by (7.66).

The robust H_∞ control problem can be solved using game theory methods (Basar & Bernhard, 1991, James & Baras, 1993), which are closely related to the above-mentioned optimal control techniques. Henceforth we shall consider the following special game problem, for fixed $t_0 < t_1$, corresponding to finite horizon disturbance attenuation.

We seek to minimise the worst-case performance index

$$J = \max_{w(\cdot)} \left\{ \int_{t_0}^{t_1} [L(x(s), u(s)) - \tfrac{1}{2}\gamma^2 |w(s)|^2] \, ds + \psi(x(t_1)) \right\}. \quad (7.67)$$

If this cost is finite, then the resulting optimal controller achieves the goal (7.66) (when $\psi = 0$). Applying DP, the value function is given by

$$V(x, t) = \min_{u(\cdot)} \max_{w(\cdot)} \left\{ \int_t^{t_1} [L(x(s), u(s)) - \tfrac{1}{2}\gamma^2 |w(s)|^2] \, ds + \psi(x(t_1)) \right\}, \quad (7.68)$$

where $x(t) = x$, as in §7.8.2. The DPE is a first-order non-linear PDE, and is called the Hamilton–Jacobi–Isaacs (HJI) equation:

$$\begin{cases} \frac{\partial V}{\partial t} + \min_u \max_w [\nabla_x V \cdot (f(x,u) + w) + L(x,u) - \frac{1}{2}\gamma^2 |w|^2] = 0 \\ \quad \text{in } \mathbf{R}^n \times (t_0, t_1), \\ \\ V(x, t_1) = \psi(x) \text{ in } \mathbf{R}^n. \end{cases}$$

(7.69)

If the DPE (7.69) has a smooth solution, then the optimal controller is given by a value $\mathbf{u}^{\gamma*}(x,t)$ which attains the minimum in (7.69). This optimal state feedback controller is related to the optimal state feedback controller for the simple deterministic optimal control problem in §7.8.2 by

$$\mathbf{u}^{\gamma*}(x,t) = \mathbf{u}^*(x,t) + \frac{1}{\gamma^2} \mathbf{u}_{robust}(x,t) + \dots, \qquad (7.70)$$

for large γ, and for some $\mathbf{u}_{robust}$ (valid under certain conditions). The term $\mathbf{u}_{robust}$ depends on the worst-case disturbance energy, and contributes to the controller appropriately.

The infinite horizon problem can also be discussed, yielding stabilising robust feedback controllers, assuming the validity of a suitable verification theorem, etc.

7.12 Output feedback

The optimal feedback controllers discussed above all feed back the state x. This makes sense since no restriction was placed on the information available to the controller, and the state is by definition a quantity summarising the internal status of the system. As mentioned earlier, in many cases the state is not completely available, and the controller must use only the information gained from observing the outputs y – the controller can only (causally) feed back the output:

$$u(t) = \mathbf{u}(y(s), 0 \le s \le t). \qquad (7.71)$$

This *information constraint* has a dramatic effect on the complexity of the optimal solution, as we shall soon see. We begin with a discussion of a useful although suboptimal approach which is less computationally demanding.

7.12.1 Deterministic suboptimal control

For deterministic systems without disturbances, a state feedback controller is equivalent to an open loop controller, since in the absence of

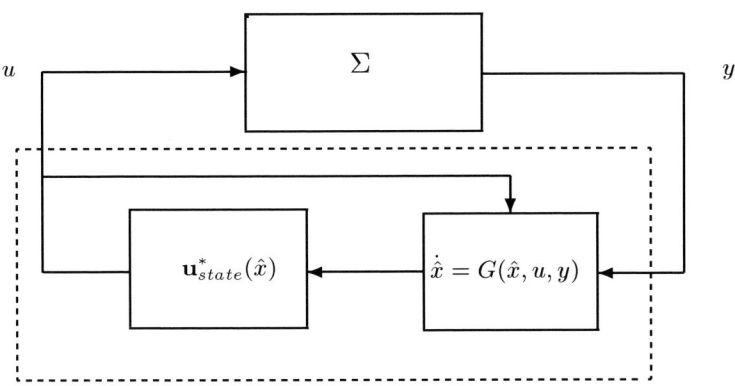

Fig. 7.13. Suboptimal controller/observer structure.

uncertainty the trajectory is completely determined by the initial state and the control applied. In the presence of disturbances (either deterministic or stochastic) this is not the case.

Because of the error correcting nature of feedback, it is still desirable to implement a feedback controller even if the design does not specifically account for disturbances. In the output feedback case, the following (suboptimal) procedure can be adopted. We suppose the output is

$$y = h(x). \qquad (7.72)$$

First, solve a deterministic optimal control problem for an optimal stabilising state feedback controller $\mathbf{u}^*_{state}(x)$, as in §7.53. Then design an *observer* or *filter*, say an ODE of the form

$$\dot{\hat{x}} = G(\hat{x}, u, y) \qquad (7.73)$$

to (asymptotically) estimate the state:

$$|\hat{x}(t) - x(t)| \to 0 \ \text{ as } \ t \to \infty. \qquad (7.74)$$

Note that this filter uses only the information contained in the outputs and in the inputs applied. Finally implement the output feedback controller obtained by substituting the estimate into the optimal state feedback controller, see fig. 7.13:

$$u(t) = \mathbf{u}^*_{state}(\hat{x}(t)).$$

Under certain conditions, the closed-loop system will be asymptotically stable (Vidyasagar, 1980). A key advantage of this procedure is that it can yield practically useful controller designs, provided one has found a filter and computed or approximated the state feedback control.

7.12.2 Optimal control

Stochastic or deterministic H_∞ optimal output feedback control is a complex subject undergoing much research at present. The optimal controller feeds back a new state variable p_t, called an *information state*:

$$u(t) = \mathbf{u}^*(p_t^*, t). \tag{7.75}$$

The information state is typically a function $p_t = p_t(x)$, which solves a filter type equation

$$\dot{p} = F(p, u, y), \tag{7.76}$$

a PDE, not an ODE. In general, it is not possible to compute p_t using a finite set of ODEs, and so it is *infinite dimensional* in this sense. There is a value function $V(p,t)$ which is a function of the information state p (not x), satisfying an infinite dimensional DPE. This determines the optimal controller (7.75). The nature of this controller is illustrated in fig. 7.14. For further information, see Kumar & Varaiya (1986), Fleming & Pardoux (1982), Hijab (1990), Lions (1988), Zakai (1969), James *et al.* (1993, 1994), James & Baras (1993).

7.13 Computation and complexity

The practical utility of optimisation procedures depends on how effective the methods for solving the optimisation problem are. Linear programming is popular in part because of the availability of efficient numerical algorithms capable of handling real problems (simplex algorithm). The same is true for optimal design of linear control systems (matrix Riccati equations). For non-linear systems, the computational issues are much more involved. To obtain candidate optimal open loop controllers using the Pontryagin minimum principle, a non-linear two-point boundary value problem must be solved. There are a number of numerical schemes available (Jennings *et al.*, 1991, Teo *et al.*, 1991,

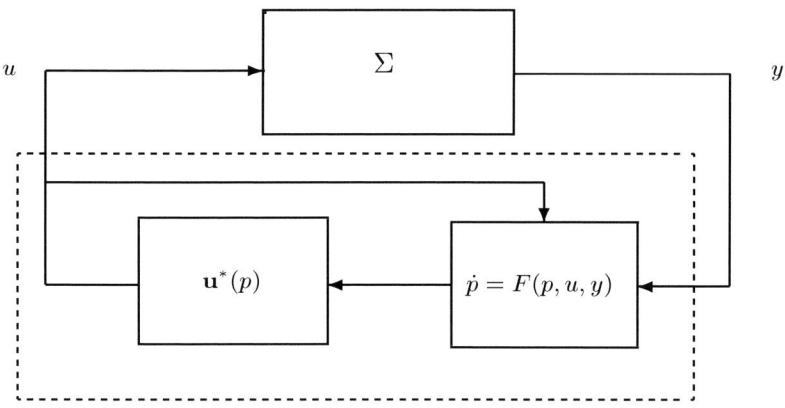

Fig. 7.14. Optimal output feedback control.

Hindmarsh, 1983). The dynamic programming method, as we have seen, requires the solution of a PDE to obtain the optimal feedback controller. In the next section, we describe a finite difference numerical method, and in the section following, make some remarks about its computational complexity. In summary, effective computational methods are crucial for optimisation-based design. It will be interesting to see non-linear dynamic optimisation flourish as computational power continues to increase and as innovative optimal and suboptimal algorithms and approximation methods are developed.

7.13.1 Finite-difference approximation

Finite difference and finite element schemes are commonly employed to solve both linear and non-linear PDEs. We present a finite difference scheme for the stabilisation problem discussed in §7.9. This entails the solution of the non-linear PDE (7.54). The method follows the general principles in the book (Kushner & Dupuis, 1993).

For $h > 0$ define the grid $h\mathbf{Z}^n = \{hz : z_i \in \mathbf{Z}, i = 1, \ldots, n\}$. We wish to construct an approximation $V^h(x)$ to $V(x)$ on this grid. The basic idea is to approximate the derivative $\nabla_x V(x) \cdot f(x, u)$ by the finite difference

$$\sum_{i=1}^{n} f_i^{\pm}(x, u) \left(V(x \pm h e_i) - V(x) \right) / h, \qquad (7.77)$$

where $e_1, \ldots, e_n$ are the standard unit vectors in $\mathbf{R}^n$, $f_i^+ = \max(f_i, 0)$

and $f_i^- = -\min(f_i, 0)$. A finite difference replacement of (7.54) is

$$V^h(x) = \min_u \left\{ \sum_z p^h(x, z|u) V^h(z) + \Delta t(x) L(x, u) \right\}. \tag{7.78}$$

Here,

$$p^h(x, z|u) = \begin{cases} \dfrac{\| f(x, u) \|_1}{\max_u \| f(x, u) \|_1} & \text{if } z = x \\ \dfrac{f_i^\pm(x, u)}{\max_u \| f(x, u) \|_1} & \text{if } z = x \pm he_i \\ 0 & \text{otherwise,} \end{cases} \tag{7.79}$$

$\| v \|_1 = \sum_{i=1}^n |v_i|$, and $\Delta t(x) = h/\max_u \| f(x, u) \|_1$. Equation (7.78) is obtained by substituting (7.77) for $\nabla_x V \cdot f$ in (7.54) and some manipulation. In Kushner & Dupuis (1993), the quantities $p^h(x, z|u)$ are interpreted as transition probabilities for a controlled Markov chain, with (7.78) as the corresponding DPE. The optimal state feedback policy $\mathbf{u}^{*h}(x)$ for this discrete optimal control problem is obtained by finding the control value attaining the minimum in (7.78). Using this interpretation, convergence results can be proven. In our case, it is interesting that a deterministic system is approximated by a (discrete) stochastic system. See fig. 7.15. One can imagine a particle moving randomly on the grid trying to follow the deterministic motion or flow, on average. When the particle is at a grid point, it jumps to one of its neighbouring grid points according to the transition probability.

In practice, one must use only a finite set of grid points, say a set forming a box D^h centred at the origin. On the boundary ∂D^h, the vector field $f(x, u)$ is modified by projection, so if $f(x, u)$ points out of the box, then the relevant components are set to zero. This has the effect of constraining the dynamics to a bounded region. From now on we take (7.78) to be defined on D^h.

Equation (7.78) is a non-linear implicit relation for V^h, and iterative methods are used to solve it (approximately). There are two main types of methods, viz. *value space iteration*, which generates a sequence of approximations

$$V_k^h(x) = \min_u \left\{ \sum_z p^h(x, z|u) V_{k-1}^h(z) + \Delta t(x) L(x, u) \right\} \tag{7.80}$$

to V^h, and *policy space iteration*, which involves a sequence of approx-

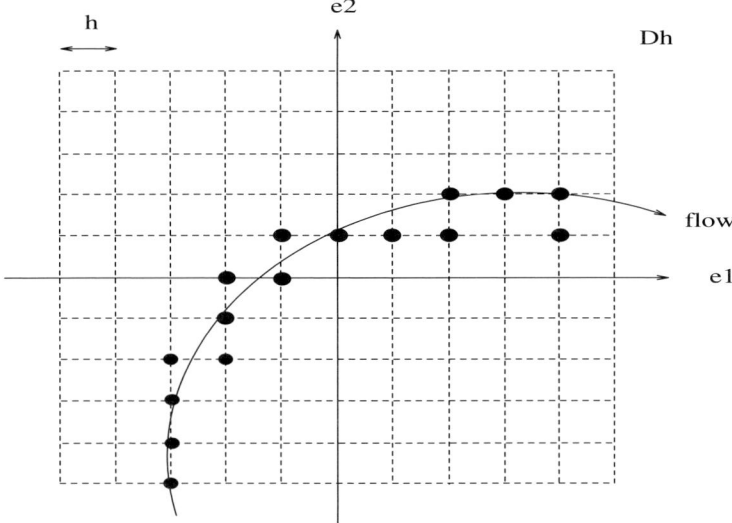

Fig. 7.15. The finite difference grid showing a randomly moving particle following the deterministic flow.

imations $\mathbf{u}_k^{*h}(x)$ to the optimal feedback policy $\mathbf{u}^{*h}(x)$. The are many variations and combinations of these methods, acceleration procedures, and multigrid algorithms. Note that because of the local structure of equation (7.78), *parallel computation* is natural.

Numerical results for the stabilisation of the inverted pendulum are shown in §7.9. The non-linear state feedback controller $\mathbf{u}^{*h}(x)$ (fig. 7.11) was effective in stabilising the pendulum (fig. 7.12).

7.13.2 Computational complexity
7.13.2.1 State feedback

One can see easily from the previous subsection that the computational effort demanded by dynamic programming can quickly become prohibitive. If N is the number of grid points in each direction e_i, then the storage requirements for $V^h(x)$, $x \in D^h$ are of order N^n, a number growing exponentially with the number of state dimensions. If $N \sim 1/h$, this measure of computational complexity is roughly

$$O(1/h^n).$$

The complexity of a discounted stochastic optimal control problem was analysed by (Chow & Tsitsiklis, 1989). If $\varepsilon > 0$ is the desired accuracy and if the state and control spaces are of dimension n and m, then the compuational complexity is

$$O(1/\varepsilon^{2n+m}),$$

measured in terms of the number of times functions of x and u need be called in an algorithm to achieve the given accuracy.

Thus the computational complexity of state feedback dynamic programming is of exponential order in the state space dimension, n. This effectively limits its use to low dimensional problems. For $n = 1$ or 2, it is possible to solve problems on a desktop workstation. For $n = 3$ or 4, a supercomputer such as a Connection Machine CM-2 or CM-5 is usually needed. If n is larger, then in general other methods of approximation or computation are required, see, e.g. Campillo & Pardoux (1992).

7.13.2.2 Output feedback

In the state feedback case discussed above, the DPE is defined in a finite dimensional space $\mathbf{R}^n$. However, in the case of output feedback, the DPE is defined on an *infinite dimensional* (function) space (in which the information state takes values). Consequently, the computational complexity is vastly greater. In fact, if n is the dimension of the state x, and a grid of size $h > 0$ is used, then the computational complexity is roughly

$$O(1/h^{1/h^n}),$$

a number increasing doubly exponentially with n. In the output feedback case, it is particularly imperative that effective approximations and numerical methods be developed.

7.14 Concluding remarks

In this chapter we have surveyed a selection of topics in non-linear control theory, with an emphasis on *feedback design*. As we have seen, feedback is central to control systems, and techniques from differential equations, differential geometry and dynamic optimisation play leading roles in their design and analysis. Feedback is used to stabilise and regulate a system in the presence of disturbances and uncertainty, and the main problem of control engineers is to design feedback controllers. In practice, issues of robustness and computation are crucial, and in situations with partial information (output feedback), the complexity can be very high.

Bibliography

Anderson, B. D. O., & Moore, J. (1990). *Optimal Control: Linear Quadratic Methods*. Prentice Hall, Englewood Cliffs, N. J.

Athans, M., & Falb, P.L. (1965). *Optimal Control*. McGraw-Hill, New York.

Basar, T., & Bernhard, P. (1991). *H_∞-Optimal Control and Related Minimax Design Problems*. Birkhauser, Boston.

Basar, T., & Olsder, G. J. (1982). *Dynamic Noncooperative Game Theory*. Academic Press, New York.

Bellman, R. (1957). *Dynamic Programming*. Princeton Univ. Press, Princeton, N.J.

Bertsekas, D.P. (1987). *Dynamic Programming: Deterministic and Stochastic Models*. Prentice Hall, Englewood Cliffs, N. J.

Boothby, W. M. (1975). *An Introduction to Differentiable Manifolds and Riemannian Geometry*. Academic Press, New York.

Brockett, R. W. (1976). Control Theory and Analytical Mechanics. Martin, C., & Hermann, R (eds), *The 1976 Ames Research Center (NASA) Conference on Geometric Control Theory*. Math Sci Press, Brookline, Mass.

Brockett, R. W. (1983). Asymptotic Stability and Feedback Stabilization. Brockett, R.W., Millman, R.S., & Sussmann, H.J. (eds), *Differential geometric control theory*. Birkhauser, Basel-Boston.

Campillo, F., & Pardoux, E. (1992). Numerical Methods in Ergodic Optimal Stochastic Control and Application. *Pages 59–73 of:* Karatzas, I., & Ocone, D. (eds), *Applied Stochastic Analysis, LNCIS 177*. Springer-Verlag, New York.

Chow, C. S., & Tsitsiklis, J. N. (1989). The computational complexity of dynamic programming. *J. Complexity*, **5**(1), 466–488.

Clarke, F. H. (1983). *Optimization and Non-Smooth Analysis*. Wiley-Interscience, New York.

Craig, J. J. (1989). *Introduction to Robotics: Mechanics and Control*. Addison-Wesley, New York.

Crandall, M. G., & Lions, P. L. (1984). Viscosity solutions of hamilton-jacobi equations. *Trans. AMS*, 183–186.

Fleming, W. H., & Pardoux, E. (1982). Optimal control for partially observed diffusions. *SIAM J. Control Optim.*, **20 (2)**, 261–285.

Fleming, W. H., & Rishel, R. W. (1975). *Deterministic and Stochastic Optimal Control*. Springer-Verlag, New York.

Fleming, W. H., & Soner, H. M. (1993). *Controlled Markov Processes and Viscosity Solutions*. Springer-Verlag, New York.

Fleming, W. H. Chair. (1988). *Future Directions in Control Theory: A Mathematical Perspective Panel Report*. SIAM, Philadelphia.

Hermann, R., & Krener, A. J. (1977). Nonlinear controllability and observability. *IEEE Trans. Automatic Control*, **22-5**, 728–740.

Hijab, O. (1990). Partially observed control of markov processes. *Stochastics*, **28**, 123–144.

Hindmarsh, A. C. (1983). ODEPACK: A systemized Collection of ODE Solvers. *Pages 55–64 of: Scientific Computing*. North Holland, Amsterdam.

Hunt, E. R., & Johnson, G. (1993). Controlling chaos. *IEEE Spectrum*, **30-11**, 32–36.

Hunt, L. R. Su, R., & Meyer, G. (1983). Global transformations of nonlinear systems. *IEEE Trans. Automatic Control*, **28**, 24–31.

Isidori, A. (1985). *Nonlinear Control Systems: An Introduction*. Springer-Verlag, New York.

James, M. R., & Baras, J. S. (1993). Robust h output feedback control for nonlinear systems. *IEEE Trans. Automatic Control*, **to appear**.

James, M. R., Baras, J. S., & Elliott, R. J. (1993). Output Feedback Risk-Sensitive Control and Differential Games for Continuous-Time Nonlinear Systems. *32nd IEEE Control and Decision Conference*.

James, M. R., Baras, J. S., & Elliott, R. J. (1994). Risk-sensitive control and dynamic games for partially observed discrete-time nonlinear systems. *IEEE Trans. Automatic Control*, **AC-39-4**, 780–792.

Jennings, L. S., Fisher, M. E., Teo, K. L., & Goh, C. J. (1991). Miser3: Solving optimal control problems–an update. *Advances in Engineering Software and Workstations*, **13**, 190–196.

Kailath, T. (1980). *Linear Systems*. Prentice Hall, Englewood Cliffs, N. J.

Khalil, H. K. (1992). *Nonlinear Systems*. Macmillan, New York.

Kumar, P. R., & Varaiya, P. (1986). *Stochastic Systems: Estimation Identification, and Adaptive Control*. Prentice Hall, Englewood Cliffs, N. J.

Kushner, H. J., & Dupuis, P. G. (1993). *Numerical Methods for Stochastic Control Problems in Continuous Time*. Springer-Verlag, New York.

Lions, P. L. (1988). Viscosity solutions of fully nonlinear second order equations and optimal stochastic control in infinite dimensions, Part II: Optimal control of zakai's equation. *Lecture Notes Math*, **1390**, 147–170.

Nijmeijer, H., & van der Schaft, A. J. (1990). *Nonlinear Dynamical Control Systems*. Springer-Verlag, New York.

Pomet, J. B. (1992). Explicit design of time-varying stabilizing control laws for a class of controllable systems without drift. *Systems and Control Letters*, **18**, 147–158.

Pontryagin, L. S., Boltyanskii, V. G., Gamkriledze, R. V., & Mischenko, E. F. (1962). *The Mathematical Theory of Optimal Processes*. Wiley-Interscience, New York.

Sontag, E. D. *Some Topics in Neural Networks*. Tech. rept. Technical Report, Siemens Corporate Research, N.J.

Sontag, E. D. (1990). *Mathematical Control Theory: Deterministic Finite Dimensional Systems*. Springer-Verlag, New York.

Sontag, E. D. (1992). Feedback stabilization using two-hidden-layer nets. *IEEE Trans. Neural Networks*, **3**, 981–990.

Teo, K. L., Goh, C. J., & Wong, K. H. (1991). *A Unified*

Computational Approach to Optimal Control Problems. Longman Scientific and Technical, London.

Vidyasagar, M. (1980). On the stabilization of nonlinear systems using state detection. *IEEE Trans. Automatic Control,* **25 (3)**, 504–509.

Vidyasagar, M. (1993). *Nonlinear Systems Analysis.* Prentice Hall, Englewood Cliffs, N. J.

Zakai, M. (1969). On the optimal filtering of diffusion processes. *Z. Wahr. Verw. Geb.,* **11**, 230–243.

8
Parallel computers and complex systems

GEOFFREY FOX & PAUL CODDINGTON,
Northeast Parallel Architectures Center,
Syracuse University,
Syracuse,
NY 13244, USA

8.1 Introduction

The power of high performance computing1 is being used in an increasingly wide variety of applications in the physical sciences, and in particular in the study of complex systems. The performance of supercomputers has increased by roughly a factor of two every 18 months since electronic computers were first developed. A number of supercomputer manufacturersaare aiming to deliver Teraflop (10^{12} floating point operations per second) performance well before the end of the decade.

Hardware trends imply that all computers, from PCs to supercomputers, will use some kind of parallel architecture by the end of the century. Until recently parallel computers were only marketed by small start-up companies (apart from Intel Supercomputer Systems Division), however recently Cray, Hewlett-Packard and Convex, IBM, and Digital have all begun marketing commercial parallel computers. Software for these systems is a major challenge, and could prevent or delay this hardware trend toward parallelism. Reliable and efficient systems software, high level standardised parallel languages and compilers, parallel algorithms, and applications software all need to be available for the promise of parallel computing to be fully realised.

A characteristic feature of the research on parallel computing at the Caltech Concurrent Computation Program (C3P), and more recently the Northeast Parallel Architectures Center (NPAC) at Syracuse University, is that many of the people who have worked in these groups (including ourselves) have a background in physics, so much of this research has made use of ideas from both physics and computer science. The goal of this work has been to make parallel computers more effective and easier to use for a wider variety of applications (Fox, 1987, 1988b, 1992a, Fox *et al.*, 1988b, Angus *et al.*, 1990).

Parallel computers are complex entities used to simulate complex problems. While the physical sciences have developed several qualitative and quantitative methods to understand complex systems, other fields, in particular computer science, have not. Thus, it is not surprising that physics concepts, especially those related to complex systems, are helpful in developing a theory of computation and indeed may become more important as the computers and the problems they simulate get larger and more complicated. Here we present a review of these concepts. Several references contain more detailed discussions such as Fox *et al.*, (1985), Fox and Otto, (1986), Fox and Furmanski, (1988a, 1988b), Fox *et al.*, (1988a, 1988b, 1989, 1994, 1990b, 1991d, 1991e, 1992a).

In §8.2 we give an overview of the state of the art and future trends in parallel computing, concentrating on the use of parallel computers for simulation, particularly of complex systems. We describe recent progress in defining a standardised, portable, high level parallel language called High Performance Fortran, an extension of Fortran 90 designed for efficient implementation of data parallel applications on parallel, vector and sequential high performance computers. An outline of the language is presented, and we discuss its ability to handle different applications in computational science, particularly the difficulties of implementing irregular problems. We also discuss other problems such as software integration and the use of concepts such as visualisation, virtual reality and metacomputing to enhance the useability of high performance computers. Further discussions of issues concerning parallel computing can be found in Fox et al.,(1988b), Fox (1989, 1991c, 1992b, 1992a), Messina (1991).

We have found that when trying to understanding the use of parallel computers it is often very helpful to view the application, the software and the computer as complex systems. We have used these concepts to develop a theory of computation for parallel computers. In §8.3 we lay the foundations for this theory by presenting the view of computation as a set of maps from one of these complex systems to another, and introducing the concepts of space and time for these complex systems. §8.4 describes spatial properties – size, topology, dimension, and a physical analogy for data partitioning of slowly-varying problems leading to concepts of temperature and phase transitions.

In §8.5, we discuss temporal properties – a string model for very adaptive problems and a duality between the temporal structure of problems and the memory hierarchy of computers. Just as in physics, locality is a critical issue in high performance computing. We need to ensure that the data needed for a computation is readily available for the arithmetic unit. Delays increase as the data is placed in memory which is further away from the processor. Locality underlies the design and use by compilers of caches in 'ordinary' sequential computers and the nature of the networks used to link the individual computer nodes in a parallel system. Matching the problem locality to the computer locality is a key to good performance.

In §8.6, we briefly discuss the concept of problem architecture and its relation to the better understood computer architecture, in order to understand which problems are suitable for which computers. We also apply these ideas to compilers, which are viewed as mapping one space-time system into another.

Finally, in §8.7, we discuss the idea of *physical computation*, or adapting techniques from the physical sciences to create useful computational algorithms for such general problems as optimisation. Such techniques are also applied to problems in parallel computing, such as balancing the computational load between processors.

8.2 Parallel computers and simulation

8.2.1 Parallel computing

Carver Mead of Caltech in an intriguing public lecture once surveyed the impact of a number of new technologies, and introduced the idea of 'headroom' – how much better a new technology needs to be for it to replace an older, more entrenched technology. Once the new technology has enough headroom, there will be a fairly rapid crossover from the old technology, in a kind of phase transition. For parallel computing the headroom needs to be large, perhaps a factor of 10 to 100, to outweigh the substantial new software investment required. The headroom will be larger for commercial applications where programs are generally much larger, and have a longer lifetime, than programs for academic research. Machines such as the nCUBE and Thinking Machines CM-2 were comparable in price/performance to conventional supercomputers, which was enough to show that 'parallel computing works' (Fox *et al.*, 1988b, Fox, 1992a), but not enough to take over from conventional machines. It will be interesting to see whether the new batch of parallel computers, such as the CM-5, Intel Paragon, IBM SP-2, Maspar (DECmpp) MP-2, Cray T3D, etc. have enough headroom.

Parallel computers have two different models for accessing data:

- shared memory – processors access a common memory space;
- distributed memory – data is distributed over processors and accessed via message passing between processors;

and two different models for accessing instructions:

- SIMD (Single Instruction Multiple Data) – processors perform the same instruction synchronously on different data,
- MIMD (Multiple Instruction Multiple Data) – processors may perform different instructions on different data.

Different problems will generally run most efficiently on different computer architectures, so a range of different architectures will be available for the some time to come, including vector supercomputers, SIMD and MIMD parallel computers, and networks of RISC workstations. The user would prefer not to have to deal with the details of the different hardware, software, languages and programming models for the different classes of machines. So the aim of supercomputer centers is transparent distributed computing, sometimes called 'metacomputing' – to provide simple, transparent access to a group of machines of different architectures, connected by a high speed network to each other and the outside world, and to data storage and visualisation facilities. Users should be presented with a single system image, so they do not need to deal with different systems software, languages, software tools and libraries on each different machine. They should also be able to run an application across different machines on the network.

Parallel computing implies not only different computer architectures, but different languages, new software, new libraries, and will open up computation to new fields and new applications. It also offers a different way of viewing problems. Virtually all complex real-world problems are inherently parallel, in that many different elements of the problem domain interact with one another at any given time. In a sequential language, the structure of the problem must be artificially broken up to fit within the confines of the sequential computer, for which only one computation can occur at a time. Using parallel computers and parallel languages allows the programmer to better preserve the problem structure in the software, and perhaps also in the algorithm.

Over the last 10 years we have learned that parallel computing works – the majority of computationally intensive applications perform well on parallel computers, by taking advantage of the simple idea of 'data parallelism', which means obtaining concurrency by applying the particular algorithm to different sections of the data set concurrently (Hillis, 1985, Hillis & Steele, 1986, Fox *et al.*, 1988b). Data parallel applications are scalable to larger numbers of processors for larger amounts of data.

Another type of parallelism is 'functional parallelism', where different processors (or even different computers) perform different functions, or different parts of the algorithm. Here the speed-ups obtained are usually more modest and this method is often not scalable, however it is important, particularly in multidisciplinary applications.

Surveys of problems in computational science (Angus *et al.*, 1990, Denning & Tichy, 1990, Fox, 1992b) have shown that the vast majority (over 90%) of applications can be run effectively on MIMD parallel computers, and approximately 50% on SIMD machines (probably less for commercial, rather than academic, problems). Currently there are many different parallel architectures, but only one – a distributed memory MIMD multicomputer – is a general, high performance architecture that is known to scale from one to very many processors.

8.2.2 Parallel languages

Using a parallel machine requires rewriting code written in standard sequential languages. We would like this rewrite to be as simple as possible, without sacrificing too much in performance. Parallelising large codes involves substantial effort, and in many cases rewriting code more than once would be impractical. A good parallel language therefore needs to be portable and maintainable, that is, the code should run effectively on all current *and* future machines (at least those we can anticipate today). This means the language should be scalable, so that it can work effectively on machines using one or millions of processors. Portability also means that programs can be run in parallel over different machines across a network (distributed computing).

There are some completely new languages specifically designed to deal with parallelism, for example `occam`, however none are so compelling that they warrant adoption in precedence to adapting existing languages such as Fortran, `C`, `C++`, Ada, Lisp, Prolog, etc. This is because users have experience with existing languages, good sequential compilers exist and can be incorporated into parallel compilers, and migrating existing code to parallel machines is much easier. In any case, to be generally suable, especially for scientific computing, any new language would need to implement the standard features and libraries of `C` and Fortran (Fox, 1991b, Fox, 1991d).

The purpose of software, and in particular computer languages, is to map a problem onto a machine, as described in §8.3.1. A drawback of current software and languages is that they are often designed around the *machine* architecture, rather than the *problem* architecture. This can make it very difficult to port the code from one machine to another, and in particular from a sequential computer to a parallel computer. It is possible for compilers to extract parallelism from a dependency analysis of sequential code (such as Fortran 77), however this is not usually very effective. In many cases the parallelism inherent in the problem will be obscured by the use of a sequential language or even a sequential algorithm. A particular application can be parallelised efficiently if, and only if, the details of the problem architecture are known. Users know the structure of their problems much better than compilers do, and can create their algorithms and programs accordingly. If the data structures are explicit, as in Fortran 90, then the parallelism becomes much clearer.

Each class of problem architectures requires different general constructs from the software, and a study of problem architectures is helpful in formulating the requirements for parallel languages and software (this is described in more detail in §8.6). Currently there are two language paradigms for distributed memory parallel computers: message passing and data parallel languages. Both of these have been implemented as extensions to Fortran and C. Here we will concentrate on Fortran.

8.2.2.1 Message passing Fortran

Message passing is a natural model of programming distributed memory MIMD computers, and is currently used in the vast majority of successful applications using MIMD machines. The basic idea is that each *node* (processor plus local memory) has a program that controls, and performs calculations on, its own data (the 'owner-computes' rule). Non-local data may need to be obtained from other nodes, which is done by communication of messages.

In its simplest form, there is one program per node of the computer. The programs can be different, although they are usually the same. However they will generally follow different threads of control, for example different branches of an IF statement. Communication can be asynchronous, but in most cases the algorithms are *loosely synchronous* (Fox et al., 1988b), meaning that they are usually controlled by a time or iteration parameter and there is synchronisation after every iteration, even though the communications during the iteration process may not be synchronous.

If parallelism is obtained from standard domain decomposition, then the parallel program for each node can look very similar to the sequential program, except that it computes only on local data, and has a call to a message passing routine to obtain non-local data. Schematically, a program might look something like the following:

>   CALL COMMUNICATE (*required non-local data*)
>   DO i *running over local data*
>   CALL CALCULATE (*with i's data*)
>   END DO

Note that it is more efficient to pass all the non-local data required in the loop as a single block before processing the local data, rather than pass each element of non-local data as it is needed within the loop. The advantages of this style of programming are:

- It is portable to both distributed and shared memory machines.
- It should scale to future machines, although to achieve good efficiencies schemes to overlap communication with itself and with calculation may be required.
- Languages are available now and are portable to many different MIMD machines. Current message passing language extensions include Express, PICL, PVM, and Linda.
- There will soon be an industry standard Message Passing Interface (Message Passing Interface Forum, 1994).
- All problems can be expressed using this method.

The disadvantages are:

- The user has complete control over transfer of data, which helps in creating efficient programs, but explicitly inserting all the communication calls is difficult, tedious, and error prone.
- Optimisations are not portable.
- It is only applicable to MIMD machines.

8.2.2.2 Data parallel Fortran

The goal of the Fortran 90 standard is to 'modernise Fortran, so that it may continue its long history as a scientific and engineering programming language'. Although Fortran 90 is a sequential language, some of its major new features are the array operations to facilitate vector and data parallel programming.

Data parallel languages have distributed data just as for the message passing languages, however the data is explicitly written as a globally addressed array. As in the Fortran 90 array syntax, the expression

```
DIMENSION A(100,100), B(100,100), C(100,100)
A = B + C
```

is equivalent to

```
DO i = 1, 100
   DO j = 1, 100
      A(i, j) = B(i, j) + C(i, j)
```

END DO
END DO

The first expression clearly allows easier exploitation of parallelism, especially as a DO loop of Fortran 77 can often be 'accidentally' obscured, so a compiler can no longer see the equivalence to Fortran 90 array notation. Migration of data is also much simpler in a data parallel language. If the data required to do a calculation is on another processor, it will be automatically passed between nodes, without requiring explicit message passing calls set up by the user. For example, a program fragment might look something like the following, using either an array syntax with shifting operations to move data (as in Fortran 90)

$$A = B + SHIFT\ (C,\ \textit{in i direction})$$

or explicit parallel loops in a FORALL statement using standard array indices to indicate where the data is to be found (FORALL is not in the Fortran 90 standard, but is present in many dialects of data parallel Fortran)

FORALL i,j
$A(i,j) = B(i,j) + C(i-1,j)$

The advantages of this style of programming are:

- it is relatively easy to use, since message passing is implicit rather than explicit, and parallelism can be based on simple Fortran 90 array extensions;
- it is scalable and portable to both MIMD and SIMD machines;
- it should be able to handle all synchronous and loosely synchronous problems, including ones that only run well on MIMD;
- data parallel languages such as CM Fortran and MasPar Fortran are available now that are based on Fortran 90 array syntax;
- an industry standard, High Performance Fortran (HPF), has been adopted, which is an extension of Fortran 90 that builds on existing data parallel languages (High Performance Fortran Forum, 1993, Koelbel et al., 1994).

The disadvantages are:

- the need to wait for good HPF compilers;
- not all problems can be expressed in this way.

8.2.2.3 High Performance Fortran

A major hindrance to the development of parallel computing has been the lack of portable, industry standard parallel languages. Currently, almost all parallel computer vendors provide their own proprietary parallel languages which are not portable even to machines of the same architecture, let alone between SIMD and MIMD, distributed or shared memory, parallel or vector architectures. This problem is now being addressed by the High Performance Fortran Forum (HPFF), a group of over 40 organisations including universities, national laboratories, computer and software vendors, and major industrial supercomputer users. HPFF was created to discuss and define a set of extensions to Fortran called High Performance Fortran. The goal was to address the problems of writing portable code that would run efficiently on any high performance computer, including parallel computers of any architecture (SIMD or MIMD, shared or distributed memory), vector computers, and RISC workstations. Here 'efficiently' means 'comparable to a program hand-coded by an expert in the native language of a particular machine'.

The HPF standard was finalised in May 1993. HPF is designed to support data parallel programming. It is an extension of Fortran 90, which provides for array calculations and is therefore a natural starting point for a data parallel language. HPF attempts to deviate minimally from the Fortran 90 standard, while providing extensions that will enable compilers to provide good performance on a variety of parallel and vector architectures. While HPF was motivated by data parallel languages for SIMD machines, it was developed to enable such languages to be portable to any computer architecture, including MIMD, vector and sequential machines (Fox, 1991b, Choudhary *et al.*, 1992b, Bozkus *et al.*, 1993).

HPF has a number of new language features, including:

- new directives that suggest implementation and data distribution strategies to the compiler; they are structured so that a standard Fortran compiler will see them as comments and thus ignore them;
- new language syntax extending Fortran 90 to better express parallelism;
- standard interfaces to a library of efficient parallel implementations of useful routines, such as sorting and matrix calculations;
- access to extrinsic procedures which can be defined outside the language, for example by using Fortran with message passing, in order

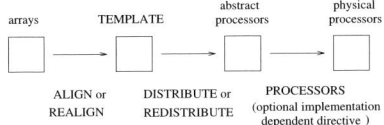

Fig. 8.1. Mapping data onto processors in High Performance Fortran.

to handle certain operations that cannot be expressed very well (or at all) in HPF.

The strategy behind HPF is that the user writes in an SPMD (Single Program Multiple Data) data parallel style, with conceptually a single thread of control and globally accessible data. The program is annotated with assertions (compiler directives) giving information about desired data locality and distribution. The compiler then generates code implementing data and work distribution.

In the HPF model, the allocation of data to processors is done using a two-level mapping of data objects to processor memories, referred to as *abstract processors*. This is shown in fig. 8.1. First the data objects (typically array elements) are *aligned* relative to one another, using an abstract indexing space called a *template*. A template is then *distributed* onto a rectilinear arrangement of abstract processors. The final mapping of abstract processors to the same or a smaller number of physical processors is not specified by HPF, and is implementation dependent.

8.2.2.4 HPF compilers and Fortran 90D

HPF is defined to be portable between computers of different architectures. As its name suggests, a major goal of High Performance Fortran is to have efficient compilers for all these machines. Vectorising compilers work by analysing data dependencies within loops, and identifying independent data sets that can be processed simultaneously. However, obtaining parallelism solely through dependency analysis has not proven to be effective in general, so for all commands in HPF the dependencies are directly implied, enabling the compiler to generate more efficient code.

Compilers will implement HPF differently on different computer architectures, for example:

SIMD computers – parallel code with communication optimised by compiler placement of data;

MIMD computers – a multi-threaded message passing code with local data and optimised send/receive communications;

vector computers – vectorised code optimised for the vector units;

RISC computers – pipelined superscalar code with compiler generated cache management.

A subset of HPF has been defined to enable early availability of compilers. The first implementation of HPF is the Fortran 90D compiler being produced by NPAC (Bozkus *et al.*, 1993). The alpha version of the compiler was demonstrated at Supercomputing 92, and the beta release is now available. The compiler currently runs on MIMD parallel computers: the nCUBE/2, Intel iPSC/860, and a network of Sun workstations. The next target architecture is heterogeneous networks, and in the future the compiler will be optimised for specific architectures and released as a commercial product by the Portland Group. An example of the performance of the current Fortran 90D compiler compared to a hand-coded message passing program is shown in fig. 8.2 for a Gaussian elimination problem. The HPF code for the main routine is shown in fig. 8.3.

Fortran 90D will continue to be developed as a superset of HPF, in order to research new functionality that may be added to the HPF standard in the future. For example, language facilities for handling parallel I/O are being investigated (Bordewaker & Choudhary, 1994), which is a major area of concern that was not addressed by the initial HPF standard.

8.2.3 Systems integration and visualisation

Recent advances in parallel programming languages such as High Performance Fortran are expected to improve the useability of parallel processing for the simulation of complex problems, especially in industry. However, complex 'real world' computationally intensive applications in areas such as fluid dynamics, product design, or concurrent engineering require even more powerful and versatile tools. Such applications typically contain several modules with varying degrees of inter-modular interaction. Some modules such as digital signal processors, 3D renderers or partial differential equation (PDE) solvers map naturally onto the HPF programming model, while some others are inherently sequen-

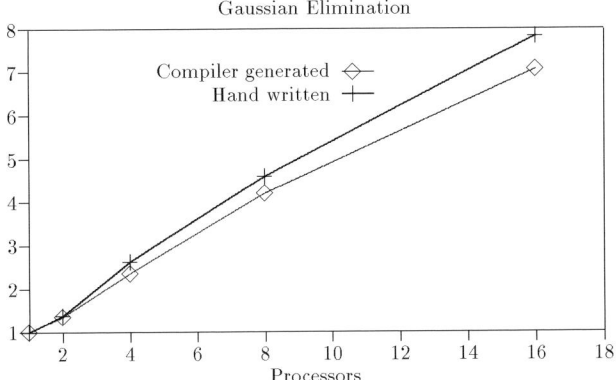

Fig. 8.2. Performance of the Fortran 90D compiler versus hand-written message passing code on the Intel iPSC/860 for Gaussian elimination.

tial. Also, a realistic application contains typically several data parallel modules, some of them interacting in the data parallel mode as well, for example different layers of a multigrid PDE algorithm, or subsequent filters in machine vision systems. Finally, the process of integrating individual components into the full application is a complex task itself, and so is the process of module synchronisation, interactive debugging and fine-tuning the parameters of a prototype. New generation High Performance Distributed Computing (HPDC) software integration tools, required to handle this type of computational complexity, are currently being constructed. We summarise here recent activities in this area at NPAC.

Currently a popular approach is based on dataflow visualisation systems such as AVS (Advanced Visualisation System). This model supports a network of computational modules, implemented as individual UNIX processes, and interacting via the RPC (Remote Procedure Call) protocol under control of the AVS kernel. Individual modules can be placed on different machines and hence the model provides support for heterogeneous distributed computing. Some of these modules can also be installed on parallel platforms, thereby extending the paradigm to the HPDC level. Visual editing tools for such a network are also offered by the system which facilitate application prototyping, integration, monitoring and fine-tuning. Finally, several default visualisation modules

```
              PROGRAM gaussian

              PARAMETER (N = 100)
              PARAMETER (NN = 100)
              INTEGER index(N), iTmp(1)
              INTEGER indexRow, i, j, k
              REAL a(N,NN), row(NN), fac(N)
              REAL maxNum
!HPF$ PROCESSORS p(4)
!HPF$ TEMPLATE templ(100)
!HPF$ DISTRIBUTE templ (BLOCK) ONTO p
!HPF$ ALIGN a(*,i) WITH templ(i)
!HPF$ ALIGN row(i) WITH templ(i)
              index = -1
              DO k = 1, N
                 iTmp = MAXLOC(a(:,k), MASK = index .EQ. -1)
                 indexRow = iTmp(1)
                 maxNum = a(indexRow,k)
                 index(indexRow) = k
                 fac = a(:,k) / maxNum
                 row = a(indexRow,:)
                 FORALL (i=1:N, j=k:NN, index(i) .EQ. -1)
        &           a(i,j) = a(i,j) - fac(i) * row(j)
              END DO
              END
```

Fig. 8.3. Gaussian elimination programmed in High Performance Fortran.

come with the system and allow for sophisticated data visualisation and rendering tasks.

AVS is adequate only for relatively static scientific visualisation tasks. There is no support for system-wide synchronisation nor for real-time interactive services required for advanced simulation tasks such as virtual reality (VR). Also, while parallel (e.g. HPF) modules can be included in the AVS network, there is no support for parallel I/O or parallel dataflow between individual HPF tasks. All communication must be mediated by the corresponding host programs, which causes substantial bottlenecks.

We are currently developing a set of tools that will allow us to extend AVS functionality in these areas and to provide support for virtual reality simulations as well as for televirtuality services providing remote VR user interfaces. The underlying software model is provided by the

MOVIE (Multi-tasking Object-oriented Visual Interactive Environment) model (Furmanski, 1992, Furmanski et al., 1993). A MOVIE system is a network of MOVIE servers – interpreters of a high-level object-oriented programming language, MovieScript. MovieScript extends , in areas such as graphical user interface (GUI) prototyping, Fortran 90 style array syntax and operating support for real-time multi-threading.

A specific design of a MOVIE network can be adapted to a particular computational domain. In particular, all dynamic features of the AVS model in the heterogenous distributed mode can be reproduced in terms of MOVIE tasks or threads, but the model also offers support for multi-tasking data parallel processing and interactive real-time programming. In the early development stage is the next level tool, which will allow for concurrent execution of and parallel dataflow between several HPF modules. An AVS-like visual network editor will also be provided to facilitate application editing tasks. Finally, we also plan to provide support for World Wide Web (WWW) services in terms of CGI (Common Gateway Interface) scripts. Such scripts, passed from the Web browser to the Web server to the MOVIE host server will allow for interactive control of simulations running on remote high performance computers. This architecture will also enable the development of prototype televirtuality (TVR) services. Current hypermedia browsers such as Mosaic are not adequate to support two-way interactivity but there are ongoing VR-oriented activities in the WWW community, and several new consumer level VR products will soon offer remote support as well.

8.3 Complex systems and a theory of parallel computing

8.3.1 Mapping problems onto computers

For this section, we shall consider a *complex system* as a large collection of, in general, disparate members. Those members have, in general, a dynamic connection between them (see chapter 3). A dynamic complex system evolves by a probabilistic or deterministic set of rules that relate the complex system at a later 'time' to its state at an earlier 'time'. Complex systems studied in chemistry and physics, such as a protein or the universe, obey rules that we believe we understand more or less accurately. The military play war games, which is the complex system formed by a military engagement. This and more general complex systems found in society obey less clear rules.

One particular important class of complex systems is that of the *com-*

plex computer. In the case of a hypercube such as the nCUBE, or other multicomputers such as the Intel Paragon or Thinking Machines CM-5, the basic entity in the complex system is a conventional computer and the connection between members is a communication channel implemented either directly in VLSI, on a PC board, or as a set of wires or optical fibres. In another well-known complex computer, the brain, the basic entity is a neuron and an extremely rich interconnection is provided by axons and dendrites (see chapter 10).

In many situations, one is concerned with mapping one complex system into another. Solving a problem consists of using one complex system, the *complex processor*, to 'solve' another complex system, the *complex problem*. In building a house, the complex processor is a team of masons, electricians, and plumbers, and the complex problem is the house itself. In this article, we are mainly interested in the special case where the complex processor is a complex computer and modelling or simulating a particular complex problem involves mapping it onto the complex computer.

Simulation or modelling begins with a map

$$\text{Nature (or system to be modelled)} \xrightarrow{\text{map theory}} \text{Idealization or Model}$$

(8.1)

This map would often be followed by a computer simulation, which consists of mapping the model onto the computer. This whole process can be broken up into several maps, as shown in fig. 8.4. We illustrate the procedure using the example of a computational fluid dynamics study of airflow around an airplane, where the complex systems used are:

S_0 is nature – the actual flow of air around the airplane.

S_1 is a (finite) collection of molecules interacting with long-range Van der Waals and other forces. This interaction defines a complete interconnect between all members of the complex system S_1.

S_2 is the infinite degree of freedom continuum with the fundamental entities as infinitesimal volumes of air connected locally by the partial differential operator of the Navier–Stokes equation.

$S_3 = S_{\text{num}}$ could depend on the particular numerical formulation used. Multigrid, conjugate gradient, direct matrix inversion and alternating gradient would have very different structures in the direct numerical solution of the Navier–Stokes equations. The more radical cellular automata approach would be quite different again.

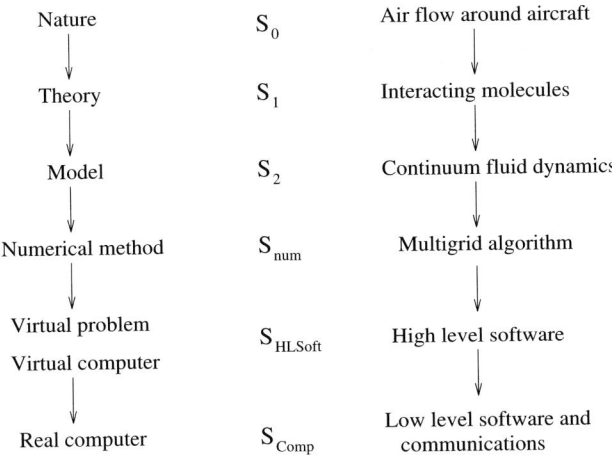

Fig. 8.4. Computation and simulation as a series of maps.

$S_4 = S_{\text{HLSoft}}$ would depend on the final computer being used and division between high and low level in software. The label HLSoft denotes 'High Level Software'.

$S_5 = S_{\text{comp}}$ would be S_{HLSoft} embroidered by the details of the hardware communication (circuit or packet switching, wormhole or other routing). Further, we would often need to look at this complex system in greater resolution and expose the details of the processor node architecture.

Nature, the model, the numerical formulation, the software, and the computer are all complex systems, and they can be quite different. We are interested in the structure of all these complex systems and the maps between them. Note that each of the successive maps in fig. 8.4 results in a loss of information. As reviewed in §8.6, we can discuss key problems in the design of software systems in terms of minimising information loss.

Typically, one is interested in constructing the maps in fig. 8.4 to satisfy certain goals, such as minimising the execution time of the computer simulation (the main focus of the high performance computing community), minimising the time required to write the computer program (the main focus of the computer science and software engineering community), and obtaining the best agreement of the model with the effects seen in nature (the main focus of the scientific community). We therefore get a class of optimisation problems associated with the different complex systems and the mappings between them. Parallel computing can therefore be looked at as 'just' an optimisation problem, even if we can't agree on exactly what to optimise – there are obvious tradeoffs between fidelity of the model and the amount of computation required to solve it, the speed of the program and the ease of implementation (for example using assembler versus a high level language), and so on.

One approach to solving these optimisation problems is the use of methods developed from the study of complex physical systems, such as simulated annealing, genetic algorithms, or neural networks. These are used to minimise a cost function that expresses the goals described above. Typically, in studying performance, the cost function would be the execution time of the problem on a computer. For software engineering, the cost function would also reflect user productivity. These physical optimisation methods were originally developed as ways of minimising the energy of a physical system. In the rest of this section, and in §8.4 and 8.5, we will show that computational problems can be looked at using a space–time analogy to a physical system, so that the cost function for optimising the map from complex problem to complex computer does in fact resemble the energy function of a physical system. This motivates the use of these physical optimisation techniques for solving these problems, an approach we refer to as *physical computation*, which is discussed in §8.7.

In this chapter we will concentrate on the mappings in fig. 8.4 that take us from the model of the world (the complex problem) to the simulation of that model on a parallel computer (the complex computer). Mapping a complex problem onto a complex computer involves *decomposition*. We can consider the simulation of the complex problem as an *algorithm* applied to a *data domain*. We divide the data domain into pieces that we call *grains* and place one grain in each node of the concurrent computer.

If we consider a typical matrix algorithm such as multiplication

$$a_{ij} = \sum_k b_{ik} c_{kj} \qquad (8.2)$$

we have a data domain formed by the matrix elements, which we generally call *members*. The algorithm (8.2) defines a graph connecting these members and these connected members form a complex system. The standard *decomposition* involves submatrices stored in each node. Edges of the graph connect members in different submatrices (i.e. members of the complex system stored in separate nodes of the complex computer). To be precise, in the map

Complex Problem	$\longrightarrow$	Complex Computer
Members	map into	memory locations
Internal connections	map to	arithmetic operations
Internode or 'cut' connections	map to	communication followed by arithmetic operations

In §8.4, we will be considering topological properties of complex systems which correspond to the map

Complex Problem	$\longrightarrow$	Topological Structure
Members	map into	points in a geometric space
Connections	map into	(nearest neighbour) structure

In the optimal decomposition studies in §8.4 and §8.5, we will be considering dynamic properties of complex systems for which it will be useful to consider the map

Complex Problem	$\longrightarrow$	Discrete Physical System
Members	map to	particles or strings
Connections	map to	force between particles or strings

We see that different classes of complex systems realise their members and interconnections in different ways. We find it very useful to map general systems into classes having a particular choice for members and interconnects. To be precise, complex systems have interconnects that can be geometrical, generated by forces, electrical connection (e.g. wire), structural connection (e.g. road), biological channels or symbolic relationships defined by the laws of arithmetic. We map all these interconnects into electrical communication in the multicomputer implementation. On the other hand, in the simulated annealing approach to load balancing, we map all these interconnects to forces.

8.3.2 The space–time picture of computation

The above discussion was essentially static and although this is an important case, the full picture requires consideration of dynamics. We now 'define' space and time for a general complex system.

We associate with any complex system a *data domain* or *'space'*. If the system corresponds to a real or simulated physical system, then this data domain is a typically three-dimensional space. In such a simulation, the system consists of a set of discrete objects labelled by an index i and is described as a function of the positions $\underline{x}_i(t)$ of the objects at each time t.

For example, seismic exploration for oil fields involves measuring echoes of sound waves that are reflected off various underground strata. Using these measurements to reconstruct the strata formation involves solving the wave equation, a standard second order differential equation that describes the propagation of the sound waves. The equation is discretised in space (a three-dimensional grid representing some part of the earth's crust) and time (the time-step) to give a finite difference equation that can be solved on a computer. Only local data (nearest-neighbour points in the grid) is required to solve the difference equations at each time-step.

Other complex systems have more abstract data domains:

(i) In a computer chess program, the data domain or 'space' is the pruned tree-like structure of possible moves.
(ii) In matrix problems, the data domain is either a regular two-dimensional grid for full matrices or an irregular subset of this for sparse matrices.
(iii) In a complex computer (defined in §8.3.1), 'space' is the set of nodes of the parallel computer, or at a finer resolution, the set of memory locations.

The data domain will have certain dependencies contingent on the model, for example a dependence on nearest-neighbour grid points for problems with local interactions, or a dependence on all other data points for N-body simulations where forces are long-ranged. The data domain can thus be viewed as a set of interconnected nodes (the data elements) connected by edges (the data dependencies), which form what we call the *computational graph* (see chapter 3). This is defined by a time slice of the full complex system.

Note that we can examine the data domain of the complex computer hardware in terms of a computational graph, just as we can for the computer software or algorithms. The computational graph of a multicomputer is formed by the individual computer nodes with the edges of the graph determined by the interconnection topology (or architecture) of the multicomputer.

In a physical simulation, the complex system evolves with time and is specified by the nature of the computational graph at each time-step. If we are considering a statistical physics or Monte Carlo approach, then we no longer have a natural time associated with the simulation. Rather, the complex system is evolved iteratively or by Monte Carlo sweeps. We will find it useful to view this evolution or iteration label similarly to time in a simple time-stepped simulation. We thus consider a general complex system defined by a data domain, which is a structure given by its computational graph. This structure is extended in 'time' to give the 'space–time' cylinders. For our previous examples:

(i) *chess*: time labels depth in tree;
(ii) *matrix algebra*: time labels iteration count in iterative algorithms or 'eliminated row' in a traditional full matrix algorithm such as Gaussian elimination;
(iii) *complex computer*: the time dependence is just the evolution given by either the cycle time of the nodes or the executed instructions. SIMD machines give an essentially static or synchronous time dependence, whereas MIMD machines can be very dynamic.

We expand the discussion of temporal properties in §8.5. We will also discuss in §8.6 an interesting class of problems and a corresponding way of using MIMD machines, called loosely synchronous. These are microscopically dynamic or temporally irregular but become synchronous when averaged over macroscopic time intervals.

Domain decomposition for data parallel computing is just the mapping of the spatial domain (data) of the complex problem onto the spatial domain (nodes) of the complex computer. This differs from the computational model for a sequential computer, where all aspects of the problem are mapped to the time domain of the computer. In contrast, another type of data processor, a seismograph, maps the time dependence of an earthquake onto the spatial extent of a recording chart. The general problem of computation is to map the space–time domain of the problem onto the space–time domain of the computer in an effective way.

8.4 Spatial properties of complex problems and complex computers

8.4.1 System size and geometry

The *size* N of the complex system is an obviously important property. Note that we think of a complex system as a set of *members* with their spatial structure evolving with time. Sometimes the time domain has a definite 'size', but often one can evolve the system indefinitely in time. However, most complex systems have a natural spatial size with the spatial domain consisting of N members. In the matrix example, Gaussian elimination has n^2 spatial members (matrix elements) evolving for a fixed number n of 'time' steps. As usual, the value of the spatial size N will depend on the granularity or detail with which one looks at the complex system. One could consider a parallel computer at the level of transistors with a very large value of N, but usually we look at the processor node as the fundamental entity and define the spatial size of a parallel computer viewed as a complex system, by the number N_{proc} of processing nodes.

Consider mapping a finite difference simulation with N_{num} grid points, such as solving the wave equation for a seismic exploration simulation as described in §8.3.2, onto a parallel machine with N_{proc} processors. An important parameter is the *grain size* n of the resultant decomposition. We can introduce the *problem* grain size $n_{\text{num}} = N_{\text{num}}/N_{\text{proc}}$, and the *computer* grain size n_{mem} as the memory contained in each node of the parallel computer. Clearly we must have $n_{\text{num}} < n_{\text{mem}}$ if we measure memory size in units of seismic grid points. More interestingly, we will later relate the performance of the parallel implementation of the seismic simulation to n_{num} and other problem and computer characteristics. We find that in many cases, the parallel performance only depends on N_{num} and N_{proc} in the combination $N_{\text{num}}/N_{\text{proc}}$ and so grain size is a critical parameter in determining the effectiveness of parallel computers for a particular application.

Another set of parameters describes the topology or structure of the spatial domain associated with the complex system. The simplest parameter of this type is the geometric dimension d^{geom} of the space. Our early parallel computing used the binary hypercube of dimension d, which has $d_{\text{comp}} = d$ as its geometric dimension. This was an effective architecture because it was richer than the topologies of most problems. For many physical grid-based simulations such as seismic exploration, the geometric dimension of the problem d_{num} is just the dimension of the physical space being simulated (3 for this example). The performance of the simulation also depends on whether the software system preserves the spatial structure of the problem, in which case $d_{\text{HLSoft}} = d_{\text{num}}$.

8.4.2 Performance model for a multicomputer

The performance of a multicomputer is usually defined in terms of parallel *speedup* S and *efficiency* ε. Speedup is just how much faster a multicomputer executes the parallel program on N nodes compared to the sequential program on one node. Efficiency measures what fraction of the maximum speedup N is actually achieved, so that

$$S = \varepsilon N. \tag{8.3}$$

Efficiency will usually be less than 1 since there are overheads involved in parallel computing, such as the cost of communicating data between processors. Let us try to quantify these costs by defining the following parameters for a multicomputer:

- t_{calc} – the typical time required to perform a generic calculation. For scientific problems, this can be taken as a floating point operation.
- t_{comm} – the typical time taken to communicate a single word between two nodes connected in the hardware topology.

The definitions of t_{comm} and t_{calc} are imprecise above. In particular, t_{calc} depends on the nature of the node and can take on very different values depending on the details of the implementation; floating point operations are much faster from registers than from slower parts of the memory hierarchy. On systems built from processors like the Intel i860 chip, these effects can be large – t_{calc} could be 0.0125μ s from registers (80 Megaflops) and a factor of ten larger when the variables a, b are fetched from dynamic RAM. Communication speed t_{comm} depends on internode message size (a software characteristic) and the latency (startup time) and bandwidth of the computer communication subsystem. It will also generally be slower for communications between nodes that are not directly connected in the multicomputer topology, so that messages have to be routed between intermediary nodes. This effect will depend on the problem being solved – it can be negligible for grid-based problems with only local data dependencies (such as the seismic simulation), or a factor of 2 or more for problems with a lot of non-local communication (such as a parallel Fast Fourier Transform).

The overhead f_C due to communication can be expressed as

$$f_C = \frac{\text{total time for communication}}{\text{total time for calculation}}. \tag{8.4}$$

It is easy to see that if the parallel overhead is due solely to communication, then

$$S = \frac{N}{1 + f_C}. \tag{8.5}$$

Let us examine the communication overhead for a simple grid-based problem, such as our seismic simulation example. We use standard domain decomposition to map the problem domain (a $d_{\text{num}} = 3$ grid of points) onto the computer (a hypercube, for example) so that every processor has a cubic section of the grid. The grain size n_{num} will be $L^{d_{\text{num}}}$, where L is the length of a side of the grain (the cube of grid points on each processor).

The total amount of computation on each node of the computer will be proportional to t_{calc} times the grain size, which is the volume $L^{d_{\text{num}}}$ of the grain. For this problem, the data dependencies are all local (nearest-neighbour), so only data at the edge of the grain needs to be communicated. The total amount of communication is thus proportional to t_{comm} times the surface area $L^{d_{\text{num}}-1}$ of the grain. So from equation (8.4) we have that

$$\begin{aligned} f_C &\propto \frac{1}{L} \frac{t_{\text{comm}}}{t_{\text{calc}}} \\ &\propto \frac{1}{n_{\text{num}}^{1/d_{\text{num}}}} \frac{t_{\text{comm}}}{t_{\text{calc}}}. \end{aligned} \tag{8.6}$$

It can be shown (Fox et al., 1994) that in general the overhead due to internode communication can be written in the form

$$f_C \propto \frac{N_{\text{proc}}^{\alpha}}{n_{\text{num}}^{\beta}} \frac{t_{\text{comm}}}{t_{\text{calc}}}. \tag{8.7}$$

The term $t_{\text{comm}}/t_{\text{calc}}$ indicates that communication overhead depends on the relative performance of the internode communication system and node (floating point) processing unit. A real study of parallel computer performance would require a deeper discussion of the exact values of t_{comm} and t_{calc}. More interesting here is the dependence on the number of processors N_{proc} and the problem grain size n_{num}. As described above, grain size $n_{\text{num}} = N_{\text{num}}/N_{\text{proc}}$ depends on both the problem and the computer. The value of β is given by

$$\beta = \frac{1}{d_{\text{info}}} \tag{8.8}$$

where the *information dimension* d_{info} is a generalisation of the geometric dimension for problems whose structure is not geometrically based. This will be described in the next subsection. It is independent of the parameters of the computer. α is given by

$$\begin{aligned}\text{if}\quad d_{\text{num}} < d_{\text{comp}} &, \quad \alpha = 0 \\ \text{if}\quad d_{\text{num}} > d_{\text{comp}} &, \quad \alpha = \left(\frac{1}{d_{\text{comp}}} - \frac{1}{d_{\text{num}}}\right)\end{aligned} \tag{8.9}$$

which quantifies the penalty, in terms of a value of f_C that increases with N_{proc}, for a computer architecture that is less rich than the problem architecture. An attractive feature of the hypercube architecture is that d_{comp} is large and one is essentially always in the regime governed by $\alpha = 0$ in (8.9). Recently, there has been a trend away from rich topologies like the hypercube towards the view that the node interconnect should be considered as a routing network or switch to be implemented in the very best technology. The original MIMD machines from Intel, nCUBE and Ametek all used hypercube topologies as did the SIMD Connection Machine CM-1 and CM-2. The nCUBE-2 introduced in 1990 still uses a hypercube topology, but both it and the second generation Intel iPSC/2 used more sophisticated routing. The latest Intel Touchstone Delta and Paragon use a two-dimensional mesh with wormhole routing. It is not clear how to incorporate these new node interconnects into the above picture, and further research is needed here. Presumably, we would need to add new complex system properties and perhaps generalise the definition of dimension d_{comp}, as we will now do for d_{num} in order for equation (8.7) to be valid for problems whose structure is not geometrically based.

8.4.3 Information dimension

Returning to (8.5), (8.7), (8.8), and (8.9) we note that we have not properly defined the correct dimension d_{num} or d_{comp} to use. We have implicitly equated this to the natural *geometric dimension* but this is not always correct. This is illustrated by the complex system S_{num} consisting of a set of particles in three dimensions interacting with a long-range force such as gravity or electrostatic charge. The geometric

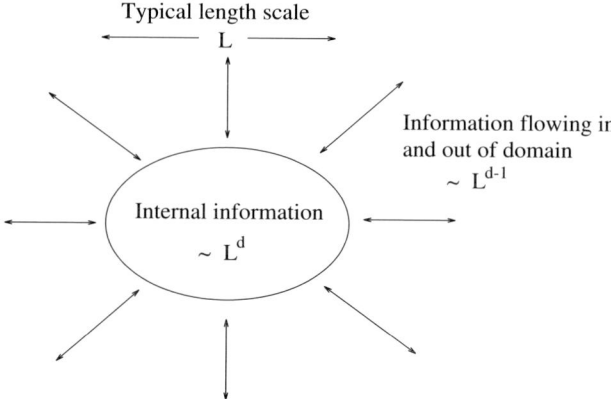

Fig. 8.5. The information density and flow in a general complex system of length L.

structure is local with $d_{\text{num}}^{\text{geom}} = 3$ but the complex system structure is quite different; all particles are connected to all others.

We define the *information dimension* d^{info} for a general complex system to reflect the system connectivity (Fox et al., 1998b, Fox, 1992a). This is analogous to the fractal dimension introduced by Mandelbrot (Mandelbrot, 1979) (see chapter 5), in that it may not be equal to the geometric dimension, and need not be an integer. Consider fig. 8.5 which shows a general domain D in a complex system. We define the volume V_D of this domain by the information in it. Mathematically, V_D is the computational complexity needed to simulate D in isolation. In a geometric system

$$V_D \propto L^{d^{\text{geom}}}, \tag{8.10}$$

where L is a geometric length scale. The domain D is not in general isolated and is connected to the rest of the complex system. Information I_D flows into D, and again in a geometric system I_D is a surface effect with

$$I_D \propto L^{d^{\text{geom}}-1}. \tag{8.11}$$

If we view the complex system as a graph (i.e. the computational graph), V_D is related to the number of edges of the graph with at least one of the nodes in D, and I_D is related to the number of edges cut by the surface of D. (8.10) and (8.11) are altered in cases like the long-range force problem where the complex system connectivity is no longer geometric. We define the information dimension to preserve the surface versus volume interpretation of (8.11) compared to (8.10). Thus, generally we define

$$I_D \propto V_D^{1-1/d^{\text{info}}}. \qquad (8.12)$$

With this definition of information dimension d^{info}, we find that (8.5), (8.7), (8.8), and (8.9) essentially hold in general. For simple problems, the information dimension will be approximately equal to the geometric dimension. However the information dimension will in general be larger for systems with complex structure, which have non-geometric (or 'hidden') dimensions of complexity.

An interesting example of non-trivial information dimension comes from the simulation of electronic circuits. Rent's rule (Landman & Russo, 1971, Donath, 1979) is a phenomenological rule that is used in the packaging of circuits. It relates the number of output lines (pinouts) to a power ($\approx 0.5 \to 0.7$) of the number of internal components. This implies a non-integer information dimension $d^{\text{info}} \approx 3$, which is greater than the geometric dimension $d^{\text{geom}} = 2$ for an electronic circuit. Rent's rule is approximately independent of the size of the circuit, which is analogous to the self-similarity and scaling properties of systems with non-trivial fractal dimension (see chapter 5).

For the long-range force problem, it can be shown that $d^{\text{info}} = 1$ independent of d^{geom} (Fox et al., 1994, Fox, 1992a). One might naively expect that the information dimension of such a problem would be infinite, rather than 1, since all objects interact with all other objects. However infinite information dimension applies to systems such as the telephone network, for which everyone is connected to everyone else, but *different* information is communicated to every different person. In contrast, the Voice of America radio broadcast has an information dimension of 1, since the *same* information is communicated to everyone. In the long-range force problem, the same information is broadcast by an object to every other object (e.g. the mass and position of the object for an N-body gravitational interaction problem), so the information dimension is 1.

8.4.4 A physical analogy for domain decomposition

In the previous three subsections, we described static spatial properties of complex systems that are relevant for computation. These included size, topology (geometric dimension) and the information dimension.

We will find new ideas when we consider problems that are spatially irregular and perhaps vary slowly with time. A simple example would be a large-scale astrophysical simulation where the use of a parallel computer requires that the universe be divided into domains that, due to the gravitational interactions, will change as the simulation evolves.

The performance of a computation executing on a parallel machine is crucially dependent on *load balance*. This refers to the amount of CPU idling occurring in the processors of the concurrent computer: a computation for which all processors are continually busy (and doing useful, non-overlapping work) is considered perfectly load balanced. This balance is often not easy to achieve, however.

As described in the previous section, a key to parallel computing is to split the underlying spatial domain into grains, each of which correspond to a process as far as the operating system is concerned. We will take a naive software model where there is one process associated with each of the fundamental members of the simulated system, i.e. with each 'particle' in the astrophysical simulation. This is not practical with current software systems as it gives high context switching and other overheads. However, it captures the essential issues.

The processes will need to communicate with one another in order for the computation to proceed. Assume that the processes and their communication requirements are changing with time – processes can be created or destroyed, communication patterns will move. This is the natural choice when one is considering timesharing the parallel computer, but can also occur within a single computation. It is the task of the operating system to manage this set of processes, moving them around if necessary, so that the parallel computer is used in an efficient manner.

The operating system performs two primary tasks. First, it must monitor the ongoing computation so as to detect bottlenecks, idling processors and so on. Secondly, it must modify the distribution of processes and also the routing of their associated communication links so as to improve the situation. In general, it is very difficult to find the optimum way of doing this – in fact, this is an NP-complete problem. Approximate solutions, however, will serve just as well. We will be happy if we can realise a reasonable fraction (say 80%) of the potential computing power of the parallel machine for a wide variety of computations. An example of a non-trivial domain decomposition of an irregular data domain onto processors configured in a hypercube is given in fig. 8.6.

One can usefully think of a parallel computation in terms of a physical analogy. Treat the processes (or the data elements) as 'particles' free to

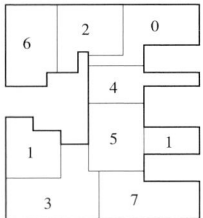

Fig. 8.6. A mapping of an irregular data domain onto processors in a hypercube.

move about in the 'space' of the parallel machine. Minimising the total execution time of the parallel computation formally requires that one minimise:

$$\max_{\text{nodes } i} C_i \qquad (8.13)$$

where C_i is the total computation time for calculation and communication. We choose to replace this mini–max problem by a least squares minimisation (Fox, 1988a) of

$$E = \sum_i C_i^2. \qquad (8.14)$$

Let m label the nodal points and (m, m') the edges of the computational graph. Then

$$C_i = \sum_{m \in i} \left[\sum_{(m,m')} Comm\,(m, m') + Calc\,(m) \right] \qquad (8.15)$$

where it takes time $Calc\,(m)$ to simulate m and time $Comm\,(m', m)$ to communicate necessary information from m' to m. If we consider the case where we can neglect the quadratic communication terms, then

$$C_i^2 \sim \sum_{\substack{(m,m') \\ m \in i}} Comm\,(m, m')$$

$$+ \sum_{m,\,m' \in i} Calc\,(m)\,Calc\,(m'). \qquad (8.16)$$

In this physical analogy, the above equation describes a 'Hamiltonian' (or 'energy function') for parallel computation, that the operating system must try to minimise, and if possible find the 'ground state' (the lowest energy state), which corresponds to the most efficient decomposition of data onto nodes of the parallel computer.

The last term in the Hamiltonian (8.16) is zero unless particles m and m' are at the same place, i.e. in the same node. In the physical analogy, this is like a short-range 'potential', where range is measured by distance between nodes in the space of the complex computer. This provides a short-range, repulsive 'force', causing the particles, and thereby the computation, to spread throughout the parallel computer in an evenhanded, balanced manner, corresponding to the requirement of load balancing.

A conflicting requirement to that of load balancing is shown in the first term of the Hamiltonian as interparticle communications – the various parts of the overall computation need to communicate with one another at various times. If the particles are far apart (distance being defined as the number of communication steps separating them) large delays will occur, slowing down the computation. This represents a long-range, attractive force between those pairs of particles which need to communicate with one another. This force is proportional to the amount of communication traffic between the particles, so that heavily communicating parts of the computation will coalesce and tend to stay near one another in the computer.

Exact minimisation of the function in (8.16) is not necessary – we have already 'wasted' some computational power using convenient high level languages, and we can surely afford to lose another 10% to load imbalance. The problem of distributing a computation onto a parallel machine in an efficient manner can therefore be fruitfully attacked using simulated annealing (Kirkpatrick *et al.*, 1983) and other 'physical' optimisation methods such as neural networks and genetic algorithms – see Flower *et al.* (1987), Fox and Furmanski (1988a), Fox (1988a, 1988a, 1991), Williams (1991), and Mansour and Fox (1993). The physical analogy described above makes this plausible, since these methods are highly appropriate minimisation techniques for (8.14). For example, simulated annealing is a standard Monte Carlo technique that was originally devised to find the ground states of spin models of magnetism that have competing interactions, such as spin glasses (see chapters 10, 6, 2). In this case, the competing interactions are the attractive 'communication' force and the repulsive 'load balance' force. We have used these methods routinely for load balancing a variety of simulations including finite element and particle dynamics simulations. Physical optimisation methods are described in more detail in §8.7.

8.4.5 System temperature and dynamic load balancing

(8.16) holds for the case of static load balancing, that is, where the data structures are static, so that the domain decomposition is done only once, at the beginning of the computation. However in general, problems and the data structures and computational graphs that describe them will be changing. In this case the data will have to be redistributed throughout the computation in order to keep the load balanced. For dynamic load balancing there will be an extra attractive 'force' in (8.16), corresponding to the penalty for moving data or processes to different nodes. High Performance Fortran provides a mechanism for the redistribution of data at runtime, which is particularly important for dynamic and irregular problems. The user can either specify the distribution, or specify the computational graph, in which case the compiler will find a good distribution using various optimisation techniques such as those described above (Choudhary et al., 1992a, Ponnusamy et al., 1993).

Using the physical analogy we introduced in §8.4.4, we can think of the operating system as a 'heat bath' that keeps the computation 'cool' and therefore near its 'ground state' (optimal solution). Most scientific simulations change slowly with time and redistribution of processes by the operating system can be gradual. Thus, we can think of the computation as being in *adiabatic* equilibrium at a complex system *temperature* T_{problem} which reflects the ease of finding a reasonable minimum. T_{problem} will be larger for problems that change more rapidly and where the operating system does not have 'time' to find as good an equilibrium (Fox et al., 1985, Fox & Otto, 1986, Fox et al., 1994). Redistribution of the data takes time, and for some problems this time may be significant, perhaps even longer than the simulation time between data redistribution. However there is a simple static data distribution called *scattered decomposition* that can work quite well for very irregular and dynamic problems.

Standard static domain decomposition works by splitting the data space into large connected partitions and mapping the partitions onto nodes in the computer space in a way that preserves data locality and balances the computational load. However this static decomposition will produce substantial load imbalance for problems with dynamic data structures that produce 'hot spots', or data intensive regions, that change position during the simulation, for example in N-body simulations of galactic collisions, where particles clump together due to gravitational attraction.

Scattered decomposition works by going to the opposite extreme, that is, breaking up the data into very small partitions and then 'scattering' the partitions among the nodes of the computer, so that each node receives some data from all regions of the problem space – see Morrison and Otto (1986) and Fox *et al.* (1988b, 1994). In this case any data hot spots will also get distributed fairly evenly among the processors. As the partitions are made smaller, the load balancing will improve. The price paid, of course, is increased communication overhead. The scattered decomposition will require much more communication traffic than the standard domain decomposition. Often, however, communication between nodes is relatively cheap compared to the computation required, and so the scattered decomposition becomes an attractive possibility. This corresponds to ignoring the first term in (8.16) (the communications cost) and trying to minimise only the second term, which we have seen requires spreading data connected in the computational graph to different nodes, which is exactly what scattered decomposition attempts to achieve.

One of the outstanding features of the scattered decomposition is its *stability*, meaning that as the computation changes with time (e.g. particles move, clumping occurs, etc.), the scattered decomposition is quite insensitive to these changes and will continue to load balance rather well. So it is possible to get good load balance without having to use time-consuming optimisation techniques such as simulated annealing to obtain good data distribution. Scattered decomposition has proven to be very effective for problems such as adaptive mesh finite element simulations, where the grid is much finer in regions that are changing more rapidly (Morrison & Otto, 1986, Flower *et al.*, 1987), growing a cluster in spin models of magnetism (Coddington & Ko, 1994), and certain matrix problems (Fox *et al.*, 1988b).

For a dynamic problem, the Hamiltonian of (8.16) will vary with (computer) time, and so will its minimum value (the optimal domain decomposition). The operating system will need to redistribute the data periodically to try to keep the system close to the global minimum value. In contrast, the static scattered decomposition presumably corresponds to a stable local minimum of the Hamiltonian that does not change much with time. For highly dynamic problems, the operating system may not be able to 'keep up' perfectly with the computation. In this case the

Hamiltonian that actually matters is not the instantaneous version in (8.16), but a time-averaged Hamiltonian, $\overline{H}$:

$$\overline{H}(t, t_{\text{av}}) = \int_{t}^{t+t_{\text{av}}} H(u)\, du \qquad (8.17)$$

where the averaging time, t_{av} is the time scale for the operating system to find a good domain decomposition. In the earlier terminology, t_{av} is related to the 'temperature' T_{problem} of the complex system. Note that t_{av} and the temperature are in fact characteristics of the problem, not the computer. t_{av} will be smaller for a faster computer; however what is important here is the *relative* time scale of the operating system – the time taken to compute a new domain decomposition relative to the time to do the computations between redistributing the data. Increasing the speed of the computer will decrease the time for both these tasks in roughly equal proportion, so the relative time scale will remain about the same.

An interesting point is that, in terms of $\overline{H}$, the better decomposition may actually be the scattered one. Because of the rapid shifting of the optimal decomposition as a function of time, the minimum of $\overline{H}$ corresponding to this will be raised upwards, while the scattered minimum will remain approximately the same. Two possible scenarios develop – the minima may or may not cross. Depending upon the parameters of the problem and upon the hardware characteristics of the parallel machine, a 'phase transition' may occur whereby the scattered decomposition actually becomes the better decomposition for $\overline{H}$.

The relative importance of the two terms in (8.16) is governed by the ratio $t_{\text{comm}}/t_{\text{calc}}$ introduced in §8.4.2. This plays the role of a 'coupling constant' or 'interaction strength' J, such as occurs in Hamiltonians for spin glasses and other spin models of magnetism. J increases in size as the communication performance of the hardware decreases. The scattered decomposition is favoured as either the coupling J decreases, which means communications are relatively fast so the non-locality of the data is not a problem, or as the averaging time t_{av} increases. Large t_{av} corresponds to rapidly varying problems which the operating system finds hard to equilibrate. In the earlier terminology, large t_{av} means high 'temperature' complex systems.

Thus, as we increase J or decrease problem temperature, we transition from a high temperature phase, where scattered decomposition is optimal, to a low temperature phase where standard domain decomposition is optimal. This is of course analogous to a statistical mechanics system having a phase transition separating a high temperature disordered state and a low temperature ordered state. Which phase or data decomposition is relevant depends on the properties of both the computer architecture ($J \sim t_{\text{comm}}/t_{\text{calc}}$) and the problem architecture ($t_{\text{av}} \sim T_{\text{problem}}$).

8.5 Temporal properties of complex problems and complex computers

8.5.1 The string formalism for dynamic problems

In the previous section, we thought of a problem (the complex system S_{num} or S_{HLSoft}) as a graph (the computational graph) with vertices labelled by the system member m and edges corresponding to the linkage between members established by the algorithm. This is a good picture for what we called 'adiabatic' problems that change slowly with time. In this case, it makes sense to think of slicing the 'space–time' cylinder formed by the complex system and just consider the computational graph – the spatial structure at fixed time. However, this is not appropriate for asynchronous problems or for loosely synchronous problems that are rapidly varying or *dynamic* – those with high temperature T_{problem} in the language of §8.4.4. For such problems, the operating system cannot 'keep up' with the variation of the computational graph – the graph changes significantly over the time period that the operating system takes to partition the computational graph.

In adiabatic problems, our physical analogy was that of members mapped to particles interacting by forces given by the member interconnect. One might imagine that a reasonable analogy for *dynamic* problems would be to add a 'kinetic energy' term to give time dependence to the member positions, however it is not clear how to do this. Rather, we change the analogy so that members are mapped to 'strings' representing their *world lines*, that is, their path through the space–time of the complex computer. At computer time t, the complex system member m is located at position $\underline{x}_m(t)$. $\underline{x}$ is a position in the complex computer space. At its simplest $\underline{x}$ is just a node number, but we can look at a finer resolution and consider $\underline{x}$ as a position in the global computer memory. This allows one, in principle at least, to set up a formalism to study the full memory hierarchy of the system including caches and register use. Each member now corresponds not just to a position $\underline{x}_m$ but to a world line $\{\underline{x}_m(t)\}$. The execution time T_{par} on a parallel machine is a functional of the world lines

$$T_{\text{par}} \equiv T_{\text{par}}\left(\{\underline{x}_0(t)\}\ldots\ldots\{\underline{x}_m(t)\}\ldots\right). \tag{8.18}$$

The structure of the original dynamic complex system leads to an expression for (8.18) that is similar to the simpler (8.16). There is a repulsive force between world lines corresponding to load balancing. There is an attractive force corresponding to the dynamic interconnection between the members m. The details of this depend on the relation between clock time t and the simulation time t_m of each member m.

The most straightforward approach to minimise T_{par} would be simulating annealing with the basic 'move' being a change $\{\underline{x}_m(t)\} \to \{\underline{x}_m(t)\}'$ which is typically local in both $\underline{x}$ and t. This gives a formalism similar to quantum chemistry or lattice gauge theories. One can also use an optimisation method based on neural networks. These points are described in greater detail in §8.7.

We have applied these ideas to message routing in a network (Fox & Furmanski, 1988b), and more generally to combining networks which implement global reduction formulae such as forming a set of sums

$$y_j = \sum_i M_{ji} x_i \tag{8.19}$$

where y_j, M_{ji}, and x_i are all distributed over the nodes of a parallel computer.

A very preliminary examination was given in (Fox & Koller, 1989) of the application of these ideas to register allocation for compilers. We have explored more deeply the application of these methods to multi-vehicle navigation (Fox et al., 1988a, Fox, 1990a). In that case $\{\underline{x}_m(t)\}$ is the path of vehicle m in a two or three dimensional space with m at position $\underline{x}_m$ at time t.

8.5.2 Memory hierarchy

Modern workstations have heirarchical memory, formed by the cache and main memory. Obtaining good performance from these computers requires minimising cache misses, so that data is referenced from cache and not main memory, which can be an order of magnitude slower. This is often referred to as the need for 'data locality'. This makes clear the analogy with distributed memory parallel computers, where data locality is needed to minimise communication between processors.

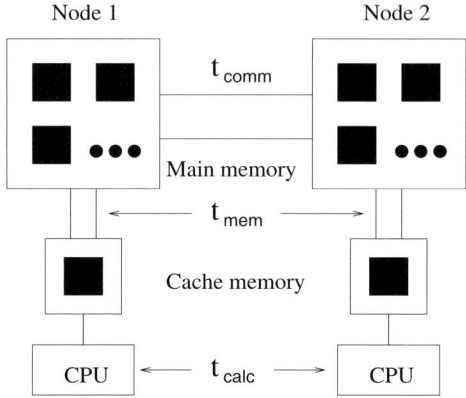

Fig. 8.7. The fundamental time constants of a heirarchical memory parallel computer.

There is one essential difference between cache and distributed memory. Both need data locality, but in the parallel case the basic data is static and fetches additional information as necessary. This gives the familiar surface-over-volume communications overheads of (8.7). However, in the case of a cache, all data must stream through it, not just the data needed to provide additional information. We can use our space–time picture of computation to view the data streaming through heirarchical memory as a distribution of data in the temporal direction.

Let us introduce a new time constant, t_{mem}, which is the time it takes to load a word into cache. This is illustrated in fig. 8.7. The cache overhead has exactly the same form as the communication overhead in (8.7), if we simply replace t_{comm} by t_{mem} and n_{num} for n_{time}, where n_{time} is the temporal blocking factor, or the number of iterations in the problem between cache flushes. The overhead is a surface-over-volume effect just as for a distributed memory machine, but now the surface is in the temporal direction, and the volume is that of a domain in both space and time.

It is remarkable that t_{mem}, time, and memory hierarchy are completely analogous to t_{comm}, space, and distributed memory. In particular, the well-known methods for improving the performance of caches and registers correspond to blocking (clumping) the problem in its time direction. This is analogous to blocking the problem in space to improve performance on a distributed memory parallel machine.

High performance computer architectures exploit data locality with a memory hierarchy implemented either as a multilevel cache and/or with distributed memory on a parallel machine. Good use of cache requires blocking in time; good use of distributed memory requires blocking in space. In general, full space–time blocking is required to give a universal implementation of data locality that will lead to good performance on both distributed and heirarchical memory machines. This strategy is used in the implementation of the BLAS-3 primitives in LAPACK (Demmel, 1991).

The directives in High Performance Fortran essentially specify data locality, so we believe that an HPF compiler can use the concepts of this section to optimise cache use on heirachical memory machines. Thus, HPF should provide good performance on all high-performance computers, not just parallel machines.

8.6 Problem architectures and parallel software

In a series of papers (Angus *et al.*, 1990, Fox, 1991b, Fox, 1991d), we have developed a qualitative theory of the architectures of problems, analogous to the well-known classification of parallel computer architectures into SIMD and MIMD. This is summarised in table 8.1, which introduces five general problem classes. Let us return to the concept of fig. 8.4 – namely, computation is a map between problem and computer, and software is an expression of this map. We have explored in depth this concept of problem architecture and its use for clarifying which problems run well on SIMD machines and which on MIMD. One can also understand which problem classes parallelise naturally on massively parallel machines. Here, we just describe the consequences for software, which are summarised in table 8.1.

We believe that successful software models will be built around problem and not machine architecture. We see that some of the current languages – both old and new – are flawed because they do not use this principle in their design. The language often reflects artifacts of a particular machine architecture and this naturally leads to non-portable codes that can only be run on the machine whose architecture is expressed by the language. On the other hand, if the language expresses properly the problem structure, then a good compiler should be able to map it into a range of computer architectures.

We can illustrate this with Fortran 77, which we can view as embodying the architecture of a sequential machine. Thus, software written in

Table 8.1. *Architectures for five problem classifications*

Synchronous: data parallel
> Tightly coupled. Software needs to exploit features of problem structure to get good performance. Comparatively easy, as different data elements are essentially identical.
> *Candidate software paradigms: High Performance Fortran, Parallel Fortran 77D, Fortran 90D, CMFortran, Crystal, APL, C++.*

Loosely synchronous: data parallel
> As above but data elements are not identical. Still parallelises due to macroscopic time synchronisation.
> *Candidate software paradigms: may be extensions of the above, however C or Fortran with message passing is currently the only guaranteed method.*

Asynchronous
> Functional (or data) parallelism that is irregular in space and time. Often loosely coupled and so need not worry about optimal decompositions to minimise communication. Hard to parallelise, not usually scalable to large numbers of processors.
> *Candidate software paradigms: PCN, Linda, object-oriented approaches.*

Embarrassingly parallel
> Independent execution of disconnected components.
> *Candidate software paradigms: Many approaches work – PVM, PCN, Linda, Network Express, ISIS, etc.*

Metaproblems
> Asynchronous collection of loosely synchronous components where the component program modules can be parallelised.
> *Candidate software paradigms: PCN, Linda, ADA, controlling modules written in synchronous or loosely synchronous fashion.*

Fortran 77 maps the space–time structure of the original complex system into a purely temporal or control structure. The spatial (data) parallelism of the problem becomes purely temporal in the software, which implements this as a sequential loop over the data (a DO loop in Fortran 77). Somewhat perversely, a parallelising compiler tries to convert the temporal structure of a DO loop back into spatial structure to allow concurrent execution on a spatial array of processors. Often parallelising compilers produce poor results as the original map of the problem into sequential Fortran 77 has 'thrown away' information necessary to reverse this map and recover unambiguously the spatial structure. The first (and some ongoing) efforts in parallelising compilers tried to directly 'parallelise the DO loops'. This seems doomed to failure in general as it does not recognise that in nearly all cases the parallelism comes from spatial and not control (time) structure. Thus, as described in §8.2.2.4, we are working on the development of a parallelising compiler for Fortran D and High Performance Fortran, where the user adds additional information to tell the compiler about the spatial structure (Fox, 1990b, Fox, 1991b, Fox, 1991d, Fox *et al.*, 1991, Wu & Fox, 1991, Bozkus *et al.*, 1993). We are optimistic that this project will be successful for the synchronous and loosely synchronous classes defined in table 8.1.

Most languages do not express and preserve space time structure. Array languages such as APL and Fortran 90 are examples of data parallel languages that at least partially preserve the space time structure of the problem in the language. Appropriate class libraries can also be used in C++ to achieve this goal. We expect that development of languages which better express problem structure will be essential to get good performance with an attractive user environment on large scale parallel computers. The results in §8.5.2 show that data locality is critical in sequential high performance (hierarchical memory) machines as well. Thus, we would expect that the use of languages that properly preserve problem structure will lead to better performance on all computers.

8.7 Physical computation and optimisation

Physical computation can be loosely classified as the use of physical analogies or methods from the physical sciences in computational studies of general complex systems (Fox, 1991e). One example is the use of simulated annealing (an idea from physics) for optimisation problems such as chip routing and placement (Kirkpatrick *et al.*, 1983). Another is the use of neural networks (an idea from biology) in learning and pattern recognition for problems in computer vision and robotics (see chapter 10).

Optimisation is a particularly important application of physical computation. Simic has introduced the term *physical optimisation* to describe the many different optimisation methods of this kind (Simic, 1990, Fox, 1991a). It is not surprising that techniques based in the physical sciences are useful for solving optimisation problems, since most laws of physics can be formulated variationally as optimisation problems, many physical systems act so as to minimise energy or free energy, and evolution in nature is also involved in optimisation.

As mentioned in §8.4.4, physical optimisation techniques such as simulated annealing, neural networks and genetic algorithms can be usefully applied to domain decomposition and load balancing, an important optimisation problem in parallel computation. However these methods can be used to tackle general optimisation problems, and indeed have successfully been applied to a wide variety of problems. Physical optimisation methods can be contrasted with other methods for optimisation: heuristics can be considered as an approach motivated by the problem; maximum entropy or information theory as approaches from electrical engineering; combinatorial optimisation methods from mathematics; and linear programming and rule-based expert systems from computer science.

There is no universally good approach to optimisation. Each method has different tradeoffs in robustness, accuracy, speed, suitability for parallelisation, and problem size dependence. For instance, neural networks do simple things on large data sets and parallelise easily, whereas expert systems do complex things on smaller data sets and are difficult to parallelise. Parallel algorithms for physical optimisation methods are not usually trivial, and are not always similar to the sequential algorithms.

The nature of the problem is very important in terms of which method is most suitable. For instance, what is the shape of the cost function? Are the local minima deep or shallow, wide or narrow, relatively few or numerous? Are the minima correlated or uncorrelated? Does the problem require the exact global minimum, or is a good approximate minimum sufficient? Physical optimisation methods try to find good approximate solutions, not necessarily the best solution. They work well for many complex real-world problems, for which approximate solutions are all that is required, and indeed all that is warranted by the usually imprecise data or models. Also, many of these problems are NP-complete, so that only approximate solutions are feasible given a limited computational resource.

Here we briefly describe some physical optimisation methods. More detailed reviews can be found in (Fox, 1991a, Fox, 1992a).

8.7.1 Simulated annealing

Simulated annealing is a very general optimisation method that stochastically simulates the slow cooling of a physical system to its ground state (Kirkpatrick, 1983, Otto, 1989 and Laarhoven and Aarts, 1987). The cost function for the problem is viewed as an energy function, and a parameter T analogous to temperature is introduced. The algorithm works by using an iterative Monte Carlo technique, that is, by proposing changes to the state of the system, and either accepting or rejecting a change using the Metropolis criterion – if the cost (energy) is decreased, the change is accepted; if the energy is increased by δE, it is accepted with a probability $\exp(-\delta E/T)$. The process is started at a high temperature where almost all proposed changes are accepted, and the temperature is gradually reduce to zero, where changes are only accepted if they decrease the energy. The zero temperature algorithm is just the *greedy* or *hill-climbing* algorithm, which works poorly in most cases since it can get trapped in local minima. Simulated annealing at non-zero T allows the system to probabilistically increase the energy and thus escape from local minima.

The rate of cooling is crucial to the performance of the algorithm. It can be shown that if the temperature is decreased slowly enough (logarithmically), then the global optimum will be found (with enough trials, since this is a probabilistic method). The basic idea is the same as real annealing (for example, of steel) – if the temperature changes are small enough, the system can maintain thermal equilibrium throughout the procedure, so it will finish up in the zero temperature equilibrium state, i.e. the ground state or global optimum. However a logarithmic cooling schedule is much too slow for most problems, so usually a faster (exponential) cooling schedule is used. For many problems this will still keep the system close enough to equilibrium that it will be very near the ground state energy at zero temperature, and thus find a near-optimal solution. One of the main problems with simulated annealing is that finding a good cooling schedule is generally a trial-and-error procedure. Some advances have been made in finding *adaptive* annealing schedules, where the temperature is reduced depending on the measured values of the energy for the particular problem. Theoretically this promises a great improvement in performance, however in practice it is often difficult to find the optimal cooling schedule given the limited number of measurements available.

Another critical part of the algorithm is the choice of the method for updating the system state. If large changes are made to the system variables, then the energy change δE will generally be large, which will result in most of the proposed changes being rejected. If the changes at each iteration are chosen to be small, so that most of them are accepted, it will take many iterations to reach a very different, uncorrelated state of the system. A tradeoff is required, since in both these cases moving through the search space will be very slow. The method can be greatly improved if a way can be found to make substantial changes to the system without changing the energy too much.

Simulated annealing is popular because it is simple and it can be easily applied to any optimisation problem. However it may not be effective unless a good update method and cooling schedule are used.

8.7.2 Deterministic annealing

This approach is similar to simulated annealing, however instead of using a stochastic (Monte Carlo) approach, a simple heuristic is used to minimise the free energy $F = E - TS$ at each temperature T, where E is the cost function (or energy) and S is analogous to the entropy of a physical system. Notice that at zero temperature the energy and free energy are equivalent. The free energy is formally defined as $F = -T \log Z$, where

$$Z = \sum_{\text{states } C} \exp\left[-E(C)/T\right] \qquad (8.20)$$

is known as the partition function in statistical mechanics. This approach is similar to methods used in quantum chemistry to find the ground state of a complex molecule.

Deterministic annealing has been used very effectively in data clustering problems (Rose et al., 1990). For the simple case of grouping points in space into clusters based on distance, it is possible to construct an energy function for which the minimum of the free energy can be computed deterministically by iteratively solving an implicit equation. This particular example has an interesting temperature dependence. At high temperature the points will all be in a single cluster. As the temperature is decreased, the points will split into two clusters, then three, and so on, at various critical temperatures. The temperature is related to the size of the clusters, or the distance scale at which the system is observed. For a given problem we will need to specify a particular minimum distance scale or temperature.

8.7.3 Neural networks

The use of neural networks for optimisation was introduced for the travelling salesman problem in (Hopfield & Tank, 1986), and although the method is not very effective for this application, the method and basic ideas are important for a range of problems. The travelling salesman problem (TSP) is the classic NP-complete discrete optimisation problem, for which the salesman has to find a tour that minimises the distance travelled in visiting a given set of cities. We introduce neural variables η_p^i that are 1 if the ith step of the tour passes through city p, and zero otherwise. The cost function can be written very simply in terms of these neural variables, however an extra penalty term needs to be added to implement the constraint that only one of the η_p^i can be non-zero for a given i or p. The cost function then looks like a statistical physics problem, with 'spins' η_p^i governed by an 'energy function'. This formulation of the TSP can in fact be solved using simulated annealing, however the neural network approach uses a faster approximate method, similar to a mean-field approximation in statistical physics. Unfortunately this approximate method does not work well for even a modest number of cities (Wilson & Pawley, 1988).

This general approach has however been very effective for a number of other problems, including load balancing a parallel computer (Fox & Furmanski, 1988a). In this case, we can introduce neural variables just as for the TSP, except that η_p^i is now 1 if data element i is assigned to processor p. However this will give the same problems as found for the TSP. Instead, we define the neural variables by the binary decomposition of the processor number

$$p = \sum_{k=0}^{d-1} 2^k \eta_k^i \qquad (8.21)$$

where there are M data elements and $N = 2^d$ processors.

The $Md = M \log_2 N$ neural variables provide a non-redundant specification of the data decomposition, compared to the MN redundant variables in the TSP-like formulation. This approach was obviously motivated by parallel computers with a hypercube topology, however it can be used for an arbitrary topology.

Using non-redundant variables allows us to construct an energy (cost) function for data decomposition along the lines of (8.16)

$$E = E_{\text{calc}} + E_{\text{comm}} \quad \text{where}$$

$$E_{\text{calc}} = \frac{1}{N} \sum_{m,m'} Calc\,(m)\, Calc\,(m') \prod_{k=0}^{d-1} [1 + s_k(m)\, s_k(m')]$$

$$E_{\text{comm}} = \frac{1}{4} \sum_{m,m'} Comm\,(m, m') \sum_{k=0}^{d-1} [1 - s_k(m)\, s_k(m')] \qquad (8.22)$$

with 'spins' $s_k(m) = 2\eta_k^m - 1$ taking the values ± 1. In this case, the energy function has no constraint (penalty) terms, and the Hopfield–Tank method works extremely well, being comparable in quality to simulated annealing results but much faster, since a deterministic mean-field approximation is used (Williams, 1991). This indicates that the problems found in using neural networks for optimisation lie not with the method, but with the choice of variables and the necessity of introducing penalty terms.

We have also used neural network optimisation successfully for optimising compilers (Fox & Koller, 1989, Fox *et al.*, 1989), and robotic vehicle navigation (Fox *et al.*, 1988a, Fox, 1990a).

8.7.4 Elastic net

The elastic net was introduced as a physically based approach that outperformed the neural network method for TSP (Durbin & Willshaw, 1987). The basic idea behind the elastic net is to 'invent' a physical system whose equilibrium state is the desired optimum. In the case of the TSP, we consider an elastic string with beads for each city. The beads are attracted to each other with a simple elastic force that will try to shrink the length of the path to zero, and thus drive the system towards the minimum path. There is also an attractive force between the beads and the cities that drives the system towards enforcing the constraint that the tour must pass through each city. The comparative strength of these two competing forces is a parameter similar to temperature in annealing. We start with the elastic force being dominant, and then slowly change the forces until finally the bead-city force is dominant so that we end up with a valid (hopefully near-optimal) tour.

Simic has shown an interesting relation between neural networks and elastic nets for the TSP (Simic, 1990). Both correspond to deterministic annealing using similar mean field approximations, but with different choices of degrees of freedom and thus different constraints.

8.7.5 Genetic algorithms

Genetic algorithms are based on evolutionary processes in biology. The basic idea is to encode the system parameters as 'genes' which make up a set of 'chromosomes', each describing a different state of the system. For example, in load balancing a multicomputer, each gene would specify the processor number for a specific node in the computational graph. We start with a base population of chromosomes that undergo changes due to the application of genetic operators such as crossover and mutation. Crossover is like mating, in that the genes of two individual chromosomes are randomly combined to form a new individual. Mutation occurs by randomly changing a gene in the chromosome. Once a new population is formed by these operations, it is compared with the old population, with each chromosome being assigned a 'fitness' (the cost for the optimisation problem). Only the fittest individuals are retained for the next generation ('survival of the fittest').

Although genetic algorithms provide an interesting and often very effective optimisation method, obtaining good performance usually requires very careful mapping of the problem variables onto the genes, a good choice of genetic operators, and a lot of tuning of parameters in the algorithm. This is analogous to the necessity for choosing good problem-specific update moves and cooling schedules for simulated annealing. Another problem with genetic algorithms is that there is no natural way to decide when a good solution has been reached and the process should be stopped, so it is usually followed by some post-processing using a hill-climbing technique or other fast heuristic.

8.7.6 Simulated tempering

A variation of simulated annealing, known as *simulated tempering*, has recently been introduced and used to study certain types of lattice spin models (Marinari & Parisi, 1992). Tempering differs from annealing in two ways: it allows both heating and cooling of the system; and it keeps the system in equilibrium when the temperature is changed. Both of these changes are beneficial for optimisation. The first allows for 'reheating' when the annealing gets stuck in local minima at low temperatures. This is currently often used as an ad-hoc addition to standard simulated annealing. The second takes care of one of the main difficulties in using simulated annealing, which is coming up with a cooling schedule that is not too slow but keeps the system close to equilibrium as the temperature goes to zero. We are currently working on applying this new method to general optimisation problems.

8.8 Conclusion

We have found that using ideas and techniques from the realms of complex systems and the physical sciences can provide useful and powerful insights into parallel computing and computer science, and in particular the efficient use of parallel computers for problems in computational science. However many of the ideas presented here are well out of the mainstream of research in computer science, and have attracted little attention. This is perhaps due to the general unfamiliarity of the computer science community with many of these methods and concepts from the physical sciences.

There are interesting synergies between computer science, computational science, the physical sciences, and the theory of complex systems. The development of models of computation and general optimisation techniques seem particularly suited to an interdisciplinary approach drawing from all these areas.

Acknowledgements

This work has been collaborative with many people, especially W. Furmanski and S. Otto. We would like to thank them and our colleagues in the Caltech Concurrent Computation Program (C3P) and the Northeast Parallel Architectures Center (NPAC) for their insight and help.

This work was supported by the Center for Research on Parallel Computation (CRPC) with National Science Foundation Cooperation Agreement No. CCR-9120008. The US Government has certain rights in this material.

Bibliography

Angus, I. G., Fox, G. C., Kim, J. S., & Walker, D. W. (1990). *Solving Problems on Concurrent Processors.* Vol. 2. Prentice Hall, Englewood Cliffs, NJ.

Bordewaker, R., & Choudhary, A. (1994). *HPFio: HPF with Parallel I/O Extensions.* Tech. rept. NPAC SCCS-613.

Bozkus, Z., Choudhary, A., Fox, G., Haupt, T., & Ranka, S. (1993). *Fortran 90D/HPF Compiler for Distributed Memory MIMD Computers: Design, Implementation, and Performance Results.* Tech. rept. NPAC SCCS-498.

Choudhary, A., Fox, G., Ranka, S., Hiranandani, S., Kennedy, K., Koelbel, C., & Saltz, J. (1992a). Software Support for Irregular and Loosely Synchronous Problems. *Pages 43–52 of: Computing Systems in Engineering*, vol. 3. Pergamon Press, Oxford. NPAC Technical Report SCCS-297b, CRPC-TR92258.

Choudhary, A., Fox, G., Ranka, S., & Haupt, T. (1992b). Which Applications Can Use High Performance Fortran and FortranD — Industry Standard Parallel Languages? *Proc. Fifth Australian Supercomputer Conference.* NPAC Technical Report SCCS-360.

Coddington, P. D., & Ko, S.-H. (1994). *Parallel Wolff Cluster Algorithms*. Tech. rept. SCCS-619. NPAC.

Demmel, J. (1991). LAPACK: A portable linear algebra library for high-performance computers. *Concurrency: Practice and Experience*, **3**, 655–666.

Denning, P. J., & Tichy, W. F. (1990). Highly parallel computation. *Science*, **250**, 1217.

Donath, W. E. (1979). Placement and average interconnection lengths of computer logic. *IEEE Trans. Circuits and Systems*, **16**, 272.

Durbin, R., & Willshaw, D. (1987). An analogue approach to the traveling salesman problem using an elastic net method. *Nature*, **326**, 689–691.

Flower, J., Otto, S., & Salama, M. (1987). Optimal mapping of irregular finite element domains to parallel processors. *Proc. Symposium on Parallel Computations and Their Impact on Mechanics, ASME Winter Meeting.* Caltech Technical Report C3P-292b.

Fox, G., & Otto, S. (1986). Concurrent computation and the theory of complex systems. *Hypercube Multiprocessors*, 244–268. Caltech Technical Report C3P-255.

Fox, G., Otto, S. W., & Umland, E. A. (1985). Monte Carlo physics on a concurrent processor. *J. Stat. Phys.*, **43**, 1209. Caltech Technical Report C3P-214.

Fox, G. C. (1987). Questions and unexpected answers in concurrent computation. *Pages 97–121 of:* Dongarra, J. J. (ed), *Experimental Parallel Computing Architectures.* Elsevier Science, North-Holland, Amsterdam. Caltech Technical Report C3P-288.

Fox, G. C. (1988a). A review of automatic load balancing and decomposition methods for the hypercube. *Pages 63–76 of:* Schultz, M. (ed), *Numerical Algorithms for Modern Parallel Computer Architectures.* Springer-Verlag, Berlin. Caltech Technical Report C3P-385b.

Fox, G. C. (1988b). The hypercube and the Caltech Concurrent Computation Program: A microcosm of parallel computing. *Pages 1–40 of:* Alder, B. J. (ed), *Special Purpose Computers.* Academic Press, New York. Caltech Technical Report C3P-422.

Fox, G. C. (1989). Parallel Computing. *The Encyclopedia of Physical Science and Technology 1991 Yearbook.* Academic Press, New York. Caltech Technical Report C3P-830.

Fox, G. C. (1990a). *Applications of the Generalized Elastic Net to Navigation.* Caltech Technical Report C3P-930.

Fox, G. C. (1990b). Hardware and software architectures for irregular problem architectures. *Pages 125–160 of:* Mehrotra, P., Saltz, J., & Voigt, R. (eds), *Unstructured Scientific Computation on Scalable Microprocessors.* Scientific and Engineering Computation Series. MIT Press, Cambridge, Mass. NPAC Technical Report SCCS-111, CRPC-TR91164.

Fox, G. C. (1991a). Approaches to physical optimization. *Pages 153–162 of:* Dongarra, J., & et al. (eds), *Proc. 5th SIAM Conference on Parallel Proccesses for Scientific Computation.* SIAM, Philadelphia. Caltech Technical Report C3P-959, Technical Report

SCCS-92, CRPC-TR91124.

Fox, G. C. (1991b). Fortran D as a portable software system for parallel computers. *Proc. Supercomputing USA/Pacific 91, Santa Clara, Calif.* NPAC Technical Report SCCS-91, CRPC-TR91128.

Fox, G. C. (1991c). Achievements and prospects for parallel computing. *Concurrency: Practice and Experience*, **3**, 725–739. Caltech Technical Report C3P-927b, NPAC Technical Report SCCS-29b, CRPC-TR90083.

Fox, G. C. (1991d). *The Architecture of Problems and Portable Parallel Software Systems.* Tech. rept. NPAC SCCS-134, Revised SCCS-78b.

Fox, G. C. (1991e). Physical computation. *Concurrency: Practice and Experience*, **3**, 627–653. Caltech Technical Report C3P-928b, NPAC Technical Report SCCS-2b, CRPC-TR90090.

Fox, G. C. (1992a). Lessons from Massively Parallel Applications on Message Passing Computers. *Proc. 37th IEEE International Computer Conference, San Francisco, Calif.* NPAC Technical Report SCCS-214.

Fox, G. C. (1992b). The use of physics concepts in computation. *Chap. 3 of:* Huberman, B. A. (ed), *Computation: The Micro and the Macro View.* World Scientific Publishing Co., River Edge, NJ. Caltech Technical Report C3P-974, NPAC Technical Report SCCS-237, CRPC-TR92198.

Fox, G. C., & Furmanski, W. (1988a). Load balancing loosely synchronous problems with a neural network. *Pages 241–278 of:* Fox, G. C. (ed), *Proc. Third Conference on Hypercube Concurrent Computers and Applications*, vol. 1. ACM Press, New York. Caltech Technical Report C3P-363b.

Fox, G. C., & Furmanski, W. (1988b). The physical structure of concurrent problems and concurrent computers. *Pages 55–88 of:* Elliott, R. J., & Hoare, C. A. R. (eds), *Scientific Applications of Multiprocessors.* Prentice Hall, Englewood Cliffs, NJ. Caltech Technical Report C3P-493.

Fox, G. C., & Koller, J. G. (1989). Code generation by a generalized neural network: General principles and elementary examples. *Parallel and Distributed Computing*, **6**(2), 388–410. Caltech Technical Report C3P-650b.

Fox, G. C., Furmanski, W., Ho, A., Koller, J., Simic, P., & Wong, Y. F. (1988a). Neural networks and dynamic complex systems. *Proc. 1989 SCS Eastern Conference, Tampa, Florida.* Caltech Technical Report C3P-695.

Fox, G. C., Johnson, M. A., Lyzenga, G. A., Otto, S. W., Salmon, J. K., & Walker, D. W. (1988b). *Solving Problems on Concurrent Processors.* Vol. 1. Prentice Hall, Englewood Cliffs, NJ.

Fox, G. C., Furmanski, W., & Koller, J. (1989). The use of neural networks in parallel software systems. *Mathematics and Computers in Simulation*, **31**, 485–495. Caltech Technical Report C3P-642b.

Fox, G. C., Hiranandani, S., Kennedy, K., Koelbel, C., Kremer, U., Tseng, C.-W., & Wu, M.-Y. (1991). *Fortran D Language Specification.* Tech. rept. NPAC SCCS-42c, CRPC-TR90079. Unpublished.

Fox, G. C., Messina, P. C., & Williams, R. D. (1994). *Parallel

Computing Works. Morgan Kaufmann Publishers, San Francisco.
Furmanski, W. (1992). *MOVIE – Multitasking Object-oriented Visual Interactive Environment.* Tech. rept. NPAC SCCS-353.
Furmanski, W., Faigle, C., Haupt, T., Niemiec, J., Podgorny, M., & Simoni, D. A. (1993). Movie model for open systems based high performance distributed computing. *Concurrency: Practice and Experience*, **5**, 287–308.
High Performance Fortran Forum. (1993). *High Performance Fortran Language Specification.*
Hillis, W. D. (1985). *The Connection Machine.* MIT Press, Cambridge, Mass.
Hillis, W. D., & Steele, G. (1986). Data parallel algorithms. *Commun. ACM*, **29**, 1170.
Hopfield, J. J., & Tank, D. W. (1986). Computing with neural circuits: a model. *Science*, **233**, 625.
Kirkpatrick, S., Gelatt, C. D., & Vecchi, M. P. (1983). Optimization by simulated annealing. *Science*, **220**, 671–680.
Koelbel, C. H., Loveman, D. B., Schreiber, R. S., Steele Jr., G. L., & Zosel, M. E. (1994). *The High Performance Fortran Handbook.* MIT Press, Cambridge, Mass.
Landman, B. S., & Russo, R. L. (1971). On a pin versus block relationship for partitions of logic graphs. *IEEE Trans. Comp.*, **C20**, 1469.
Mandelbrot, B. (1979). *Fractals: Form, Chance, and Dimension.* Freeman, San Francisco.
Mansour, N., & Fox, G. C. (1993). Allocating data to multicomputer nodes by physical optimization algorithms for loosely synchronous computations. *Concurrency: Practice and Experience.* NPAC Technical Report SCCS-350.
Marinari, E., & Parisi, G. (1992). Simulated tempering: A new Monte Carlo scheme. *Europhys. Lett.*, **19**, 451. NPAC Technical Report SCCS-241.
Message Passing Interface Forum. (1994). *MPI: A Message Passing Interface Standard.* Available via anonymous ftp from ftp.mcs.anl.gov in the directory pub/mpi.
Messina, P. (1991). Parallel computing in the 1980s – one person's view. *Concurrency: Practice and Experience*, **3**, 501–524. Caltech Technical Report CCSF-4-91.
Morrison, R., & Otto, S. (1986). The scattered decomposition for finite element problems. *Journal of Scientific Computing*, **2**, 59–76. Caltech Technical Report C3P-286.
Otten, R. H. J. M., & van Ginneken, L. P. P. P. (1989). *The Annealing Algorithm.* Kluwer Academic Publishers, Dordrecht.
Ponnusamy, R., Saltz, J., & Choudhary, A. (1993). *Runtime Compilation Techniques for Data Partitioning and Communication Schedule Reuse.* University of Maryland Technical Report UMIACS-TR-93-32.
Rose, K., Gurewitz, E., & Fox, G. C. (1990). A deterministic annealing approach to clustering. *Pattern Recognition Letters*, **11**, 589–594. Caltech Technical Report C3P-857.
Simic, P. (1990). Statistical mechanics as the underlying theory of

'elastic' and 'neural' optimizations. *Network*, **1**, 89–103. Caltech Technical Report C3P-787.

van Laarhoven, P. J. M., & Aarts, E. H. L. (1987). *Simulated Annealing: Theory and Applications*. Kluwer Academic Publishers, Dordrecht.

Williams, R. D. (1991). Performance of dynamic load balancing algorithms for unstructured mesh calculations. *Concurrency: Practice and Experience*, **3**, 457–481. Caltech Technical Report C3P-913b.

Wilson, G. V., & Pawley, G. C. (1988). On the stability of the travelling salesman problem algorithm of hopfield and tank. *Biol. Cybern.*, **58**, 63–70.

Wu, M., & Fox, G. C. (1991). *Fortran 90D Compiler for Distributed Memory MIMD Parallel Computers*. NPAC Technical Report SCCS-88b, CRPC-TR91126.

9
Are ecosystems complex systems?

ROGER H. BRADBURY[1], DAVID G. GREEN[2] & NIGEL SNOAD[3]

[1] National Resource Information Centre,
PO Box E11 Queen Victoria Terrace ACT 2600, AUSTRALIA,
[2] Charles Sturt University, Albury NSW 2640, AUSTRALIA
[3] Australian National University, ACT 0200, AUSTRALIA

9.1 Introduction

It is the living world that first confronts us with real complexity. From coral reefs to rainforests, complex patterns are evident everywhere in the living world. Their intricacy and beauty has always astounded us.

Ecology – the study of plants and animals in their environment – emerged as a distinct discipline at the beginning of the twentieth century. Its subject is nothing less than the complexity of the living world itself. Ecology attempts to link patterns to processes. The hope is that hidden within the complexity living systems lie patterns that will enable us to understand the biological processes involved.

In this brief introduction to complexity in ecology we introduce some of the key issues, especially those that relate most closely to the subject of complexity in general.

9.2 What is ecological complexity?

Ecological complexity arises from several sources. These include:

- the sheer abundance of plants and animals – both in numbers and kinds;
- the richness of the interactions between individuals and populations;
- the wide variations in environmental conditions that affect organisms and
- the many different kinds of processes that occur in nature.

Ecologists have taken a variety of approaches to dealing with complexity. Below we look briefly at some of these.

9.2.1 Why is taxonomy relevant?

Our first response to complexity is taxonomy. Whatever the subject we attempt to order and classify phenomena so that we may see a pattern. It is not surprising that taxonomy grew out of biology. Systematic biology – the science of classifying species – represents perhaps the oldest tradition in our attempts to understand the complexity of the living world. The founders of systematic biology, from Aristotle to Linnaeus, were really the first to study complexity in its own right.

Taxonomy addresses two main issues. By naming and describing species, taxonomy defines the units that we have to deal with in understanding – and managing – complexity in the living world. Taxonomy also makes statements about the processes by which biological complexity arose. By placing species within higher groups (genera, families, etc.) it expresses theories about evolutionary history.

In recent times, growing world-wide concern about endangered species and other environmental issues has placed taxonomy in the conservation spotlight. Simply naming and listing the world's species is a huge task in itself. The total number of species in the world is perhaps 10 million (estimates range from 2 to 100 million) (World Resources Institute, 1992). Only about 1.4 million of these species have been described. The critical issue is that the task of finding and naming all the world's species could take literally hundreds of years to complete, whereas the conservation problems that require those names need to be solved in a matter of decades at most.

Taken in a more general sense, taxonomy is the process of identifying objects of study. When viewed this way taxonomy is usually the first stage in the development of any science. In particular as ecology began to separate itself from other parts of biology at the turn of the century, it did so by describing and defining ecological phenomena – a taxonomic process. This taxonomy of structure – *phytogeography* – classified (say) vegetation into broad ecological units (McIntosh, 1985). A major part of this work, which continues today, concerns the description, dissection and definition of ecological complexity. We can see a continuous tradition from the early phytogeographers, through the first major synthetic work in ecology – Charles Elton's (Elton, 1927) *Animal Ecology* – to modern work using multivariate statistics to classify biological communities (Pielou, 1984).

9.2.2 What is succession?

Succession is the process of orderly change in the composition of an ecosystem. In a cleared patch of forest, for example, trees that are adapted to disturbed conditions soon reinvade. As they grow the trees create shade and change the soil and moisture regime as well. In doing so they allow other tree species to establish and eventually replace them.

Clements (1905) theory of succession was the first logical system that could be described as ecological theory. The Clementsian theory sought to explain the complexity of ecological phenomena in terms of a holistic super-organism that developed to a climax state controlled by the climate. Clements proposed that any ecosystem has a natural state – the *climax*. This is a community of species that is in balance with its environment. Following any disturbance, such as a fire, an ecosystem passes through an orderly sequence of community types (*succession*) until its equilibrium *climax* state is restored (Clements, 1905) (see fig. 9.1, after Whitaker, 1975).

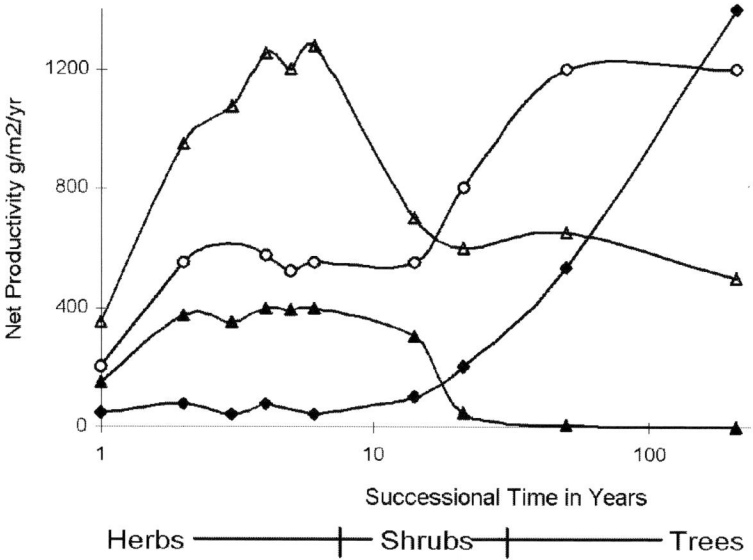

Fig. 9.1. Changes in productivity, biomass, and species diversity during forest succession.

9.2.3 Paradigms for dealing with complexity

Ecologists adopt two main approaches in dealing with the complexity of ecological systems: autecologyand synecology. *Autecology* is the study of how a population of a single species interacts with its environment (fig. 9.2, based on Hutchinson, 1978). *Synecology* is the study of the interactions among communities of species and their environment. This distinction was first made by Schroter in 1896 (Chapman, 1931). In simplistic terms, autecology generally adopts a reductionist approach to ecological systems, whereas synecology takes a holistic approach.

Synecology, exemplified by Clements' idea of the 'super-organism', looks at organisms and populations in relation to whole ecosystems; autecology is more in the naturalist tradition. For instance, the autecological tradition stresses the search for the simpler constituents or

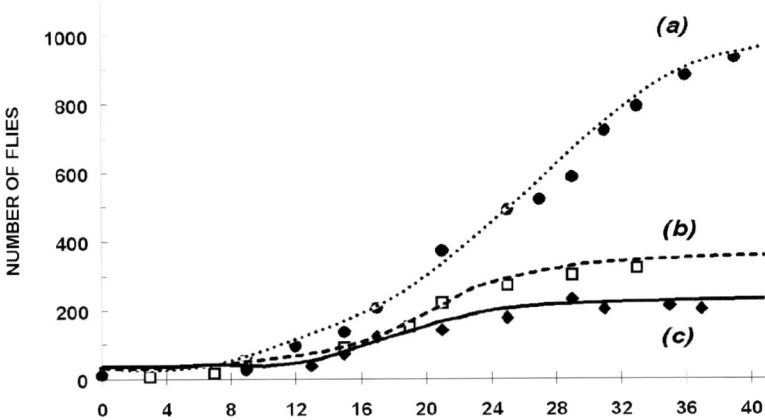

Fig. 9.2. An early example of single species ecology: Measurements of population growth for one species of fruit fly under different environmental conditions, establishing the classic sigmoid growth curve.

components that make up the observed complexity. The holist, synecological tradition stresses the search for simple patterns and processes underlying the observed complexity.

In reviewing Gell-Mann's (1994) book *The Quark and the Jaguar*, Marschall (1994) neatly states the dilemma of those who would study complexity:

Complexity is what financial markets, mammalian immune systems and ecological communities have in common. The ability to interact with the environment, to recognise patterns in the world and to apply acquired knowledge to the modification of future behaviour is easily detected, yet the definition of complexity remains elusive (Marschall, 1994, p. 45).

The idea of ecology as a study of natural complexity has not changed much over the years. Nor has the issue of how to do ecology. We can see this in the startlingly modern thesis of Forbes (1887) in his famous article 'The lake as a microcosm':

The lake is an old and relatively primitive system, isolated from its surroundings. Within it matter circulates, and controls operate to produce an equilibrium comparable with that in a similar area of land. In this microcosm nothing can be fully understood until its relationship to the whole is clearly seen.

Forbes's description could still serve today as a clear statement of a holistic approach to ecology.

There is a continuing debate within ecology about the relative merits of the reductionist and holistic approaches to understanding ecological phenomena. Examples include the controversy over Clements' theory of the community as a super-organism at the turn of the century, the issue of the role of competition in the 1920s and 1930s (Scudo & Ziegler, 1978), the issue of density dependent regulation of population in the middle years of the century (Andrewartha & Birch, 1954), and the multispecies modelling of the 1960s and 1970s (Cody & Diamond, 1975). The common thread in each of these debates is the nature of the proper object of study in ecology.

By the middle of the twentieth century, a strongly empirical single-species ecology made a reductionist autecology the norm. According to this paradigm, the interesting ecological questions could be reduced to five (Andrewartha & Birch, 1954): how was the distribution and abundance of animals affected by the weather, other animals of the same kind, other organisms of different kinds, food, and a place in which to live?

The above paradigm was later challenged by a synecological view, championed first by Macarthur and Wilson (1967) and later by Robert May (1974, 1976b). This view holds that fundamental ecological questions have more to do with species interactions than with the physical determinants of a species' distribution. The manifest complexity of ecosystems is seen as the result of a small set of ecological interactions among species – e.g. predation, competition and symbiosis – and it is assumed that the dynamics of these interactions can be described by relatively simple mathematical models (fig. 9.3).

9.2.4 Food chains and food webs

An important theme in ecology is the cycling of material and energy within ecosystems. Elton, the first ecologist to dissect out the component processes of ecological complexity, identified the *food chain*, the *food web* (which he called the *food cycle*, the *ecological niche*, and the *pyramid of numbers* (fig. 9.4). Elton's focus on what eats what in the community also paved the way for the trophodynamic approach (Lindeman, 1942). For instance, we can view ecosystems as a *trophic pyramid* in which primary producers (e.g. grass) support consumers (e.g. antelope), who in turn support secondary consumers (e.g. lions). Elton's empirical work on the distribution and abundance of animals such as voles and lem-

Are ecosystems complex systems? 345

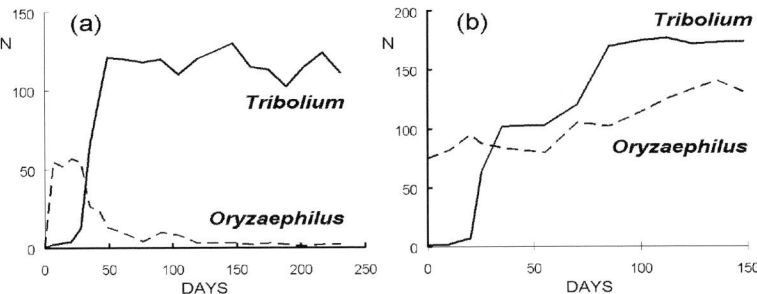

Fig. 9.3. Demonstrating the effects of competition: Crombie (1945) showed that the outcome (a) of competition between two species of flour beetle, *Tribolium*, cultivated together changed (b) when the environment was made mode complex by introducing short lengths of capillary tubing (thus providing a refuge for the weaker competitor) in to the flour (based on Hutchinson,1978).

mings, whose populations often fluctuate violently, established a tradition of patient, long-term field work and did much to dispel then-current ideas of a *balance of nature* (see §9.4). It was thus a precursor to the complexity/stability debate which followed later.

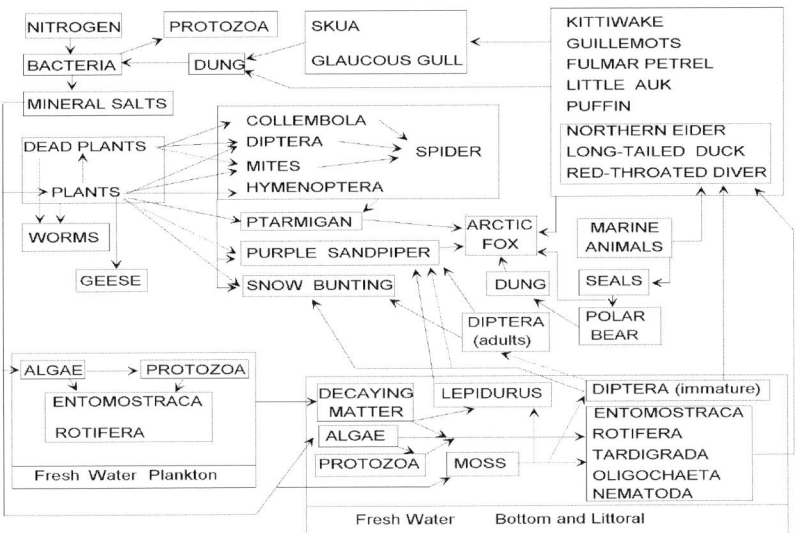

Fig. 9.4. Real-life complexity. Elton's (1927) original diagram for the food web on Bear Island in the Barents Sea (after Summerhayes, 1923).

9.2.5 What is diversity?

Diversity is the idea that ecologists use to capture the complexity of communities. In general, the more varied the mix of species in a community, the greater the potential for complex interactions and processes to occur.

The first, and most obvious measure of diversity is the number of species – *species richness*. However this measure does not always express the variability in species composition. Biological communities are made up of many species: some are common; some rare. In intuitive terms a community with many rare species but dominated by a single species, is less diverse than one with fewer, but more common species (fig. 9.5).

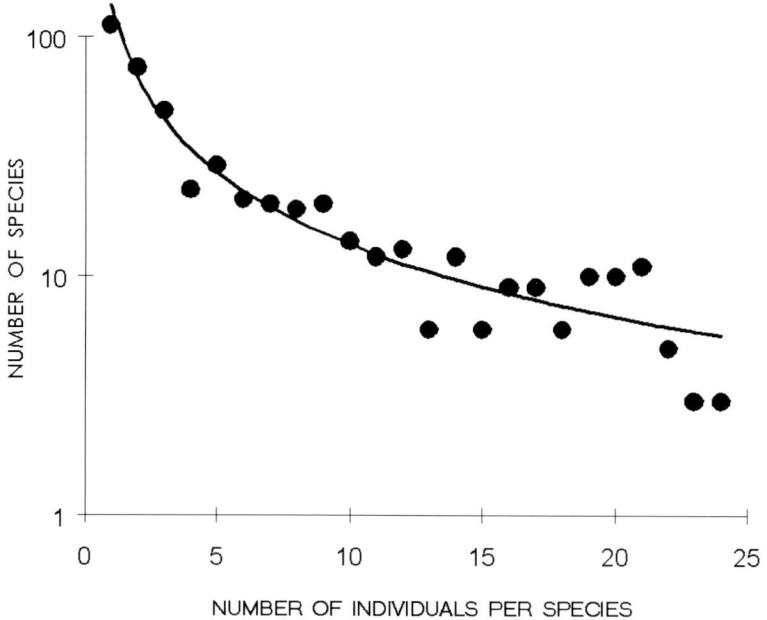

Fig. 9.5. The early measurement of diversity: Fisher, Corbet and Williams (1943) fitted a log-series curve to the distribution of individuals in to species, and used the curve's parameters as an index of diversity (after Fisher *et al.*, 1943).

Another way of thinking about diversity is to imagine picking individuals at random from the community. What is the probability that any two individuals will be from the same species? The higher this probability, the lower the diversity of the whole community, and vice versa. It follows that, other things being equal, diversity will be higher in communities in which the abundances of the various species are much the same; that is, no species dominates the community. This is because the probability that two randomly selected individuals will be from the same species is lower in communities where there is less dominance. This aspect of diversity is called *evenness* sometimes also *beta diversity*. Analogies with thermodynamics and information theory led ecologists to adopt Shannon's index ($-N \log N$) as a measure of evenness. Many robust partitionable measures based on information theory and probability theory have been proposed and used over the years. Pielou (1975) provides an excellent introduction to them.

Diversity thus confounds two separate issues: evenness and richness. As Pielou (1975) notes in her classic work :

To describe a community's diversity merely in terms of its diversity index is to confound these two factors; a community with a few, evenly represented species can have the same diversity index as one with many, unevenly represented species (pp. 14–15).

Besides measuring overall diversity, ecologists have also tried to describe and understand the overall patterns of species abundance. For example, one approach concentrated on establishing the distribution of species abundances in natural communities. Fisher *et al.* (1943) proposed the log-series distribution while Preston (1948, 1962) proposed the log-normal distribution. This work led to the idea of biological diversity as a measurable trait of ecosystems. Parameters from these distributions have been used as diversity indices. Work in this vein continued with the development of information-theoretic measures of diversity. These were first used in ecology by Margalef (1957) and rapidly evolved into a sophisticated suite of measures of community structure (1975) which dissected out the components of diversity and related them to measures of niche width and overlap.

In more recent times, the focus of this ecology has returned to the concerns of the *fin de siècle* phytogeographers – the objective, empirical description of biological communities – using multivariate techniques of classificationand ordination(Pielou, 1984). This work was pioneered by W. T. Williams (Williams & Lambert, 1959) who was the first to use the power of modern computers to analyse and describe the complexity of biological communities.

9.2.6 Do ecological communities exist?

An important simplifying idea in ecology is the notion of *community* This refers to groupings of populations that occur together naturally. Here again we see a debate between autecology and synecology. The question of whether the population mixtures that we observe are *natural* is extremely contentious. Does a particular collection of species occur because they 'belong' together or are they simply random assemblages? By 'belong' we understand that the species concerned have adapted to have a suite of mutual interactions and dependencies. Either they cannot survive without the other species, or the community itself is more stable than a random collection of species would be.

The above question is far from hypothetical. Much of the conservation debate centres around the question of whether to conserve whole communities or to focus on saving individual species. If *communities* are simply random associations of species, then we need not preserve communities; instead we could (at least in principle) conserve species by relocating them in areas not needed for other purposes. On the other hand, if ecosystems do contain suites of species that exhibit strong mutual dependencies, then the only way to conserve those species is to set aside representative tracts of land where their natural communities occur.

The 'sharp end' of the above debate rests with classifications of landscape cover types. *Classification* is the process of grouping together sets of similar sites. The assumption is that all sites classified as belonging to the same group are examples of a single community type. Now classification algorithms can form groups as large or small as you like – it simply depends on where you draw the cutoff level when deciding whether sites are similar enough to belong to the same group. The problem is when to stop the algorithm. Taken to its logical extreme the conservation argument would be that *every* site constitutes a unique group and should be conserved. The exploitation argument would be that all sites belong to a single group, so it does not matter how many sites are destroyed. To preserve the community we need to retain only a single example.

In trying to understand ecological communities, we need to distinguish between *pattern* and *process*. Much of the above discussion has been concerned with patterns – especially species composition or species distribution. In at least some ecosystems, the essential feature is not the pattern, but the process involved. The most complex and diverse ecosystems are rainforests and coral reefs. In both cases the complexity that we see is the end product of a long process. In rainforests it can take many centuries to form rich soils and multi-layered canopies that enable so many species to flourish. Likewise coral reef communities can come into being only when corals build the reefs that provide protected and varied environments for a host of shallow-water species.

9.3 Is nature predictable?

The notion of community, as discussed above, is strongly related to issues of predictability and scale. If (say) the tree Antarctic Beech were always to occur in temperate rainforest, then the location of temperate rainforests would be a strong predictor of the distribution of that tree species. In practice, this kind of predictability is sometimes possible, but often not. Species can be very plastic in their associations with other species.

Scale is important too. Mutual dependencies are often scale-related: the predictability of a species' distribution often increases with spatial scale (Green *et al.*, 1987). Also, different views and methodologies can give vastly different results for the same system. For instance, within reef ecosystems, studies of coral populations have often used ordination and classification, which tend to emphasise patterns and reduce *noise* whereas studies of the associated fish populations have more often analysed data using analysis of variance and related techniques that emphasise random variations.

Scale inhibits our ability to predict nature in other ways too. Many ecological processes act on time or spatial scales that are too vast to permit experiments. To make matters worse, environmental managers often have to deal with altered systems in which (say) introduced species lead to new and unique situations about which nothing is known and for which correct management decisions are vital to the survival of entire species. For these reasons ecologists are turning increasingly to simulation, and other models, with which they can perform *virtual experiments* that would be impossible to carry out in the real world.

9.4 Is there a balance of nature?

The expression 'the balance of nature' sums up one of the most all-pervasive and influential ideas to emerge from ecology. The notion that equilibrium is the natural state of ecosystems has captured the public imagination and achieved the status of dogma within some sections of the conservation movement.

Many early ecologists accepted the idea of a balance of nature implicitly. For example Clements' theory assumed that succession leads to an equilibrium *climax* state (Clements, 1905). This theory dominated plant ecology for most of the twentieth century, until evidence accumulated for other kinds of dynamics, such as chronically disturbed ecosystems (Noble & Slatyer, 1979) and long-term instabilities in vegetation history (Davis, 1976, Green, 1982).

The equilibrium assumption underlies a lot of mathematical work in theoretical ecology (Scudo & Ziegler, 1978). For instance, Macarthur and Wilson's theory of island biogeography suggested that for any island there is an equilibrium number of species that it can sustain (Macarthur & Wilson, 1967). This number depends on many factors, especially the island's area. Furthermore, rates of species turnover depend on the island's distance from the mainland (or other source of immigrants). The most contentious aspect of the theory has been attempts to extend the term 'island' to any type of isolated habitat, such as patches of native vegetation within farmland, isolated mountain tops and lakes in a landscape.

An important argument in support of environmental equilibrium is the idea of *carrying capacity*. This concept proposes that in any ecosystem, certain environmental variables act as *limiting factors* that restrict the size of the population that can be supported. For particular populations the limiting factor may be (say) the amount of food or the area available for breeding animals to form territories. The *carrying capacity* is the maximum population size that the environment can support, given the factors that limit the population's size. It seems a reasonable assumption that populations will tend to increase in size until they are close to carrying capacity.

In the latter part of the century, equilibrium assumptions in ecology have been increasingly called into question. Discoveries about the long-term role of human disturbance in shaping environmental history (Ponting, 1991) cast doubt on the very meaning of the term 'natural'. Likewise the increasingly urgent need to address environmental management has led ecologists to search for non-equilibrium models. As early as 1930, Elton stated his firm opposition to the concept:

The balance of nature does not exist, and, perhaps, never has existed. The numbers of wild animals are constantly varying to a greater or lesser extent,

Are ecosystems complex systems? 351

and the variations are usually irregular in period and always irregular in amplitude (Elton, 1930).

Ecologists are also coming to terms with the sheer time-scale of many ecological processes. A single stand of trees, for instance, may live for hundreds of years, giving a false appearance of constancy. Likewise rainforests rely for their persistence on continuation of nutrient cycles. There is no certainty that broken cycles will re-establish themselves.

9.4.1 What is ecological stability?

The idea of *stability* in ecology encompasses a wide variety of different ways in which systems resist change. As a property of biological communities, however, stability has proved much easier to allude to than to define. Ecologists are confident that they can recognise instances of a biological community in different places, say a particular type of rainforest in different valleys. Likewise, they can recognise the same community at the same place but at different times. But defining and identifying stability in these systems is difficult (cf. fig. 9.6).

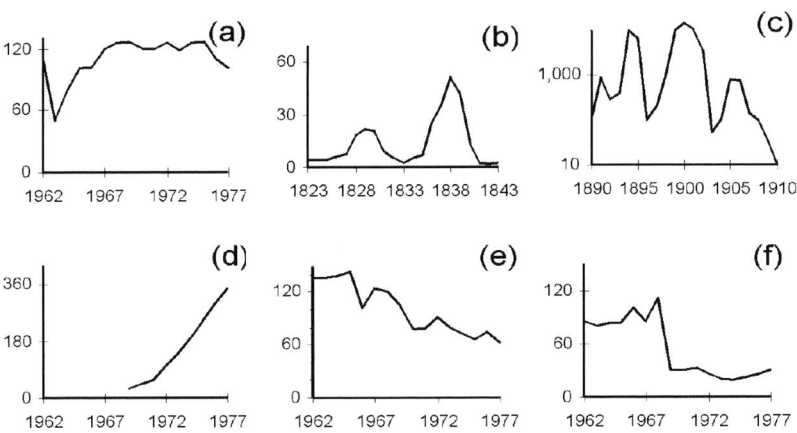

Fig. 9.6. Fluctuations in populations: Pimm (1982) assembled a set of population time series data to show the wide variety of dynamics exhibited by real populations (after Pimm, 1982).

Strictly speaking, 'stability' refers to the nature of an equilibrium point for some variable x. An equilibrium point $x = x^*$ is said to be *stable* if the system tends to return to the equilibrium after being displaced from it. Mathematically, this condition means that the rate of displacement away from the equilibrium point must be opposite in sign from the direction of the displacement. One way to summarise this behaviour is to refer to a variable S, where

$$S = \frac{d}{dx}\frac{d(x-x^*)}{dt}. \tag{9.1}$$

Then the equilibrium at x^* is

- *stable* if $S < 0$ (that is, x eventually returns to x^* after being displaced);
- *unstable* if $S > 0$ (that is, displacements of x away from x^* are self-perpetuating);
- *neutral* if $S = 0$ (that is, displacements of x away from x^* are independent of time).

Note that the above strict interpretation of stability implies a pre-existing equilibrium.

The central problem is to identify exactly what 'stability' means in an ecological context. For individual populations stability has the meaning described above. It refers to situations in which population size possesses equilibrium states to which it returns after a disturbance.

Another problem is how to generalise the above notion to the whole suite of populations that comprise an ecosystem. A more general notion of stability is captured by interpreting a system's *state* to mean its species composition, rather than exact abundances. Holling (1973) coined the term *resilience* for this notion. A community is *resilient* if it returns to the same species composition after a disturbance. An even more elementary type of stability is captured by the term *viability*. We can call a community *viable* if it retains its initial species composition in the absence of disturbance. That is, feedback and other interactions within the system do not drive some species to local extinction. Roberts (1979) used the term *feasible* to describe model systems displaying this property.

A more accessible, and more practical, approach to stability is to consider an ecosystem's components and how they interact. This approach places less emphasis on a putative equilibrium and more emphasis on community dynamics that are measurable in the field and tractable in theory. In reviewing this issue, Pimm (Pimm, 1984) concluded that stability is a qualitative attribute that has four components: *resilience, persistence, resistance* and *variability*:

- *Resilience* is how fast the variables return to their equilibrium after perturbation.
- *Persistence* is the time a variable lasts before it is changed.
- *Resistance* is the degree to which a variable is changed, following a perturbation.
- *Variability* is the variance of the variables over time.

Complicating the above is the issue of scale. Perturbations operate over different space and time scales. A spatial disturbance of 1 m^2 is not only a smaller perturbation than a disturbance of 1 km^2, it is qualitatively different. The boundary is smaller relative to the total area, so processes across the boundary (such as migration) will be different. Considering time, long lived organisms such as trees will be less resilient and more persistent in numbers than short lived organisms such as annual plants.

9.4.2 Chaotic dynamics in ecological interactions

May's landmark analyses (1972, 1974, 1976a) of the development of chaos in simple dynamic population models introduced the concepts of complex non-linear dynamics to ecological systems (chapter 6). Many population models exhibit non-linear dynamics.

In unrestricted population growth the rate of change $\frac{dN}{dt}$ in population size N with time t is a simple function of the population size and the reproduction rate r:

$$\frac{dN}{dt} = rN. \qquad (9.2)$$

This relationship leads to exponential population growth $N(t) = N_0 e^{rt}$. An important constraints on growth occurs if an environment's carrying capacity K inhibits the rate of reproduction. This *logistic growth* leads to the classic sigmoid pattern of population growth (cf. fig. 9.2). We can express logistic growth by the equation

$$\frac{dN}{dt} = rN(1 - \frac{N}{K}). \qquad (9.3)$$

If reproduction occurs periodically, not continuously, then population change deviates increasingly from the sigmoid pattern as the reproduction rate r increases, leading eventually to chaos (see fig. 9.7).

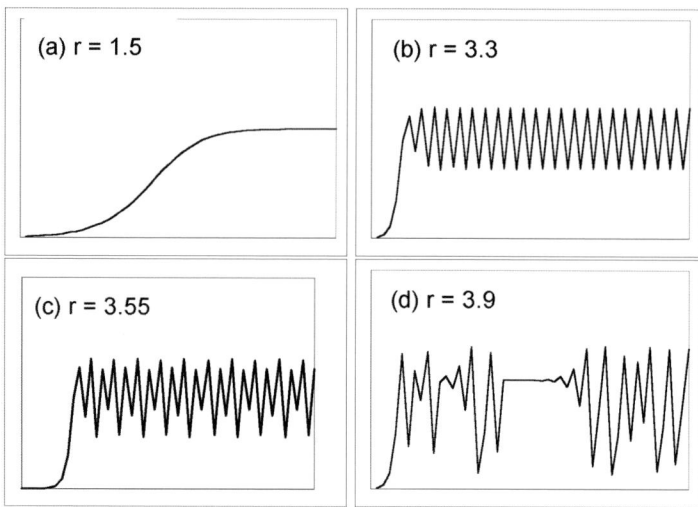

Fig. 9.7. The onset of chaos in a model of logistic growth. In each case the model settles onto an attractor after a brief period of transient behaviour. The nature of the attractor changes as the population's growth rate increases: (a) a stable equilibrium; (b) cycles of period 2; (c) cycles of period 4 (as period doubling occurs); (d) a strange attractor. Because the population consists of organisms, rather than being simply a mathematical variable, any crash that sends its size to zero means that it becomes locally extinct.

Interactions between populations provide another source of non-linearity in ecology. These interactions can take many forms, such as predation and competition. For instance the famous Lotka-Volterra equations for predator–prey interactions can be written as follows (for foxes and rabbits):

$$\frac{dF}{dt} = r_F RF - m_F F, \qquad (9.4)$$
$$\frac{dR}{dt} = r_R R - m_R RF.$$

Here r_R and r_F denote the reproductive rates of rabbits and foxes, respectively. Likewise m_R and m_F denote the mortality rates.

Feedback loops are especially common in multi-species systems. In populations with seasonal reproduction delays arising from feedback tend to produce cyclic behaviour (fig. 9.8).

For real biological systems the gap between theoretical analyses and observations has proved difficult to bridge. For instance, it is difficult to

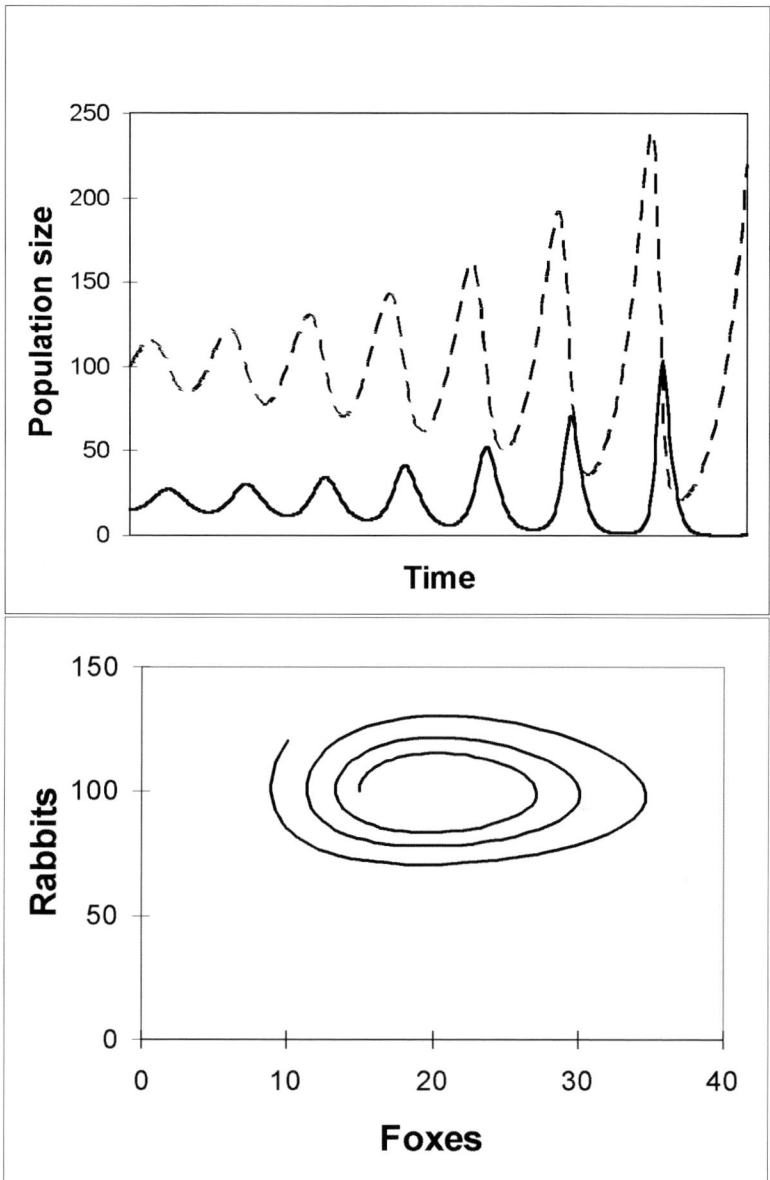

Fig. 9.8. Cyclic behaviour arising from feedback in the foxes and rabbits population model. The first graph shows population changes with time. The second graph is a phase diagram showing the two populations plotted against one another (for the first three cycles only). In the scenario shown here, positive feedback produces forced oscillations that eventually lead to local extinction of the fox population.

accurately determine the values of control parameters such as survival rates. Also it is difficult to distinguish experimentally between complex dynamics and stochastic behaviour.

As a recent example of a more complete study of biological non-linear dynamics Costantino et al. (1995) have shown a full range of stable-state, periodic and aperiodic behaviour in flour beetle populations. Their experiment involved the careful manipulation of laboratory survival rates to shift between these behaviours. They expressed the relationships between larval, pupal and adult populations as difference equations. Population changes depended on the rate of cannibalism of pupae by adults, rates of adult death from other causes and maturation rates. Most importantly they directly compared their models with experimental time-series data by converting the equations to a stochastic model with added noise. Experimentally, adult beetles were added or removed at each census to give a number of fixed death rates. Their results were in complete agreement with the dynamics of their model – as the adult mortality rate increased from zero to complete, the population dynamics made transitions from stable equilibria to stable two- cycles, to stable equilibria again and eventually to aperiodic oscillations (chaos).

9.5 Environments and interactions

9.5.1 What is the role of landscapes?

During the 1980s increasing concern with conservation and land management resulted in a new subdiscipline known as *landscape ecology*.

The traditional approach of ecologists to landscapes has been to treat them as the substrates for environmental variation. Changes in ecosystem structure and composition are seen to reflect changes in environmental conditions from place to place. This approach is embedded in standard field methodology. Sampling methods traditionally employ the *quadrat* – usually a square of fixed area – as the unit of sampling. A well-developed body of statistical theory for experimental design and interpretation has grown up around quadrat sampling (Pielou, 1974). In particular, ecologists study spatial variation by examining differences between sets of quadrats. Randomly arranged sets of quadrats are the standard way to sample uniform environments. Contiguous lines of quadrats (*transects*) provide a way to study variations along *environmental gradients* such as changes in soil moisture on a hillside or increasing salinity in an intertidal swamp.

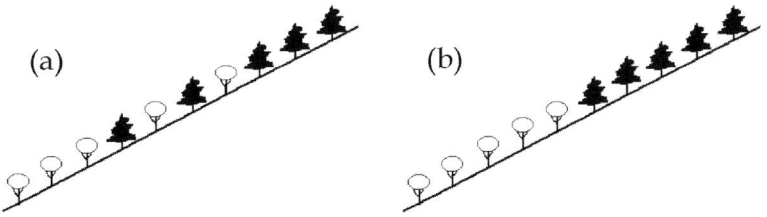

Fig. 9.9. The effect of competition on species distributions along an environmental gradient. (a) A 'smooth' transition between tree types. (b) A sharp transition produced by competition.

One of the most important results to spring from transect studies is that environmental variations alone do not suffice to explain the spatial distributions of organisms. Research by Pielou (1974) and other quantitative ecologists showed that landscape interactions, such as competitionand dispersal, play prominent roles too. For instance, competition between species often leads to truncated distributions along an environmental gradient (fig. 9.9). The significance of these findings is that ecosystems are not controlled in simple fashion by *external* (i.e. abiotic) factors. Internal, biotic interactions within a system play an important role too. These biotic interactions are likely to be non-linear and complex.

The development of ecological models, especially simulations, also reflects an increasing awareness that interactions are important. Many early models, such as those of Botkin (1972), simulated the growth of individual plants within an idealised *stand*, which was taken to typify an entire forest. To account for spatial variations, such as changes in moisture and light on a hillside, the model was run again and again, changing parameter values each time. This approach allowed ecologists to model some interactions, such as competition, and some non-equilibrium effects, such as disturbance history, but it assumed that spatial interactions were negligible.

9.5.2 Cellular automata in landscape ecology

Perhaps the main contribution of cellular automata (CA) models of ecosystems (Green, 1989, Hogeweg, 1988) has been to show that interactions within a landscape can totally alter the structure and dynamics of ecosystems.

CA models of landscapes consist of fixed arrays in which the cells represent square areas of land surface. The size of each cell fixes the scale. A cell's *state* corresponds to environmental features, such as coral type in a reef model or topography in a landscape simulation. *Neighbourhood functions* simulate landscape processes (e.g. seed dispersal, spread of epidemics). To represent time the models update the state of all cells in the grid iteratively. The time step is set by the nature of the processes being simulated.

The CA approach has many advantages. On a practical level, it is compatible both with pixel-based satellite imagery and with quadrat-based field observations. Also it enables processes that involve movement through space, e.g. dispersal, fire (Green et al., 1990), to be modelled easily. On a theoretical level, results about the behaviour of cellular automata (Wolfram, 1984, Wolfram, 1986) can be applied to ecological systems (Green, 1993). Cellular automata models have been used successfully to simulate many environmental systems, including fire spread (Green et al., 1990), starfish outbreaks (Hogeweg & Hesper, 1990, Bradbury et al., 1990, van der Laan, 1994), spread of disease (Pech et al., 1992, Green, 1993), and forest dynamics (Green, 1989, Hendry & McGlade, 1995) (see also chapter 2 on self-organisation).

Some CA models (Green et al., 1990, Pech et al., 1992) have been carefully calibrated to represent real systems and have been extensively validated using field data. Many others are abstract representations that embody, in a general way, biotic processes within a landscape (e.g. space filling, space-clearing) that are common to many different ecological systems.

9.5.3 Complexity in behaviour

A dilemma of modern behavioural ecology is the question of how complex, interactive behaviour patterns, such as honeycomb packing in bees can be genetically encoded. An increasingly popular and useful approach to such questions is the developing new field of artificial life (Alife), in which organisms and their behaviour are simulated along with their environments. Investigations into self-organisation in Alife models have given valuable clues as to how acting locally with simple behavioural rules can give rise to extremely complex behaviour on a more global scale, obviating the need for complex genetic programming.

Perhaps the most important results to emerge from Alife studies is that global properties, such as colony structure, can emerge from simple, local rules of behaviour. For example Hogeweg and Hesper showed that the organisation of bumble-bee colonies emerged out of the interactions that the bees had with their environment and with each other (Hogeweg & Hesper, 1983). Subsequent studies have shown that the same general principle holds true in many other systems, such as ant colonies and flocks of birds.

In another study of bee behaviour, Camazine has shown that simple local behavioural rules can give rise to the self-organisation of comb patterns (Camazine, 1991). The combs of feral honey bee colonies have a characteristic pattern of brood, pollen (short-term food storage) and honey (longer-term food storage) cells. The brood is kept in a central cluster (perhaps to ensure a well-regulated incubation temperature) and surrounded by a rim of pollen storage cells with honey reserves peripheral to this. Despite the continual removal and replacement of brood, honey and pollen (at different rates), with the cell content type often changing, the general structure is robust and appears well adapted for the optimal raising of the brood. A complex behavioural blueprint would be required for the bees to know the pattern and be able to adapt it to local environmental conditions and food supply fluctuations. In Camazine's model honey and pollen are deposited randomly by workers into empty cells and later used to feed the brood at a rate dependent on the number of brood cells immediately adjacent to the honey/pollen cell.

One key theme emerging from Alife behavioural studies, such as the above, is that the development of optimal foraging or even more general behavioural patterns are not necessarily difficult to evolve. In such studies individuals are represented as an encoding of behavioural rules (as neural networks or finite state automata (Jefferson et al., 1991)) which are then selected for according to a definition of success (such as foraging ability), mutated and retested (see also Taylor and Jefferson, 1994).

Sexual selection is another area of intense study and contrasting ideas. In many species males are chosen as mates on the basis of physical traits which at first seem to reduce fitness, i.e. the brilliant peacock tail displays or ever more encumbering Bird of Paradise tails (which may reach several times body-length). The locking in of female mate choice strategies for such characteristics is well established but the original development of such choices is still debated.

Enquist and Arak (1994) and Johnstone (1994) have shown how simple neural networks intrinsically develop a preference for images with more symmetric features or colour patterns. They suggest that this 'aesthetic sense' is due to the ease with which symmetric patterns can be recognised in a range of orientations and how training to recognise a range of individuals from a slightly asymmetric population will result in the easy identification of symmetric individuals (the *population average*).

Enquist and Arak (1993) had earlier shown that preferences for features such as long tails can evolve as a by-product of the need for mate-recognition. These preferences can co-evolve along with the corresponding male features, thereby prompting the development of distinct mate choice strategies. They also show how the development of longer and longer tails is possible even when disadvantageous to the individuals concerned. Thus there is no need for the display of exaggerated or symmetrical features to be linked with an individual's survival ability as is often suggested, though this has been shown to be true of some species.

9.6 Managing complexity in the real world

The acid test of ecological theory is whether it helps us to manage real ecosystems. In other words, is our science adequate to deal with the complexity of environmental processes in the real world? Managing the environment – or as many would put it saving the earth – evokes a real sense of urgency. The pace of environmental change is great; the risk of irreversible damage is real; yet our knowledge of how the system operates is still poor.

For simplicity we can summarise some of the main issues under a number of general headings (note the taxonomic approach!).

- *Conserving biodiversity.* As we saw earlier, simply listing the species to be saved is a huge job in itself. Practical steps are moving beyond *ad hoc* attempts to save threatened populations at particular localities. They now include more strategic measures, such as relocating species to safer habitats and setting up *genebanks* of seeds, embryos and other forms of germplasm. Theoretical and field studies are looking increasingly beyond simple questions of presence or absence and address issues such as the viability and genetic integrity of small populations.

- *Ecosystem management.* Most of the ideas presented in this chapter can be seen as contributing to our ability to manage ecosystems. Whereas most effort has gone into dealing with parks and other protected areas, managers are increasingly aware of the need for park management to be part of an integrated approach. It must deal with

the whole landscape, include the general public, and deal with all stages of land use, from planning to reclamation.
- *Coping with disturbance.* Perhaps the greatest need, and the greatest challenge, is to understand and control environmental disturbance. Most often 'disturbance' means alteration induced by human activity. The problems are many: clearing, fire, pollution, land degradation, fragmentation of habitats, and introduction of exotic pests, to name but a few. In general terms disturbance poses the greatest need for an understanding of environmental complexity. Interactions (non-linearities) abound wherever we have to deal (say) with changes within a landscape or with new species in an ecosystem.
- *Monitoring and planning.* Management agencies now adopting highly systematic approaches to environmental management. Whilst a lot of present work involves baseline studies to establish what is where, the need for regular monitoring is also widely recognised. Also widely recognised is the need for greater sharing of information. For example, plants and animals pay no heed to national borders, so different countries need to share data if they are to develop accurate pictures of species' distribution. Cooperative mechanisms for sharing information are now spreading rapidly, aided by the *information super highway*. To make planning possible, agencies everywhere are adopting a range of new approaches, including geographic information systems (GIS), environmental models and algorithms for land-use optimisation.

As an example, the development of an environmental decision support framework in Australia illustrates how some of the above issues are being addressed in practice. Cooperative work by a host of Australian environmental agencies has resulted in a set of data *layers* that give continental scale coverage of key environmental variables. These variables include digital elevation models from which it is possible to interpolate elevation, and hence (by combining with meteorological records) to estimate climate anywhere in the continent. For plant species, national cooperative projects have produced an Australian Plant Name Index (Chapman, 1991) and a landcover programme has produced databases of precise distribution data for most of the major landcover species. Combining all of the above data makes it possible to predict the potential distribution of many species. Overlaying all of the above with GISdata for public lands, roads, etc. and regular satellite monitoring of the entire continent makes it possible to address a wide range of issues about environmental management. One simplifying approach is to separate environmental questions into two types: *primary* or simple questions that involve single data layers and *second-order* questions that require analysis of the interactions between two separate data layers.

9.7 Conclusion

Yes, ecosystems are complex systems. In ecosystems we can easily see many of the characteristics of a *canonical* complex system (Kelly, 1992):

- *perpetual novelty*: the system continues to surprise the observer no matter how long it has been studied;
- *resilience*: the capacity for self-repair, graceful degradation and limping along instead of dying;
- *emergence of the aggregate*: a recognition that there exists an object of study, without necessarily assuming that the 'whole is greater than the sum of the parts';
- *formation of individuality*: each instance of a complex system is recognisable as an example of the system but also has a distinct individuality, which may also lead to an internal identity of self through the generation of internal models;
- *internal models*: through its architecture, the system can represent a level of abstraction internally, allowing it in a limited way to anticipate the future;
- *non-zero sum game*: gains somewhere in the system are not necessarily offset by losses elsewhere, nor is the system necessarily in equilibrium;
- *exploration tradeoff*: processes exist for optimally balancing new ways of surviving (exploration and learning behaviours) against ways of maximising what already exists (exploitation and efficiency behaviours).

Although ecosystems are clearly complex systems, the insights that flow from complexity theory have been slow to filter into ecological theory and practice. This situation arises because the researchers studying (say) artificial life have been based mostly in the physical sciences and computing, whereas most field ecologists have little contact with those fields. Conversely a criticism of artificial life research has been that most studies examine hypothetical systems with little regard as to the ecological validity or usefulness of their underlying assumptions. Notable exceptions include cellular automata simulations of coral reefs, forests and other ecosystems (see earlier). One of the great challenges at present is to join complexity theory with ecology. This is essential if we are to address some of the major environmental issues now facing the world. The future of our planet may well hinge on our ability to make this connection.

Bibliography

Andrewartha, H. G., & Birch, L. C. (1954). *The Distribution and Abundance of Animals.* Chicago University Press, Chicago.

Botkin, D. B., Janak, J. F., & Wallis, J. R. (1972). Some ecological consequences of a computer model of forest growth. *J. Ecol.*, **60**, 849–872.

Bradbury, R. H., van der Laan, J. D., & MacDonald, B. (1990). Modelling the effects of predation and dispersal on the generation of waves of starfish outbreaks. *Mathematical and Computer Modelling*, **13**, 61–68.

Camazine, S. (1991). Self-organising pattern formation on the combs of honey bee colonies. *Behav. Ecol. Sociobiol.*, **28**, 61–76.

Chapman, A. D. (1991). *Australian Plant Name Index (APNI) vols 1-4,.* Australian Flora and Fauna: Nos. 12-15. Canberra: Australian Government Publishing Service.

Chapman, R. N. (1931). *Animal Ecology.* McGrawHill, New York.

Clements, F E. (1905). *Research Methods in Ecology.* University Publishing Company, Lincoln, Nebraska.

Cody, M. L., & Diamond, J. M. (eds). (1975). *Ecology and Evolution of Communities.* Belknap Press, Cambridge, Mass.

Costantino, R. F., Cushing, J. M., Dennis, B., & Desharnais, R. A. (1995). Experimentally induced transitions in the dynamic behaviour of insect populations. *Nature*, **375**, 227–230.

Davis, M. B. (1976). Pleistocene biogeography of temperate deciduous forests. *Geoscience and Man*, **13**, 13–26.

Elton, C. S. (1927). *Animal Ecology.* London: Sidgewick and Jackson.

Elton, C. S. (1930). *Animal Ecology and Evolution.* Oxford University Press, New York.

Enquist, M., & Arak, A. (1993). Selection of exaggerated male traits by female aesthetic senses. *Nature*, **361**, 446–448.

Enquist, M., & Arak, A. (1994). Symmetry, beauty and evolution. *Nature*, **372**, 169–172.

Fisher, R. A., Corbet, A. S., & Williams, C. B. (1943). The relation between the number of species and number of individuals in a random sample of an animal population. *Journal of Animal Ecology*, **12**, 42–58.

Forbes, S. A. (1887). The lake as a microcosm. *Bulletin of the Science Association of Peoria, Illinois*, 77–87.

Gell-Mann, M. (1994). *The Quark and the Jaguar.* W. H. Freeman, San Francisco.

Green, D. G. (1989). Simulated effects of fire, dispersal and spatial pattern on competition within vegetation mosaics. *Vegetatio*, **82**, 139–153.

Green, D. G. (1993). Emergent behaviour in biological systems. *Pages 24–35 of:* Green, D.G., & Bossomaier, T. R. J. (eds), *Complex Systems – from Biology to Computation.* IOS Press, Amsterdam.

Green, D. G., Bradbury., R. H., & Reichelt, R. E. (1987). Patterns of predictability in coral reefs. *Coral Reefs,* **6**, 27–34.

Green, D.G. (1982). Fire and stability in the postglacial forests of southwest nova scotia. *Journal of Biogeography,* **9**, 29–40.

Green, D.G., Tridgell, A., & Gill, A. M. (1990). Interactive simulation of bushfire spread in heterogeneous fuel. *Mathematical and Computer Modelling,* **13**(12), 57–66.

Hendry, R. J., & McGlade, J. M. (1995). The role of memory in ecological systems. *Proc. Royal Soc. Lond. B,* **259**, 153–159.

Hogeweg, P. (1988). Cellular automata as a paradigm for ecological modeling. *Applied Math. Comput,* **27**, 81–100.

Hogeweg, P., & Hesper, B. (1983). The ontogeny of the interaction structure in bumble bee colonies: A mirror model. *Behavioural Ecology and Sociobiology,* **12**, 271–283.

Hogeweg, P., & Hesper, B. (1990). Crown crowding, an individual-oriented model of the Acanthaster phenomenon. *Pages 169–188 of:* Bradury, R. H. (ed), *Acanthaster and the Coral Reef: a Theoretical Perspective.* Lecture Notes in Biomathematics (88).

Holling, C. S. (1973). Resilience and stability of ecological systems. *Annual Review of Ecology and Systematics,* **4**, 1–23.

Hutchinson, G. E. (1978). *An Introduction to Population Ecology.* Yale University Press, New Haven.

Jefferson, D., Collins, R., Cooper, C., Dyer, M., Flowers, M., Korf, R., Taylor, C., & Wang, A. (1991). The Genysis system: Evolution as a theme in artificial life. *Pages 275–295 of:* Farmer, J. D., Langton, C. G., Taylor, C., & Rasmussen, S. (eds), *Artificial Life II.* SFI Studies in the Sciences of Complexity, vol. X. Santa Fe: Santa Fe Institute.

Johnstone, R. (1994). Female preference for symmetrical males as a by-product of selection for mate recognition. *Nature,* **361**, 172–175.

Kelly, K. (1992). A distributed Santa Fe system. *Bulletin of the Santa Fe Institute,* **7**, 4–6.

Lindeman, R. L. (1942). The trophic-dynamic aspect of ecology. *Ecology,* **23**, 399–418.

Macarthur, R. H., & Wilson, E. O. (1967). *The Theory of Island Biogeography.* Princeton University Press, Princeton.

Margalef, R. (1957). *La teori de la información en ecologia.* Mem. R. Acad. Cien. Artes, Barcelona.

Marschall, L. (1994). A world that slipped away. *The Sciences,* **34 (5)**, 45–46.

May, R. M. (1972). Will a large complex system be stable? *Nature,* **238**, 413–414.

May, R. M. (1974). *Stability and Complexity in Model Ecosystems.* Princeton: Princeton University Press.

May, R. M. (1976a). Simple mathematical models with very complicated dynamics. *Nature,* **26**, 459–467.

May, R. M. (1976b). *Theoretical Ecology: Principles and Applications.* W. B. Saunders, Philadelphia.

McIntosh, R. P. (1985). *The Background of Ecology*. Cambridge University Press, Cambridge.

Noble, I. R., & Slatyer, R. O. (1979). Concepts and models of succession in vascular plant communities subject to recurrent fire. *Pages 311–335 of:* Gill, A. M., Groves, R. H, & Noble, I. R. (eds), *Fire and the Australian Biota*. Australian Academy of Science, Canberra.

Pech, R., McIlroy, J. C., Clough, M. F., & Green, D. G. (1992). A microcomputer model for predicting the spread and control of foot and mouth disease in feral pigs. *Proceedings of the 15th Vertebrate Pest Conference, Newport*.

Pielou, E. C. (1974). *Population and Community Ecology*. Gordon and Breach, New York.

Pielou, E. C. (1975). *Ecological Diversity*. Wiley-Interscience, New York.

Pielou, E. C. (1984). *The Interpretation of Ecological Data*. Wiley-Interscience, New York.

Pimm, S. L. (1984). The complexity and stability of ecosystems. *Nature*, **307**, 321–326.

Ponting, C. (1991). *A Green History of the World*. London: Penguin.

Preston, F. W. (1948). The commonness, and rarity, of species. *Ecology*, **29**, 254–283.

Preston, F. W. (1962). The canonical distribution of commonness and rarity. *Ecology*, **43**, 185–215, 410–432.

Scudo, F. M., & Ziegler, J. R. (1978). *The Golden Age of Theoretical Ecology, 1923–1940*. Springer-Verlag, New York.

Taylor, C., & Jefferson, D. (1994). Artificial Life as a Tool for Biological Enquiry. *Artificial Life*, **1**, 1–13.

Tregonning, K., & Roberts, A. (1979). Complex systems which evolve towards homeostasis. *Nature*, **281**, 563–564.

van der Laan, J. D. (1994). *Spatio-temporal Patterns in Ecosystems – a Bioinformatic Approach*. Ph.D. thesis, University of Utrecht.

Whittaker, R. H. (1975). *Communities and Ecosystems*. Macmillan, London.

Williams, W. T., & Lambert, J. M. (1959). Multivariate methods in plant ecology. I. association-analysis in plant communities. *J. Ecol.*, **47**, 83–101.

Wolfram, S. (1984). Cellular automata as models of complexity. *Nature*, **311**, 419–424.

Wolfram, S. (1986). *Theory and Applications of Cellular Automata*. World Scientific, Singapore.

World Resources Institute, International Union for Conservation of Nature, United Nations Environment Programme. (1992). *Global Biodiversity Strategy*. WRI, IUCN, and UNEP, New York.

10
Complexity and neural networks

TERRY BOSSOMAIER

*Charles Sturt University,
Bathurst, NSW 2795,
AUSTRALIA*

10.1 Introduction

Neural networks fuse together ideas from all aspects of complex systems: they embody features of all the other chapters of the book; even the ecology of neural networks, the relationship of brain to animal environmental niche, is a rich and exciting field (Snyder *et al.*, 1990). In themselves neural dynamics are rich in complexity†. But animal neural networks, brains, play a fundamental role in the characterisation, prediction and control of many phenomena of the natural and man-made world. They evolved as the *monitors, memory and controllers of natural complex systems*. We should know how they work.

Three features of biological brains create diverse and fascinating behaviour:

- many autonomous agents: the human brain has 10^9 neurons, about the number of people on earth and considerably larger than, say, the largest ant colonies. Each agent (neuron) is adaptable.
- vast interconnectivity: the human brain contains of the order of 10^{14} synapses, the points of connection between neurons. Each neuron may connect with as many as 10^5 other neurons.
- each agent (nerve cell or neuron) *learns*. Biological brains are the most complex *adaptive* system of which we know. Their adaptability not only serves to cope with a changing environment but also to provide substantial redundancy to damage.

These factors alone would be enough to generate vast complexity, yet each neuron is, itself, a highly sophisticated biochemical system, made up of many interacting components. In many complex systems, we can *encapsulate* the interior workings of each agent, i.e. handle only external behaviour and interactions. Unfortunately, as we shall discover in this chapter, it is not obvious that we can do this with biological neural networks. Should we, in fact, treat a brain as one massive complex system?

But since neural networks are so successful, there is now huge interest in mimicking their behaviour for a whole range of applications from robotics through to analysing the contents of databases. In fact as this article goes to press, *Deeper Blue,* an IBM Chess playing computer, has recently beaten world champion Gary Kasparov. Its designers say they introduced *intuition* into this latest machine, begging the question of

---

† John Eccles, Nobel Laureate in brain science, described the human brain as the most complex system in the universe.

what neural computation actually is. We could simply say that it is the computation carried out by *wetware*, by nerve cells, but this would ignore the extensive insights gained from working on simplified artificial systems. But if we remove the wetware as a defining feature we are left with very few principles. Turing of course, provided an abstract model for digital computers. We simply do not have to concern ourselves with transistors or any of the detailed mechanisms underlying the binary representations and operators. We would like to do this with neural networks, and there are many simplifications possible which look similar at first. Taking a closer look, as we shall do in this chapter, casts doubt on such simplifications. Yet, we should also heed Walter Freeman's caution (Freeman, 1991a) that a good deal of the intricate structure of the human brain is concerned with keeping it operational for a few score years.

The Turing machine concept captures the minimal requirements of a computer and allows us to essentially define algorithmic computation and computability. A question frequently raised for biological brains is whether or not their computation is algorithmic. Since a brain is finite, the viewpoint from classical physics would be that every step in a brain can be successfully modelled. But there is a respect in which the type of computation is fundamentally more powerful than that of a Turing machine. Parberry (1994) shows that Boolean networks as they become infinite in size can represent a larger class of functions. The power is in the pattern of connections. If this pattern can be represented by a simpler algorithm than an exhaustive list, then the network reduces to the function class of Turing machines.

However philosophers such as John Searle (1992) have argued that there is some intrinsic property to human brains which is not, and may not ever, be captured by man-made computers. Searle is now probably in a minority, but Roger Penrose (1989, 1995) has recently offered another viewpoint of the supremacy of biological systems. Penrose starts from a computational perspective, arguing that biological brains do not always perform algorithmic computation and that some other mechanism is involved. However, he remains strictly within the domains of physical science and argues that these additional properties are quantum mechanical in nature, advocating a particular structure within the nerve cell, the fine microtubular structure of membrane, to supply them.

Closely linked to Penrose's arguments is work by David Deutsch and others on quantum computation. By admitting quantum mechanics to the working of a Turing machine, some problems might be soluble more quickly. At present consensus is building that this might be useful and active work on building quantum computational devices. We still have a great deal to learn about neural networks from the classical point of view alone before we admit quantum mechanics. We thus shall put the quantum mechanical issue aside.

Arguing that the brain can compute only within the Turing limits, maybe including quantum mechanics, does not imply that consciousness also falls within this description. Jackendoff (1987) and Chalmers (1996) have argued in recent books that consciousness might not be so reducible: it might be something else which tracks the physical events in the brain. These mystical issues do not concern us here, but the determination of a model for *neural computation* which, like the Turing machine, is independent of particular machines is far from over.

Abstracting the features of a biological brain which go to make up neural computation is not easy. A brain is a set of agents connected to other agents. Agents send messages to one another based on messages they themselves receive. Some agents serve as transducers: they input or output data to the real world (sensory systems and muscular control). The response of each agent is adaptive: the assembly of agents learns in some way as a result of its interaction with the environment. Stated thus, we see neural networks as just one embodiment of computation in complex systems, underlining the popular *nature as computation* metaphor.

Perhaps, neural computation is in some sense optimal for biological tasks, as opposed, say, to calculating prime numbers, although *idiot savants* can sometimes be good at this too. But in fact there are good arguments for believing that neural networks may be the best computational strategy for pattern recognition and system control in the natural world. The use of *artificial neural networks* in optimising the performance of parallel computers and control theory are touched on in chapters 8 and 7 respectively.

The volume of research in neural network applications is vast and its rapid growth continues with *IEEE Neural Network Proceedings* totalling some 6000 pages alone each year. *Connectionism* has invaded many areas from models of mental illness to the stock market. Yet much of this work has little bearing on complexity: the next section will outline some of the theoretical ideas developed for artificial systems in terms of what they can and cannot do and what they can learn in a reasonable time. With this framework we can then go on to look at the huge diversity and intricacy of a biological brain.

10.2 Neural computation

In chapter 3, Seeley discusses Kauffman's NK model and in §4.4.2, Aleksic discusses its relationship to cellular automata. In some ways the boolean networks typified by the NK model can be thought of as a more general system than canonical *artificial* neural networks. In the boolean model, each network node is a logical function of its inputs. If we make the pattern of connections regular and local, we get a cellular automaton. If we leave the connections arbitrary, but simplify the node function to a threshold sum, we get the canonical neuron:

$$y_j = S(\sum_i w_{ij} y_i - \theta) \qquad (10.1)$$

for cell y_j connected to cells y_i (possibly including y_j) with strength w_{ij}. The inputs are summed linearly then passed through a unit step function S corresponding to the generation of a spike when the summed input exceeds some threshold, θ. This equation dates back to the earliest days of neural modelling, from McCulloch and Pitts (1943) and is still used extensively in applied neural network research.

It is possible to find, within the animal kingdom, examples of just a few neurons carrying out a definite and unique function. Insects have very economical brains with, for example, just two H1 movement analysis neurons, one on each side of the head. Yet we think of brains as being networks of huge numbers of cells, and indeed, we think of the essential feature of neural networks as the massive parallelism they entail. So a fundamental characteristic is the digraph of the network. Neurons correspond to nodes of the graph and the edges to connections between them. Unlike the basic digraphs of chapter 3, each edge has a real number associated with it, the strength of the connection.

Such simple models are very powerful:

- they are capable of universal computation;
- they can learning arbitrary classifications from examples, emulating advanced statistical models as described by Ripley (1996);
- they can act as an associative memory, filling in gaps in an incomplete stimulus.

But there is one last component still fiercely debated in neuroscience circles at present: the issue of time. Does the precise timing of neural activity have a strong bearing on learning or performance? Debate rages on this issue today.

We can summarise this model of computation using a precise definition by Valiant (Valiant, 1994, p.50), which he terms a *neuroidal tabula rasa:* a quintuple $(G, W, X, \delta, \lambda)$ in which G is the graph describing the topology of the network, W is the set of weights that the edges can have, X is the set of states that neuroid can be in at any instant, δ is the update function for the node, and λ is the update function for the weights.

The neural network graph may exhibit several different properties: it may be

feed-forward: in which the graph is a tree with inputs feeding to outputs without any cycles. The sum in equation 10.1 does not include j. A huge proportion of practical artificial neural networks are of this kind.

recurrent or reentrant: where there are cycles in the graph. The vast majority of brain circuitry is of this kind.

fully-connected: where every neuron is connected to every other neuron. This characterises some of the early associative memory models (§10.2.5) but is rarely found in animal brains.

So, let us now look at the computational properties and limits of such abstract systems. We start with the relationship to computation by Turing machines.

10.2.1 Prediction and algorithmic complexity

When we write a program for a computer, we tend to have two main concerns: how fast it will go and how much memory it will require. We refer to these as *time* and *space* complexity. Many theoretical studies of algorithms focus on the time complexity: if a problem is characterised by a size N, how does the solution time increase with problem size. Some problems such as the well-known travelling salesman problem are known to be particularly nasty, perhaps taking exponentially increasing time. *Perhaps* here is not just vague. The time complexity of many practical problems is just not known.

There is, however, a third aspect to consider: the size of the program, the *algorithmic* complexity†. Normally, it is the data which takes up the space in our computers: images and video are extremely memory

---

† Three people seem to have come up with this theoretical concept at around the same time, Chaitin, Solomonoff and Kolmogorov. All three names might appear associated with it. We have adopted a neutral term!

hungry. But some applications with very complex rule sets may generate very large programs. Recognition of natural patterns can be such an application. Abu-Mostafa (1986) conjectures that they are *random* problems and require a particular kind of computational capacity and to which neural networks are well suited. Let's start with an example of how the irregularity in the natural world enforces lengthy algorithms.

Suppose we want to distinguish different aircraft. Such man-made objects have a great deal of regularity to them: jumbo jets of a particular series are all the same size and shape to a high accuracy. If we just wanted to distinguish, say, a 737 from a 747, we would simply have to look for the tell-tale bulge at the front. Birds do not present much problem to enthusiastic human observers, but they are nowhere near so regular: birds from a single species are not all the same size, their plumages are never quite what it says in the field guide and so on. Similarly telling apart different sorts of tree is a troublesome for rule systems (Abu-Mostafa, 1986): we can find all kinds of irregular shapes still passing as a snow gum or poplar: the rules for describing a tree are obviously pretty complicated. Checking through all these different rules for a bird or a tree essentially implies a long program (but it might still run very quickly). It is this notion of a long program which is at the heart of algorithmic complexity.

To take this idea further, we need some measure of program size, which seems to imply a standard language or computer. The Universal Turing machine (see chapter 4) is just such a computer and we can use the length of its program, p, as the algorithmic complexity measured in bits:

$$R = \log_2(p). \tag{10.2}$$

In fact the precise machine is not really important, since it only affects the complexity by a constant factor†.

The original formulation of the theory is based on binary strings and the programs to generate them: a string made up of just ones obviously requires a very short program. A random string presents much more of a problem, apart from making the intuitive meaning of random precise. If it is truly random then the only possible program is one in which the

---

† This constant factor is a translator from the language of some other machine. This might be quite large, so this argument is most useful when we are considering the rate at which program size grows with problem size.

string is itself embedded in the program. Thus the size of the program is at least as long as the string itself.

It's easy to see that most strings are random in this sense. There are 2^N strings of length N. But there are $1+2+2^2+2^3+...+2^{N-1} = 2^N - 1$ programs of length less than N, thus most strings require long programs to describe them. There is a story told of the English philosopher, Bertrand Russell, challenging someone who thought that ABC123 was an interesting car registration number by asserting that it had the same probability as any other number. Leaving aside the huge sums paid for special car number plates, our naive commentator did actually have a point. A sequence like this can be generated by a short algorithm which is a relatively rare event for any given string.

It is often noticed that most human brains cannot compete with the simplest computer when it comes to numerical computation, yet even the most powerful supercomputer is not in the running when it comes to recognising faces. We can express this formally with these complexity concepts: neural networks have high algorithmic capacity compared to their time capacity; conventional supercomputers score highly in the time capacity. *Thus the evolution of neural computation may not so much be a consequence of the feasibility of alternatives in biology but its greater suitability to the real world.*

10.2.2 Functional approximation

Cybenko (1989) demonstrated that a simple *feed-forward* network can in fact represent arbitrary mappings within I_p, the P-dimensional hypercupe. In this simple form of network we have a set of p inputs, x_j connected to a set of neurons h_k with connection strengths $w_{jk}^{(i)}$ and thresholds θ_k. These neurons, often referred to as hidden, since they come between input and output, are connected to the output neurons with strength $w_k^{(o)}$ such that for $\epsilon > 0 \; \exists \; N$ and

$$F(x_1 \ldots x_n) = \sum_{k=1}^{N} w_k^{(o)} \phi \left(\sum_{j=1}^{p} w_{jk}^{(i)} x_j - \theta_k \right) \qquad (10.3)$$

such that

$$|F - f| < \epsilon$$

for all x_j in I_p, the P-dimensional unit hypercube. Although the number of hidden neurons may be arbitrarily large, this is still a very simple network. There are no recurrent connections, there is no time dependency and there are only two homogeneous layers, in fact almost all the biological complexity is missing.

But given that a representation exists, can we find it by training and learning? It may not be that easy.

10.2.3 The complexity of training

Judd (1990) shows that the problem of determining the connection strengths, what he refers to as *loading the network,* is far from easy. Without going into the detail of his nomenclature and analysis, the principle might be stated as loading a task onto a given neural network graph grows very rapidly, possibly exponentially, with the size of the problem (total size of architecture and task, see Judd (1990, p. 42). A task is simply an association of *stimulus–response* pairs which the network has to learn. Judd's result is independent of task, node functions and architectures.

10.2.4 Classification

One particular stimulus response problem is classification, which has a long history independent of neural networks. Similarly learning a classification from examples can be addressed independently and recent books by Anthony and Biggs (1992) and Kearns and Vazirani (1994) fuse the two fields. What is interesting here is that we get one measure of the computing power of a neural network. It's not yet a very precise one or one for which there are results for a wide range of network architectures. Firstly we get a new tool to measure the power of a neural network to learn concepts, the *Vapnik–Chervonenkis dimension* (VCD). Secondly we get bounds on how many samples we need to effectively learn a concept, or, in other words, for the neural network to generalise from a limited set of examples. How many cats do we have to see to recognise a cat? So far these bounds are very coarse and much research remains to be done. The analysis does not concern itself with how learning takes place and considers only the simplest of neural networks. The behaviour of a biological network in the real world is on a totally different scale. We are going to assume, too, that the network is subjected first to a training phase and then subsequently to a testing, or usage phase†, again not a

† commonly referred to as offline learning.

property of a continuously adaptive biological system. We start with the example space, X, of examples, x_i. A target concept, f, will assign to each example a true or false label, dependent upon whether it is an example of the concept or not. So we want a training algorithm, which, fed a subset of X, will generate a hypothesis, f^*, from a concept class, F, which will classify all examples of X correctly, or, realistically, will minimise the error functiion

$$e(f^*) = |f(x) - f^*(x)|. \tag{10.4}$$

How big must this training subset be to get a given level of accuracy?

Now consider a training set, s, a set of N pairs of example and a classification bit (labelling the example as an example of the concept or not), b_i, i.e.

$$s = \{(x_0, b_0), (x_1, b_1), ..., (x_N, b_N)\}. \tag{10.5}$$

Now let us switch our attention to the learning machine and ask how it can cope with an ever increasing number of samples. Obviously the number of samples has to be finite. But the concept class may be representable in some sense by a finite number of samples. This notion is formalised by the VCD. Consider some set of N examples and their classification into two categories. We should consider what these examples are: they could be boolean values, real numbers or whatever, but the reader interested in the mathematical fine points should consult Anthony and Biggs (1992) or Kearns and Vazirani (1994). There are 2^N possible classifications. A concept class, F, *shatters* these examples if it can represent all such possibilities. The largest N, for which F can do this is, the VCD, which may be finite or infinite. Not surprisingly it turns out that only finite VC spaces are learnable. We can now set limits on the size of the sample set, or sample complexity, required for learning a function class, F.

The behaviour of a machine of finite VCD is expressed by the following theorem after Baum and Hausler (1989), adapted from (Bose & Liang, 1996)

The probability that there exists an error function $e(f^*) > \epsilon$ where $0 < \epsilon$ but where f^* only differs from a fraction $(1 - \gamma)\epsilon$ of the training examples is bounded by the inequality:

$$P \leq 8\Delta_F(2N)\exp\left(-\gamma^2 \frac{\epsilon N}{4}\right). \tag{10.6}$$

Vapnik's original observation was that the *growth function,* $\Delta_F(2N)$, increases exponentially with the number of samples up to some point, at which its increase falls to typically low-order polynomial. At this point the exponential term rapidly starts to dominate. This point is the VCD, a property of F.

The error function measures the capacity to *generalise,* i.e. to correctly predict the target outside of the training examples. Clearly the larger the VCD, the greater the number of samples required to get good generalisation. If the VCD is too large, then noise in the data tends to be fitted at the expense of generalisation. Thus, we do not want to try to learn a simple function with too powerful a function class (or neural network).

All that remains is to relate the size of a neural network to the VCD. Some simple bounds are already known for a particularly simple class of neural networks, feed forward networks:

$$d \leq 2W \log_2 z \qquad (10.7)$$

for W variable weights and thresholds and z computational nodes. Again, we might expect the VCD to be linearly related to the degrees of freedom in the network, which is the number of real variables and this is approximately the case.

The discussion has addressed the notion of what is learnable rather than how fast. However, concepts with a finite VCD are feasible to learn in polynomial time to within specified error tolerances according to Valiant's (1984) PAC (probably approximately correct) paradigm. But the VCD ideas trancend individiual paradigms, e.g. Kearns (1994, p. 70). So we have a bound on how well simple neural networks can classify *fixed* data sets. However, in the animal world, networks learn continually as their environment changes.

In practice, many, many application developers have pushed into classification entirely empirically. Off the shelf neural network software opens up the game to a huge number of players and the rules are simple: just try it. In fact the results achieved are often no better, sometimes a lot worse, than might be achieved by statistics – but far less mathematical understanding is required for a superficially satisfactory result. Ripley (1996) has recently redressed the balance between neural networks and statistics.

10.2.5 Associative memory

Two discoveries stand out in the 1980s as stimulating interest in artificial neural networks. One of them was a training algorithm for feed-forward networks, which does not really concern us in our pursuit of complexity. But the other is close to the very foundations themselves. John Hopfield, a Caltech physicist, introduced a neural network with a strong relationship to *spin glasses*. These so-called disordered materials had led Phillip Anderson to a Nobel prize a decade before. But he went even further along with Murray Gell-Mann, another Caltech Nobel laureate physicist, to pursue complexity itself and together they founded the Santa Fe Institute for the study of the sciences of complexity.

Amongst the first disordered materials to receive serious study, spin glasses reveal complex behaviour from large assemblies of binary entities (the atomic spins). Take a common *metal* such as copper, add about 1% manganese, another common metal, and something dramatic happens to the magnetic properties. The alloy is a *spin-glass* a state of matter which ushered in a whole new set of ideas in solid state physics. The underlying theory turned out to generalise far beyond the original studies, finding applications in other properties of matter, biological systems and even computational complexity. A qualitative overview may be found in Stein (1989).

The magnetic properties of these metals arises from the interactions of unpaired electrons and thus net electronic spin on each atom. The presence of the impurity in the spin glass material perturbs the local electronic distribution and thus affects the polarisabilty. But since the quantities are small, it should have been possible to treat the impurity as just a perturbation of the electronic properties of the metal. In fact, most of the existing solid-state theoretical methods failed dramatically and the nineteenth century edifice of statistical thermodynamics came under attack.

The weakness was *ergodicity*. Ergodic systems display the same average properties whether the averages are taken over time or over an *ensemble* of many examples. A particular ensemble, the *Gibbs canonical ensemble*, plays a special role here: the probability of each system configuration, J, is given by

$$P_J = Z^{-1}\exp(-\frac{E_J}{kT}) \qquad (10.8)$$

where E_J is the energy of the configuration, k Boltzmann's constant and

T the absolute temperature. Z is a normalising constant, referred to as the *partition function*

$$Z = \sum_J \exp(-\frac{E_J}{kT}). \quad (10.9)$$

The sum is over *all* configurations: the system can wander through the whole of state space given enough time.

A spin glass, however, can be trapped forever in particular regions of phase space. The example with which we introduced the discussion was a *random site* model: we have an impurity occupying various sites in the metal at random. Edwards and Anderson (Mézard et al., 1987)) transformed this to a *random bond* model, where the spins remain homogeneous on a regular lattice, but there are random strength interactions, or *couplings*, among them. The Hamiltonian is

$$E_J = -\frac{1}{2} \sum_{i \neq j} W_{ij} S_i^{(J)} S_j^{(J)}. \quad (10.10)$$

The coupling strengths are given by W.

First consider the case where most of these connections are zero. Then the spins can align themselves to the lowest energy configuration. Thus the overall energy will be the sum of the absolute value of the connection weights, and will obviously be unique. So far so good. But now as more and more connections are added, conflicts of interest between the spins arise and *frustration* appears.

Mezard et al. (1987) express the concept of frustration beautifully through the old adage that you can't please all of the people all of the time. Each connection adds a constraint to the direction of a spin, so it's fairly easy to see that as soon as we get above two spins there is potential conflict. Going now to the extreme of full connection, it also is apparent intuitively that almost no state will be significantly better than any other. Thus the energy minima become numerous and shallow. Exactly the same conclusion appears in Kauffman's (1993) analysis of the NK model of evolutionary fitness landscapes. These many shallow minima become the memories of the equivalent neural network.

In the sort of systems envisaged by the pioneers of statistical thermodynamics, at non-zero temperatures a system may get trapped in a local minimum but sooner or later it will shake out of it and ultimately end up in the lowest energy state. But in these frozen glass systems each of these local minima is really deep. Thus the system would take forever to get to some other configuration. This is broken ergodicity: the system is trapped forever in some region of phase space. In the partition function, we now have to recognise the restriction of phase space and sum over restricted regions.

Recognising the binary nature of the spiking neuron led physicists starting with Hopfield to explore models of neural systems based on spin systems. The spin states correspond to on/off states of neurons, the exchange couplings to the connection strengths and the energy function for the network corresponds to the Hamiltonian of the spin glass. In both cases, the time evolution of the system goes towards energy minima. Hopfield's model behaves as an associative memory and as an approximate computational device for hard problems such as the travelling salesman. The model has attracted much theoretical study of its capacity and learning properties.

Amit and others (1989) have pursued the simple physical models to derive global properties of neural dynamics in so-called attractor neural networks. Applications of statistical mechanics yields information about capacity, resistance to noise, spurious attractors and other properties. Such spin models are somewhat *ad hoc* but Kruglyak *et al.* (1993) show how to derive spin-based models directly from real biological models as the properties of a biological neuron are simplified.

There is a community, probably a growing one, of neuroscientists who believe that these large scale non-linear dynamics are *the key* functional property of brains. We shall return to the issue later in §10.4.

10.2.6 Simple oscillator models

Christian Huygens, discoverer of the wave theory of light, also gave one of the first reports of coupled oscillatory systems. He noticed that clocks on the same wall would interact through vibrations transmitted through the wall. This would synchnonize, keeping them precisely locked to the same time difference. In the simplest case the models take the form

$$\frac{d\phi_i}{dt} = \omega_i + \sum_{j \neq i}^{N} w_{ij} \sin(\phi_j - \phi_i) \quad (10.11)$$

for N oscillators of phase ϕ and frequency ω. The weights w_{ij} are often takento be a single coupling constant. In this case it is clear that all the oscillators will tend to lock to the same phase. Fig. 10.1 shows the phase difference of a population of ten oscillators. with the same frequency,

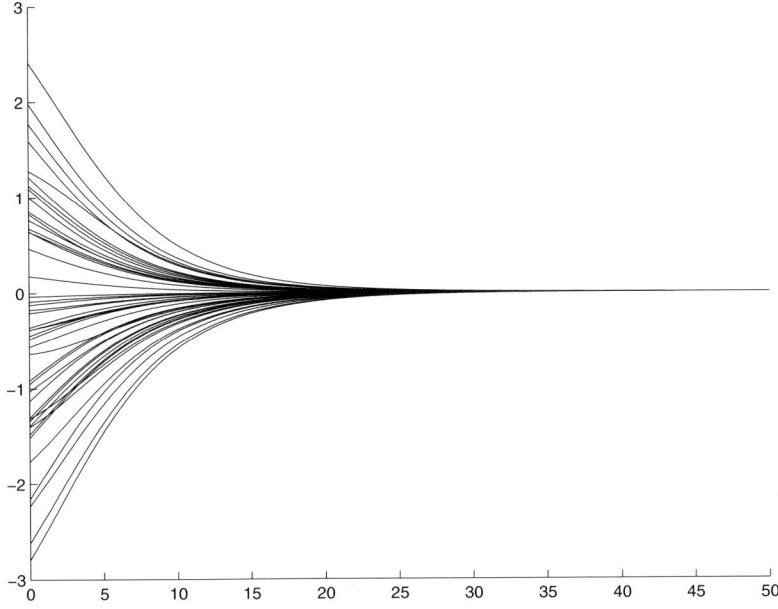

Fig. 10.1. Convergence of phase in an oscillator population.

starting from random values. If noise and some variation in frequency are added, phase locking still occurs, albeit with some error margin.

Note also that all the connections serve to encourage phase locking, if the weights are positive. Interesting results from such models are Niebur *et al.* (1991a):

- Although synchronisation occurs with nearest neighbour connections, strong robust synchronisation also occurs with sparse random connections (Niebur *et al.*, 1991b). Since connections between cortical columns are usually much less dense than intra-column connections, this has obvious implications for synchronisation between columns (§10.4.2). One might expect from the graph theory results of chapter 3 that only a small number of possible connections would be necessary to create a Hamiltonian circuit and hence the possibility of total synchronisation.
- Transmission delays reduce the period of oscillation. This is a big effect, small delays introducing large variations in frequency. Since neural transmission is not instantaneous and synchronisation, as we

shall see later, occurs over long distances in the brain, this result is extremely interesting. The implications are not yet fully elucidated.
- If one moves away from the no noise, single frequency model, more complex behaviour is observed. Simple nearest neighbour connections often lead to phase lags across the network and long delays in synchronisation.

10.3 The biological neural network

The intelligence of a brain is the result of a myriad of complex systems all interacting (successfully) together, demonstrating complex, adaptive behaviour from many components. Nerve cells do a great many jobs and are specialised in all sorts of ways. They vary in size from the tiny neurons of the human brain, crammed in at a thousand million per litre, or a quarter of a million per square millimetre of cortex†, the spinal chord neurons which run the length of the body, and the grand champion of the animal kingdom, the giant squid, whose monstrous axon underpinned much early neurophysiological research. Anatomically the morphology of neurons varies dramatically: some look like small knotted clusters, others are diffuse, as varied as bushes and trees. Not unsurprisingly their branching is often fractal-like (Smith & Lange, 1996).

Nowhere are hierarchies of complexity more evident:

- neural systems are heterogeneous with numerous different types of cells and interaction mechanisms;
- behaviour is temporally and spatially distributed; not everything happens instantaneously;
- there are practical limits imposed by metabolic and other constraints; amongst competing theories of the function of sleep is the claim that it serves to replace the energy reserves of nerve cells in the brain (Coren, 1996);
- prior knowledge (inherited as structure, connectivity, dynamical constants) is important in determining overall structure and predisposition to learning of particular tasks. The nature of this prior knowledge is subtle, however, and there is no universal agreement on the form it might take.

The extent to which all of these properties are essential to the working of a brain remains open. Neurons have to occupy space, but is the detail

† As we shall see later the cortex is a highly convoluted sheet of six layers of cells, thus the area measurement is somewhat more natural.

of their physical packing essential or intrinsic to their computational properties? Does the complex spatial organisation subserve information processing? At one extreme it might be irrelevant. At the other it might be crucial for allowing messages to arrive from one neuron to another sufficiently quickly or it might even have a control function in deliberately slowing down messages from more remote neurons.

In looking at the ways real neurons differ from the McCullough and Pitts archetype, we should start with a few defining anatomical features:

cell membrane: the encapsulating skin of the cell. But it is more than a container. It contains numerous pores, or gates, which control the flow of ions across it, leading to the electrical activity through which the cell computes and communicates. We referred earlier to Penrose's proposal that the fine structure of membrane has important computational properties, but we shall not explore this idea further here.
cell body or soma: the main metabolic part of the cell. One might think of it as functionally dividing the input from the output.
axon: the extension of the cell body allowing the output of the cell to be received some distance from the input, in the case of say spinal neurons, a comparatively huge distance away†.
dendritic tree: this is the input (afferent) part of the neuron, a vast intertwined tree of connections to other cells.
dendritic spines: are small protrusions from the dendrite on which synaptic junctions are usually located.
synapses: join input to output dendrites. Originally thought to be actual joins between the cells, effectively creating one huge cell, now known to be gaps across which cell impulses transfer information.

The cell's computation and communication is electrical. Potential differences build up between the inside of the cell and the exterior. Small differences in the extremities of the dendritic tree accumulate at the cell body. There are just two major types of output, despite the vast variation in cell morphology: graded output cells or spiking cells. Computation involves both, of course. From the vertebrate eye the *ganglion* cells

† In passing we might note that neurons specialised for long distance communication are usually *myelinated*, i.e. covered with a sheath of high electrical resistance which increases conduction speed and reduces interference between adjacent neurons. In the brain itself the grey matter is unmyelinated, distinguishing it from the white matter.

transmit visual information to the brain as spike trains while the cells intermediate between them and the photoreceptors, the amacrine and bipolar layers (Rodieck, 1973), transmit this total voltage out continuously, Spiking neurons transmit nothing until some threshold is reached at which point a discharge, or *spike,* occurs, travelling out along the axon. You can imagine this as a lightning strike: a sudden discharge of electricity developing a forked pattern as it approaches the ground‡. The output of the cell is the temporal spike sequence. For a long time the rate was thought to encapsulate all the information, but this is now recognised as simplistic.

At some points along this forked path, the active neuron will touch other neurons at so-called synaptic junctions. At these points the electrical surge can jump across to the neighbouring cell. There was some argument in the nineteenth century as to whether a tangled mass of neurons was in fact made up of individual cells, since the many points of contact look, at low resolution, like the fusing of the cell membranes. The electron microscope, however, established beyond doubt that each cell is completely closed off from its neighbours having its own cytoplasm and enclosing membrane. With hindsight this is not surprising: obviously the connection to other cells has to be highly selective, and the method nature has evolved is for the transmission across the synaptic gap to be chemical rather than electrical, thus neatly keeping the neurons separate†.

10.3.1 The single neuron

Our extremely concise survey of the single neuron will look first at the development of voltages across membranes, a fascinating example of biological control. Then we study the development of the spike discharge and finally consider how the spike influences the other cells coming into contact with the efferent dendritic tree.

10.3.2 Behaviour of membrane

The essence of the computation and communication mechanisms of the neuron is the control of the electrical potential across the membrane

‡ In fact both the dendritic trees of neurons and the patterns formed by lightning are fractal.
† But there is more than one mechanism. The most well known, and the first to be firmly identified, is this mode of chemical transmission, but electrical coupling mechanisms are also known (Kuffler *et al.*, 1988).

Table 10.1. *Ionic concentrations for a typical cell in* millimoles/litre. *The (impermeable) anion may be one of several possibilities, such as aspartate, although these details are unimportant here*

Ion	Outside	Inside
K^+	3	90
Na^+	117	30
Cl^-	120	4
A^-	0	116

of the cell. This potential arises from an ion imbalance between the interior and exterior of the cell. The membrane is electrically active: it selectively controls which ions can move across it and a powerful feedback mechanism generates the spiking activity of the cell.

Synaptic activity leads to changes in the permeability of and thus ion fluxes across the membrane. Local changes in membrane potential percolate along the dendritic tree towards the soma. The axonal discharge, or spike, is a massive change in membrane properties resulting from positive feedback (chapter 7).

To begin with we shall look at the balance of the three major ions, potassium, K^+, sodium, Na^+, and chloride, Cl^-. As indicated by the superscripts potassium and sodium each carry a single positive charge while chlorine carries a single negative charge. The concentrations of these ions inside the cell are different to the concentrations outside; illustrative figures are given in table 10.1, adapted from Kuffler (1988, p.113).

To be in equilibrium (Katz, 1966, p.42) several conditions must be satisfied:

- charge neutrality: there must be no net charge on either side of the membrane;
- osmotic balance: the total concentration of ions on each side must be the same; if the concentration were to differ, water would flow from the weaker to stronger solutions to equalise the concentrations;
- the potential gradient across the membrane exactly balances the net ionic drift; note that we talk about net drift: ions are still moving backwards and forwards.

Inspection of table 10.1 shows that these conditions are satisfied. But the membrane's permeability to different ions varies. In the resting condition the permeability to Na^+ is very low, but K^+ and Cl^- are basically free to move backwards and forwards across the membrane.

The flux of an ionic species is a balance between the concentration gradient (osmotic pressure) and the potential gradient (electrostatic force). The net flux will be zero when these two forces balance. Such an analysis is one way of deriving the ubiquitous Nernst equation (Katz, 1966, pp. 54–56):

$$E = \frac{RT}{Fz} \log_e \frac{[I_o]}{[I_i]} = 58 \log_{10} \frac{[I_o]}{[I_i]} \qquad (10.12)$$

where $[I_o]$ and $[I_i]$ denote concentrations in millimoles/litre† and the voltage, E, in millivolts, is the *equilibrium potential of the ion.*, F the Faraday constant, R the gas constant and z the charge on the ion.

With more than one ion, the membrane potential is now a balance of the movement of all ions and thus depends on the permeabilities. For diffusion against a potential gradient, we have the following expression for the ion flux, $f_i^{(i,o)}$ inward or outward,

$$f_k^i = \frac{p_k z V'[k_i]}{(e^{zV'} - 1)} \qquad (10.13)$$

where p_i is the permeability of ion i in the absence of an electrical potential and $V' = \frac{F}{RT}V$ where V is the signed potential and T the absolute temperature. We can thus derive a net flux for an ion as

$$f_i = f_i^{(o)} - f_i^{(i)} = p_i V' \frac{[I]_o - [I]_i e^{zV'}}{e^{zV'} - 1} \qquad (10.14)$$

and the current is

$$i_i = F f_i \qquad (10.15)$$

where square brackets are used to indicate ion concentration‡. Now

---

† A mole is a specific number of ions, Avogadro's number, which has the value 6×10^{23}.

‡ There is some slight electrochemical licence here. We should really use not concentration but a roughly equivalent quantity, the *activity*. In some situations, such as very high concentrations, the two are not identical, but for present purposes this is unimportant.

adding up the fluxes in each direction for all three ions, we arrive at the Goldman equation:

$$V_m = 0.58 \log \frac{[K^+]_o + \frac{p_{[Na+]}}{p_{K+}}[K^+]_o + \frac{p_{Cl-}}{p_{K+}}[K^+]_i}{[K^+]_i + \frac{p_{Na+}}{p_{K+}}[K^+]_i + \frac{p_{Cl-}}{p_{K+}}[K^+]_o} \qquad (10.16)$$

from which we can calculate the membrane potential. Using the permeability ratios, 1:0.03:0.1 for K, Na and Cl, we obtain for the table above a resting potential of minus 68 mV.

The cells signalling method depends upon these permeabilities being time and voltage sensitive. As the cell's voltage changes as a result of input to the cell, a cascade of events occur. If K^+ leaves the cell, we get *negative feedback*, tending to restore the resting state. If on the other hand, Na^+ enters, it *increases* the membrane permeability to Na^+, more enters and a rapid change in potential follows: this is the origin of the action potential or voltage spike. A rapid total depolarisation of the membrane occurs which ripples down the whole length of the axon. A slower, pumping process now restores the original ion balance: during this period the cell is relatively insensitive, giving rise to the so-called *refractory period*. Thus at this low-level analysis of the neuron we have complex behaviour controlled by feedback.

This highly sophisticated system imposes a number of constraints on wetware computation: sensitivity to extracellular ion concentrations, recovery time, energy supply for the ion pump and so on. But we have not covered one zillionth of the variables in a typical biological cell: there is greater variety of ions, there are all sorts of moderators and we still have something, maybe a great deal, to learn about the fine structure of membranes. Collective properties have to be robust to all these final details of biological function. But let us move on to consider the integration of membrane into a functioning cell.

10.3.3 Dynamics of a neuron

What we have seen so far is the behaviour of patches of membrane. How do we now integrate all these processes? Hodgkin and Huxley's seminal work, begetting them a Nobel Prize, models cell currents according to

$$i_m = C\frac{dV}{dt} + i_{active} + i_{leak} \qquad (10.17)$$

where i_m is the current travelling across the membrane, C is the membrane capacitance, V is the voltage across the membrane, i_{active} is the current due to controlled ion channels and the leakage current is simply the Ohm's law current produced by the voltage across the membrane. Hodgkin and Huxley determined the precise kinetics of the active currents, characterising the individual membrane *gates* which control the ion permeabilities.

Of the many ways of building a model of an entire neuron, the most popular at the present time is the so-called compartmental model. At this point we leave behind differential equations we might have any hope of solving as a single entity and build up the full neuron out of pieces. The simplest way of doing this is with a *cylinder* model in which we set up balances of current flow in and out of each component volume. Current in and out of a junction should sum to zero. Thus the longtitudinal current, $i_{j,j+1}$, between cylindrical segments, j and $j+1$, together with $i_{j-1,j}$ should match the membrane current, i.e.

$$i_m = i_{j-1,j} - i_{j,j+1}. \tag{10.18}$$

Now, with up to 10,000 compartments for an adequate simulation of some types of nerve cell, it should not seem surprising that this sort of level of analysis is not yet applicable to large networks. Yet it should be clear that spatial organisation and structure will clearly influence the time course of neural activity. But to examine computation in *networks* we have to discard the fine structure and *many* sources of non-linearity and noise will be pooled into a small number of composite functions (McKenna et al., 1992, chapter 4).

10.3.4 Synapses and neural communication

Chemical transmission proceeds via the transfer of a *neuro-transmitter* from across the synaptic cleft, the gap between the two neurons at the synapse. The transmitter is contained in discrete packets, or vesicles, each vesicle being referred to as a single *quantum*. The neuro-transmitter causes ion permeability and hence voltage changes in the dendrite on the other side. Each vesicle contains roughly the same number of transmitter molecules, typically a few thousand. The number of quanta rather than the vesicle size determines the strength of transmission.

Transmitters cause either a hyperpolarisation or depolarisation of the neural membrane on the other, post-synaptic, side. In the former case the synapse is inhibitory, in the latter excitatory. Strangely, any given transmitter molecule may have either effect, although it will always be the same throughout any given cell class. Acetylcholine is a widespread transmitter, which is usually excitatory. Ironically it was one of the first transmitters to be identified and, in the system in which it was found, the heart muscle of the frog, its behaviour was inhibitory.

The transmitter opens ion channels, each molecule opening many channels so giving an amplication of a thousand or so. There is some delay, typically around a millisecond, in a chemical synapse, mainly the delay in opening the ion channels. Thus this delay is reduced for new spikes arriving shortly afterwards. There are *several timescales of synaptic interaction,* ranging from this very fast facilitatory effect through to learning which is of course longer term.

Each cell makes only one kind of synapse, sign reversal being accomplished through specialised *interneurons.* There are more some 200 neuro-transmitters and a whole host of modulators which affect the transmission process. The process of learning in neural networks is the process of changing synaptic strengths, but the precise processes by which this occurs are still an active research topic.

Still more complicated is the nature of integration of signals across the spatially distributed neuron. The dendritic splines have their own somewhat non-linear properties, as reviewed by Brown *et al.* (1992).

Summarising, we see that the transmission of information between nerve cells:

- is noisy;
- has assorted time delays;
- adapts;
- occurs using a variety of transmitters and synaptic mechanisms;
- is spatially distributed.

We are still a very long way from disentangling the accidents of evolution from the compuationally necessary components. One interesting perspective on the significance of this low level detail comes from a quite different research field. Adrian Thompson at the University of Sussex has used evolutionary computing to program VLSI gate arrays. He finds the most incredible sensitivity to the individual electrical structure (i.e. manufacturing variance) of each chip, going way beyond the model of the chip as an array of logical gates. Thus evolution *is* sensitive enough to make use of the detailed organisation and dynamics of a biological brain.

10.3.5 Information in the spike train

The earliest measurements of neural information transmitter established that the firing rate, the number of spikes per second, increased with stimulus intensity. Thus firing rate became established as the measure of activity. It is still a reasonable approximation, but it is clearly not the whole story and many alternatives or refinements have been proposed. Optican and Richmond demonstrated that a spike train may carry more information than that in the firing rate alone and more recently Bialek and others have looked for quantitative measures of the information transfer (see Rieke, 1996, for a recent survey). Their results have two salient features:

(i) There is far more information in the spike train than conveyed by the firing rate alone. Individual spikes are important.
(ii) The recovery of information from the spike train proceeds through a filtering operation. The recovery algorithm is robust to loss of spikes and temporal jitter (Bialek *et al.*, 1991). The filtering operation is not unlike the time course of the generation of a voltage potential across the synapse from the arrival of a spike from another neuron.

For example, motor neurons show decidedly regular spike trains, whereas cortical neurons are closer to being Poisson distributed. Why should there be such a huge variation if the coding is based only on firing rate? In fact it is difficult to get Poisson distributions using the simple integrate and fire model (Fujii *et al.*, 1996).

In their book, *Spikes,* Rieke *et al.* (1996) effectively dispose after many years of the principle that firing rate is the dominant means of neural signalling. But, then, as soon as we admit the importance of precise timing, all sorts of new theoretical ideas open up.

Neurons could have varying integration times, may be sensitive to spike volleys only, learning might require simultaneous arrival of many spikes. *Temporal dynamics become crucial.* Thus neurons tend to behave more like spike coincidence detectors. If this turns out to be correct this will be a sea change in our thinking of how nerve cells work!

10.3.6 Learning and memory

Users of artificial neural networks are often concerned with training, providing a series of examples and correct responses. But learning in biological systems often has much more of a self-organisation character. In fact a neural learning mechanism, proposed by Donald Hebb (1949) half a century ago, still finds widespread use:

If two neurons fire at the same time, then the synaptic contacts between them are strengthened.

This simple suggestion has proved surprisingly productive, predicting spatial senstivity profiles of cell populations (Linkser, 1986). It has needed modification, and the wetware continually throws up new challenges: new chemical transmitters and modifiers are still turning up. Nitric oxide, for example, has now been clearly identified as neurologically significant, even though it is an extremely toxic gas. But unlike the precise point to point communication of the synapse, it diffuses through the network. Thus it acts as a local modifier, but thereby imposes constraints on the *topography* of a neural network as well as its *topology* defined by the network graph.

10.4 Neuronal populations

It is tempting to think, particularly in artificial intelligence applications, of simple neural networks as having some limited number of outputs representing specific things. The output might be a classification: the input is a pig or a goat; it might be a function value; or the complete representation of a partial input.

But the common notion that biological neurons are slow and noisy but operate in massively parallel arrays led to the idea of population codes. The representation of, say, a particular person's face is distributed across many neurons. There are numerous concrete examples. Georgopoulos *et al.* (1986), in recording from primate motor cortex, found cells broadly tuned to direction of movement of the arm. But the precise direction of a movement was found to be accurately represented, not in a single cell, but by a *vector sum* of the direction tunings of all the active cells. It is difficult to achieve definitive experimental evidence, however. Konishi (1991) cautions that a similar suggestion was put forward for electric fish, but eventually a cell was found which was tuned to fine stimulus discrimination. Since we are still learning about the precision of timing of individual spikes and the information capacity of an individual neuron, the story of population codes is far from over.

When we come to huge networks which make up a biological brain, we encounter continuously varying complex inputs resulting in complex behaviour. What happens in between? The last decade has brought much deeper understanding of the temporal dynamics of neural populations and there is much excitement ahead in understanding their computational properties. But, the idea that the important behaviour lies in large scale dynamics is not new by any means. Particularly seminal is Walter Freeman's pioneering work, described in a book, *Mass Action in the Nervous System* published, remarkably, in *1975*, and many papers since.

Beyond the idea of population codes we have cooperative behaviour, in which the activities of cells are *temporally* correlated. Again the idea has a long history, with Christof von der Malsberg often considered the pioneer. There are several forms of correlation:

synchronous firing: Neurons fire spikes together to represent some sort of binding among them. Many simple neural models exhibit synchronous firing with the right choice of parameters. Particularly effective is global feedback, which will produce regimented spike trains in simple models. Synchronous firing is appealing, because it fits our intuition that many neural events have got to occur rather quickly and it is also easy to see how it might be used for learning or activating cells at higher levels.

synchronous oscillation: The *firing rates* of individual neurons *oscillate* and these oscillations phase lock across populations or sub-populations. There is still some debate about the nature of these synchronies, let alone full agreement on their functional role.

large-scale dynamics: This is essentially an evolution of Freeman's idea of mass action, wherein whole populations of cells act in synergy.

It quite often happens in science that a breakthrough is imminent, perhaps with quite a few important results around, but it takes a particular perspective to make everybody sit up and take notice. In some ways this is how the cooperative behaviour movement took off: it was a paper by one of Freeman's former students, Charlie Gray, that sparked much wider interest in the idea of *linking* by temporal correlation.

10.4.1 The linking hypothesis

Anyone with any experience in image processing will know that extracting edge features from images is not that difficult. Linking them together to form coherent boundaries of objects and consequently segmenting the image is a totally different matter. How the linking takes place in a neural architecture is therefore of considerable interest and not a little puzzling.

One of the ideas, now somewhat discredited, is the idea of a grandmother cell: somewhere deep inside the brain is a cell which responds specifically to grandmother and no-one or nothing else. The trouble with this idea is that it requires rather a lot of cells, with even 10^9 is probably not enough, aside from more philosophical issues. But within the specific vision context, are we likely to find cells which respond to long borders? This would require rather a lot of cells too. Suppose we start with a token which we can join to say four other tokens. This is pretty restrictive, since in the natural world edges are hardly very smooth and linear. Even so, if we now join n tokens together, then the total number of possible contours grows exponentially in n. If we allow for this same border system to be replicated at other points in space, we end up with a vast number of cells, just for this preliminary stage of processing.

Thus we arrive at the idea that combinations of primitives might be represented not by further cells but by the temporal locking together of the cells signalling the primitives: each independent category phase locks, at a different phase to all the other oscillators. *The linking hypothesis* (§10.4.1) thus takes a very simple form in oscillator models. Thus we would have a different phase between figure and background, or between overlapping figures and so on. There is some doubt, however, as to the number of phases which can be kept distinct. This notion received direct experimental support from the work of Gray and Singer (1989) which we report in the next section.

We show in §10.4.3 that real neural equations do have oscillatory properties and see evidence for large-scale oscillatory behaviour in the whole brain (§10.4.4).

10.4.2 Linking through oscillation

Gray and Singer (1989) reported coherent oscillations in area 17 of the cat. Area 17† is the first cortical processing area for visual information.

---

† Naming of cortical structures is non-trivial and numerous historical names are mixed in with systematic classifications. Area 17 is a term used by Broca, a

In it are found cells which are tuned to be more sensitive to some orientations rather than others and frequently sensitive to movement in a particular direction. A recording from such a cell is a more or less random train of impulses and approximates a firing rate code. However, something additional occurs for an extended bar stimulus: the *firing rate* starts to oscillate at a frequency between 40 and 80 Hz. Yet more interestingly, these oscillations *phase lock* across the length of the stimulus. This phase locking occurs over a significant distance, between major cortical areas (Engel *et al.*, 1991b), and even across the corpus callosum, i.e. between the two hemispheres of the brain (Engel *et al.*, 1991a). Why was all of this so exciting? Simply because it looked like the mechanism by which the visual system encodes an extended contour. The individual tokens of the contour do not feed into a special cell, but it is their phased-locked oscillation which tells the story. Other results came in soon afterwards, but not everybody joined the club. In particular difficulties in observing the same phenomena in monkeys caused some concern.

10.4.3 From neuronal groups to chaos

A somewhat different perspective is that of Edelman's notion of neuronal groups (Edelman, 1987). Edelman talks of the *sciences of recognition*, a view originally formed from somatic mutation in the vertebrate immune system. Neuronal maps, such as we have discussed for the visual cortex, and the cells within them are formed by a combination of evolution and learning. Again, feedback between groups, *reentry*, in Edelman's terms, is crucial. Reciprocal connections between groups of say up to 10^4 neurons in different maps develop to correlate activity between the groups. Such feedback loops lead naturally to oscillatory behaviour and large-scale simulations have been used to demonstrate cooperative activity across groups, akin to the experimental results of Gray and Singer (1989, 1991).

Freeman has been exploring these mechanisms for over two decades. His emphasis is not on individual neurons, but on neuronal groups. Thus individual firing rates are averaged. The spatial temporal patterns of neural activity produced by a stimulus may be oscillatory or even chaotic, but it is this pattern which characterises the memory.

Of Freeman's very many papers just a few basic ideas can be given.

pioneer of brain anatomy (Zeki, 1993, pp. 17–21), and is still frequently used in cat physiology.

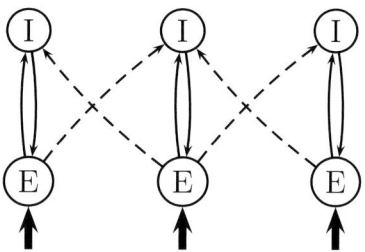

Fig. 10.2. Oscillating neuronal groups (after König and Schillen, see text).

The models come directly from physiology, mostly the olfactory bulb of the rabbit. Neurons are treated in groups, hence allowing the spiking activity to be averaged to a mean activity. Freeman's simulations are calibrated precisely with detailed parameters taken directly from physiology and reproduce the odour recognition of the bulb quite well.

His recent work has moved from oscillations to the discovery of chaotic attractors in the dynamics of the neurons. Each odour is stored as a spatio-temporal activity pattern. By this we mean the behaviour of each neuron in space as a function of time. A recent review will be found in (Freeman, 1991a) and a qualitative overview in *Scientific American* (Freeman, 1991b).

As one might imagine having read chapter 7 in this book, it is fairly easy to generate feed-back from a set of differential equations. The following model of König and Schillen, from the same group as Gray and Singer, attempted to describe the oscillatory behaviour they had discovered. It should be taken as illustrative of the mechanism rather than definitive.

In fig. 10.2 we show the basic layout of pairs of neuronal groups. Input goes to excitatory, E, which feed inhibitory cells, I. These in turn inhibit the first set after some time delay τ_d. Oscillations are coupled by additional excitation of the inhibitory cells by neighbouring excitatory cells, as shown by the dotted lines in fig. 10.2. The activity of each excitatory cell x_e and inhibitory cell x_i is governed by the equations:

$$\tau \frac{dx_e}{dt} = -\alpha x_e - w\phi(x_i(t - \tau_d)) + i_e + \eta \qquad (10.19)$$

and

$$\tau \frac{dx_i}{dt} = -\alpha x_i - w\phi(x_e(t - \tau_d)) + \eta \qquad (10.20)$$

where τ is the effective unit of time, α governs the decay of activation in the absence of input, i_e is the external input and η is introduces a small amount of noise.

The coupling strength is given by w and the output of each cell is passed through a standard sigmoid function, ϕ, in this case the Fermi function:

$$\phi(x) = \frac{1}{e^{\sigma*(\theta-x)} + 1} \qquad (10.21)$$

where σ is the slope and θ the threshold for output from the cell.

Each vertical module in the figure, a combination of excitatory and inhibitory cells, forms an oscillator. Links from the excitatory cells to adjacent inhibitory cells (dotted lines) leads to phase-locking of the oscillations. On the other hand, *desynchronising connections* in which cross-excitation occurs between the excitatory cells will break up phase locking.

There are very many variations possible on a canonical system of this kind. At this stage there are many models, and the details of how things work in biological systems is back with the experimentalists.

10.4.4 The big picture

Several models, none yet universally accepted, describe how temporal binding gets used at the next level of processing:

- fast decay of synaptic activation means that a neuron is more likely to fire if it receives a concentrated volley of spikes;
- the entrainment of a population of cells may resolve ambiguity, presenting a definite decision at higher levels;
- learning may be enhanced by activation on a short timescale;
- Edelman's neuronal groups for complex pattern recognition.

Freeman's work has emphasised throughout the significance of populations of cells acting together. One of his tools as been the EEG, a method of recording electrical fields from the scalp. EEG reacts quickly but is a very coarse tool. Its lack of precision stems from the diversity of paths and interference in passage of electrical fields from neuronal groups deep inside the brain to the skull surface. However, a newer technique retains the temporal precision but adds much better spatial resolution – the SQUID array†.

Using arrays of about 40 SQUIDS, each responsive to coherent activity in a group of about 10^4 neurons, Scott Kelso and coworkers have looked extensively at large-scale dynamics in response to simple stimuli. The dynamics of the SQUID signals turn out to be of low effective dimensionality with oscillatory processes locked to the stimulus and typical chaotic behaviour (Kelso, 1995, chapter 9).

We have entered a new phase of understanding of the way brains work. Kelso in his book *Dynamic Patterns* (1995) takes us on a tour of the non-linear dynamics in biological systems concluding with the brain experiments referred above.

Chaos, phase transitions, adaptability at the edge of chaos, all seem to be part of a vertebrate brain. It is not impossible that they are epiphenomena. But as yet the computational models, of how or why they might have evolved, are at an early stage.

10.5 Architecture and innateness

For over a hundred years, we have known that the brain is not a single homogeneous mass of neurons. It has a hierarchichal, highly modular structure. In an adult vertebrate brain we find *areas* specialised for vision and the other senses, we find motor systems, specialised memory systems, bodily control systems in a highly sophisticated dynamic whole. As we focus in on, say, vision, we again find specialisation into form, colour, stereopsis and then further specialisation into topography of visual field and so on down.

At the lowest level neurons tend to be clustered in functional groups with dense internal interconnections. Connections between groups tend to be fewer, but specialised structures for long range communication abound. In general this organisation is invariant from animal to animal for a given species, implying it is in some way coded genetically. Exactly to what level these groups are represented in the genetic code is the source of some debate which we deal with in §10.5.2. First we will use mammalian vision to illustrate something of this complex hierarchy.

† Superconducting quantum interference device.

10.5.1 Hierarchichal processing of visual information

The mammalian cortex is a highly convoluted six-layer structure. Perpendicular to the surface there is a fine scale subdivision into columns each having some distinguishing behaviour. Columns are themselves clustered into hypercolumns, and so on up to entire cortical areas. Hubel and Wiesel received the Nobel Prize for their elucidation of the properties of these columns in the area 17 of the visual cortex of the cat, one of the first stages of visual information processing in the brain). Cells were found to be sensitive to edges and bars in the visual field, each specific to a particular orientation and to direction of movement across the visual field. Cells with common properties were found in the same column at a point in the visual field. Steady changes in, say orientation, lead to clustering of columns into a so-called hypercolumn. The hypercolumns are organised topographically with each looking out into its own particular area of the visual field. Quite recently the incredible regularity of this structure has been emphasised by techniques using *in vivo* imaging. Bonhoeffer and Grinvald (1991) used optical imaging methods to watch the visual cortex of cats in action in response to visual stimuli of different orientations. Results show a clear pin-wheel structure of multiple orientations.

Jumping up the hierarchy, we have multiple *areas*. In primates there are approximately 30 areas involved in vision (Van Essen *et al.*, 1992) with over 300 interconnecting pathways. These different areas are clearly specialised for function, e.g. colour or motion, and it seems reasonably clear that some progressive refinement of visual information is occurring as the centres get further from the eye. Thus area V1 has cells responsive to simple edge/line features in images, while V2, whose main input is V1, has cells responsive to illusory contours (von der Heydt *et al.*, 1984).

Much later in the processing of the visual image in the brain, in the infero-temporal cortex, similar columnar organisation appears, but the properties of the columns are much harder to work out. How, out of the vast range of possible patterns and objects we learn to recognise, are we going to find what any particular cell does? One successful approach has been to look for visual primitives which might have overwhelming importance to the animal. Thus the importance of faces in animal behaviour is so great that physiologists, such as Rolls, conjectured that there might be special areas devoted to faces. Indeed such cells were indeed found including individual facial components, but convincingly demonstrating that they were in fact specialised for faces and not for oval objects or any other simple pattern proved at least as difficult as finding them (Perrett *et al.*, 1982). More recently, Fujita *et al.* (1992) discovered a columnar organisation for simple pattern templates in the same region, the infero-temporal cortex. This is very interesting because we do not have any sort of theory as to how we might break the world up into small object templates. Their work clearly shows systematic variation as we go down a cortical column and go from column to column, but to predict in advance what this organisation would be is far from obvious at the present time.

Suppose now that we want to build a robot vision system of the same sophistication. Can we just rely on learning and self-organisation or do we have to build the complex organisation found in animal brains? How would one build in such organisational information, or do we have a situation analogous to Kauffman's Origin of Order claim – that there are a very limited number of possible organisations which would occur? In his words (Kauffman, 1993)

'...spontaneous order lies to hand, free, as it were, for selection's further molding. In particular, asking what form such molding may take and what laws might govern it, leads us to the hypothesis that the target which selection achieves is complex systems poised in the boundary between order and chaos...'

So just what *is* innate?

10.5.2 Innateness

Self-organisation models successfully mimic quite complex cortical structures (Linkser, 1986). The cells in the first stage of processing, V1, show as we have seen particular orientation senstivities. They are also tuned to specific spatial frequencies. One popular argument was that the cells were tuned to edges and bars of different scales and orientations. But there is another possible explanation. Self-organisation will produce a very similar result by Hebbian learning (§10.3.6). Linkser demonstrated this directly, but it has also been shown that such an algorithm also produces general signal decorrelation or principal component analysis. See Haykin (1994, chapter 9) for a recent discussion.

Thus how much of the entire structure might result from learning, and could inheritance be irrelevant? Philosophically this seems unlikely. Surely there ought to be some advantage in speed of learning if nothing else by starting in the right ball park. What we do know is (Zeki, 1993):

- there are critical periods during development; if normal stimuli are not received during that period, the animal is permanently visually impaired;
- from a philosophical point of view it makes sense to encode within the genes some sort of predisposition to learning algorithms which do not change on a geological time frame, such as the mapping from two dimensional images, or image sequences, to solid three-dimensional objects, once these algorithms are in some sense optimal.

10.5.3 Genetics or self-organisation

Could we then have something akin to Piaget's idea of staged learning? Once V1 has self-organised, inputs to higher levels will now be coherent and the next level will start to develop more complex features. In some cases this self-organisation takes place prior to birth, with waves of spontaneus neural activity assisitng the correct wiring up of retinal ganglion cells (Schatz, 1992) in the cat. All the time as one looks for models of how the system is put together, the problem of evolution looms large. It is simply not possible for an animal to redesign its eye. It has to continue to survive, thus some features may remain long after their usefulness has passed. This has been demonstrated conclusively in simple animals. The crayfish, for example, has a so-called kick-reflex, for which the neural circuitry has been fully unravelled. This circuitry contains redundancy arising from *pre-adaptation:* the circuits were originally used for something else and have been co-opted (Dumont & Robertson, 1986). These biological observations have been backed up by detailed simliations of the evolutionary path (Stork *et al.*, 1991).

Significant colour vision† is only present in primates. To a rough approximation we notice two things comparing monkey with cat. Primates still use the achromatic system for tasks like stereopsis, motion perception. The colour system has almost a complete set of wiring, very much like a system just tacked on. Behaviourally the indications are that colour subserves just a few very specific functions, such as the detection of ripe fruit.

Zeki, one of the pioneers of the cortical physiology of colour, presents a superb overview of the nature of cortical structure, with many insights

---

† We have to be a little guarded here. Many mammals have some variation in photoreceptor wavelength sensitivity, but on balance do not seem to have exploited it as yet for colour.

into the slow, stumbling scientific process. He sums up the situation admirably (Zeki, 1993, p.86):

Maybe the history of neurology teaches us to ask whether the concepts that we have are broad enough to accommodate the new facts that are constantly emerging and constantly questioning our deeply held beliefs about the brain.

10.5.4 Innate knowledge in neural networks

A casual perusal of the wisdom literature will no doubt unearth all sorts of claims for prior knowledge built into our brains, be it knowledge of morality, religion or anything else. Just as we have tolerated claims of this kind for millennia, we have no difficulty in believing that the detailed knowledge possessed by a bird to build its nest or the elaborate courtship rituals across the animal kingdom should be innate. Yet, when Noam Chomsky proposed that the learning of human language by children required some innate knowledge of grammatical structure, easy acceptance was very different to the response he received. Some measure of the storm he created and the immense amount of research he stimulated may be gained from the recent book, *the Language Instinct* by Stephen Pinker (1994). Judging by Pinker, the Chomsky proposal is no longer so hotly disputed, but this issue is of interest to us in a number of ways:

The neural substrate is intriguing. We have seen already that there is nowhere near enough genetic material to encode the connections of a human brain. So how could this predisposition to learn in a particular way come about? This is one of the supreme challenges of neuroscience at the present time.

The argument for innateness was essentially an inuitive one, based on the apparent difficulty of a child learning language from examples without any prior knowledge.

10.5.5 The neural substrate

Our study of network architectures and learning gives us two broad categories for the encoding of prior knowledge. Firstly there is the architecture itself, perhaps down to a quite a low level. Alternatively there is some sort of initial setting of the connection strengths. At present this seems to be an empirical issue. Computer simulations have already been attempted for simple context-free grammars to show how innateness might be encoded. However, we can make one comment about the initial values of the connection strengths. These initial values have to take some simple algorithmic form, perhaps fractal, perhaps something quite new. But in this sense of being represented by a simple program it implies some degree of regularity to the starting conditions. It is mere speculation that this in turn might impose a constraint on regularity in the trained networks.

10.5.6 Why innateness anyway?

Convincing though this may be for language, it is not dissimilar to the arguments used by creationists against evolution. Since Chomsky's writing over two decades ago, we have learned a great deal about neural network training. Even simple feed forward networks discover regularities in the input data without anything other than random starting conditions. To make this argument precise we would need estimates of the sample and training complexity and we are very far from being able to do this for feedforward networks, let alone immeasurably more complex biological networks.

But we can turn the problem around, and say that there must be things we cannot learn. Jackendoff (1987) uses precisely this idea in the study of music. He argues that there are music-grammar structures innate to the mind, meaning that arbitrary synthetic musical grammars may in fact not be learnable. He uses serialism as an example, certainly a musical idiom difficult for the casual listener. Can we find such examples in vision? Optical illusions look like a candidate at first. Even when we know that something is an illusion, we cannot reprogram ourselves to see it in any other way. This does not show, however, that we are born with a susceptibility to illusions. They may be learnt in childhood and, like language learning ability, dismantled when no longer needed. It could also indicate a more general tendency to only allow a single possible interpretation for any given input stimulus.

Elman *et al.* (1996) take a perspective on innateness derived from studies of artificial neural networks. Training a network to perform a varied and multi-faceted task may be intractable. But staging the task in various ways, say by grading the inputs or by allowing the network to get steadily bigger, does work in selected examples. Recall our discussion of the Vapnik–Chervonenkis dimension. This is very much the way biological brains develop and should be a major theme of neural network research in coming years. Recall our discussion of VCD, §10.2.4, where we found that the number of training examples required for generalisation increases with the power of the neural network. Perhaps there is a planned staging of the growth of learning power with experience.

10.6 Conclusions

We have taken a satellite view of different aspects of neural computing and seen results from learning theory and function approximation. We have seen the complexity possible from *very* simple appoximations to biological systems. As the millennium draws to a close the mimicking and exploitation of biological information processing grows in importance. Neural networks are firmly entrenched with many and diverse successful applications.

But there is plenty of excitement left in the exploration of neural networks as complex systems. We do not have good universal models of the *edge of chaos* in brains, yet *if* it is the key to adaptability, then surely brains must exploit it. Similarly we know that evolution is capable of the most intricate mechanisms and exploitation of the structure of the world must surely be possible in the learning algorithms brains exploit.

Yet, moving upwards from this local focus, one of the great challenges is how to build a real brain: many interacting modules, some evolved directly, some adapted from elsewhere. This is the great challenge of neural computation as we approach the end of the decade of the brain. It is the great challenge faced in understanding complex systems.

Acknowledgements

The ideas in this chapter grew out of discussions with numerous people over a period of a number of years. To try to list all of them would be to risk leaving somebody out. Particularly helpful recently have been Allan Sndyer, Nigel Snoad and Daniel Osorio.

Bibliography

Abu-Mostafa, Y. (1986). Complexity of Random Problems.
  Abu-Mostafa, Y. (ed), *Complexity in Information Theory.*
  Springer-Verlag, New York.
Amit, D.J. (1989). *Modeling Brain Function: The World of Attractor Neural Networks.* Cambridge University Press.
Anthony, M, & Biggs, N. (1992). *Computational Learning Theory.*
  Cambridge University Press.
Baum, E. B., & Haussler, D. (1989). What size neural net gives valid generalization. *Neural Computation*, **1**, 151–160.

Bialek, W., Rieke, F., De Ruyter, R. R., van Steveninc, & Warland, D. (1991). Reading a neural code. *Science*, **252**, 1854–1856.
Bonhoeffer, T., & Grinvald, A. (1991). Iso-orientation domains in cat visual cortex are arranged in pinwheel-like patterns. *Nature*, **353**, 429–431.
Bose, N. K., & Liang, P. J. (1996). *Neural Network Fundamentals with Graphs Algorithms and Applications*. McGraw-Hill, New York.
Brown, T. H., Zador, A. M., Mainen, Z. F., & Claiborne, B. J. (1992). Hebbian Computations in Hippocampal Dendrites and Spines. *Chap. 4, pages 81–116 of:* McKenna, T., Davis, J., & Zornetzer, S. F. (eds), *Single Neuron Computation*. Academic Press, San Diego.
Chalmers, D. J. (1996). *The Conscious Mind*. Oxford University Press.
Coren, S. (1996). *Sleep Thieves*. Simon and Schuster.
Cybenko, G. (1989). Approximation by superpositions of a sigmoidal function. *Mathematics of Control, Signals and Systems*, **2**, 303–314.
Dumont, J. P. C., & Robertson, R. M. (1986). Neuronal circuits: an evolutionary perspective. *Science*, **233**, 849–852.
Edelman, G. M. (1987). *Neural Darwinism*. Basic Books.
Elman, J.L., Bates, E.A., Johnson, M.H., Karmiloff-Smith, A., Parisi, D., & Plunkett, K. (1996). *Rethinking Innateness*. Cambridge, Massachusetts: MIT Press.
Engel, A. K., König, P., Kreiter, A. K., & Singer, W. (1991a). Interhemispheric synchronization of oscillatory neuronal responses in cat visual cortex. *Science*, **252**, 1177–1179.
Engel, A. K., Kreiter, A. K., König, P., & Singer, W. (1991b). Synchronization of oscillatory neuronal responses between striate and extrastriate visual cortical areas of the cat. *Proc. Natl. Acad. Sci.*, **88**, 6048–6052.
Freeman, W. J. (1991a). Colligation of Coupled Cortical Oscillators by the Collapse of the Distributios of Amplitude-Dependent Characteristic Frequencies. *Pages 69–103 of:* Gear, C. W. (ed), *Computation and Cognition*. NEC Research Symposia. SIAM, USA.
Freeman, W. J. (1991b). The physiology of perception. *Scientific American*, **264**, 34–41.
Fujii, H., Ito, H., Aihara, K., Ichinose, N., & Tsukada, M. (1996). Dynamical cell assembly hypothesis – theoretical possibility of spatio-temporal coding in the cortex. *Neural Networks*, **9**(8), 1303–1350.
Fujita, I., Tanaka, K., Ito, M., & Cheng, K. (1992). Columns for visual features of objects in monkey inferotemporal cortex. *Nature*, **360**, 343–346.
Georgopoulos, A. P., Schwartz, A. B., & Kettner, R. E. (1986). Neuronal population coding of movement direction. *Science*, **233**, 1416–1419.
Gray, C. M., König, P., Engel, A. K., & Singer, W. (1989). Oscillatory responses in cat visual cortex exhibit intercolumnar synchronization which reflects global stimulus properties. *Nature*, **338**, 334–337.
Haykin, S. (1994). *Neural Networks*. Macmillan College Publishing Co. Inc., Englewood Cliffs, N.J.
Hebb, D. O. (1949). *The Organization of Behaviour*. John Wiley, New York.

Jackendoff, R. (1987). *Consciousness and the Computational Mind.* MIT Press, Cambridge, Mass.
Judd, J. S. (1990). *Neural Network Design and the Complexity of Learning.* MIT Press, Cambridge, Mass.
Katz, B. (1966). *Nerve, Muscle and Synapse.* McGrawHill, New York.
Kauffman, S. A. (1993). *Origins of Order: Self-Organization and Selection in Evolution.* Oxford University Press, Oxford.
Kearns, M. J., & Vazirani, U. V. (1994). *An Introduction to Computational Learning Theory.* MIT Press, Cambridge, Mass.
Kelso, J. A. S. (1995). *Dynamic Patterns.* MIT Press, Cambridge, Mass.
Konishi, M. (1991). Deciphering the brain's codes. *Neural Comp.*, **3**, 1–18.
Kruglyak, L., & Bialek, W. (1993). Statistical mechanics for a network of spiking neurons. *Neural Computation*, **5**, 21–31.
Kuffler, S., Nicholls, J. G., & Martin, A. R. (1988). *From Neuron to Brain.* Sinauer Associates, Sunderland, Mass.cd.
Linkser, R. (1986). From basic network principles to neural architectures (series). *Proc. Natl. Acad. Sci. USA*, **83**, 7508–7512, 8390–8394, 8779–8783.
McCulloch, W.S., & Pitts, W. (1943). A logical calculus of the ideas immanent in nervous activity. *Bulletin of Mathematical Biophysics*, **5**, 115–133.
McKenna, T., Davis, J., & Zornetzer, S. F. (eds). (1992). *Single Neuron Computation.* Academic Press, San Diego.
Mézard, M., Parisi, G., & Virasoro, M. A. (1987). *Spin Glass Theory and Beyond.* World Scientific, Teaneck, N. J.
Niebur, E., Schuster, H. G., & Kammen, D. M. (1991a). Collective frequencies and metastability in networks of limit-cycle oscillators with time delay. *Phys. Review Letters*, **67**, 2753–2756.
Niebur, E., Schuster, H. G., Kammen, D. M., & C., Koch. (1991b). Oscillator phase coupling for different two dimensional network connectivities. *Phys. Review A*, **44**, 6895–6904.
Parberry, I. (1994). *Circuit Complexity and Neural Networks.* MIT Press, Cambridge, Mass.
Penrose, R. (1989). *The Emperor's New Mind.* Oxford University Press, Oxford.
Penrose, R. (1995). *Shadows of the Mind.* Vintage.
Perrett, D. I., Rolls, E. T., & Caan, W. (1982). Visual neurones responsive to faces in the monkey temporal cortex. *Exp. Brain Res.*, **47**, 329–342.
Pinker, S. (1994). *The Language Instinct.* Penguin Press, London.
Rieke, F., Warland, D., de Ruyter van Steveninck, R., & Bialek, W. (1996). *Spikes. Exploring the Neural Code.* Cambridge, Massachusetts: Bradford Book, MIT Press, Cambridge, Mass.
Ripley, B.D. (1996). *Pattern Recognition and Neural Networks.* Cambridge University Press.
Rodieck, R. W. (1973). *The Vertebrate Retina.* W. H. Freeman, San Francisco.
Schatz, C. J. (1992). The developing brain. *Scientific American*, **267**(3), 60–67.

Searle, J. (1992). *The Rediscovery of the Mind*. MIT Press, Cambridge, Mass.

Smith, T.G., & Lange, G.D. (1996). Fractal Studies of Neuronal and Glial Cellular Morphology. Iannaccone, P.M., & Khokha, M.K. (eds), *Fractal Geometry in Biological Systems*. CRC Press, Florida.

Snyder, A. W., Bossomaier, T. R. J., & Hughes, A. (1990). Theory of Comparative Eye Design. *Pages 45–52 of:* Blakemore, C. (ed), *Vision: Coding and Efficiency*. Cambridge University Press.

Sporns, O., Gally, J. A., Reeke Jr., G. N., & Edelman, G. M. (1989). Reentrant signalling among simulated neuronal groups leads to coherency in their oscillatory activity. *Proc. Natl. Acad. Sci.*, **86**, 7265–7269.

Sporns, O., Tonini, G., & Edelman, G. M. (1991). Modelling perceptual grouping and figure ground separation by means of active reentrant connections. *Proc. Natl. Acad. Sci.*, **88**, 129–133.

Stein, D. L. (1989). Spin glasses. *Scientific American*, **261**(1), 36–43.

Stork, D. G., Jackson, B., & Walker, S. (1991). Non-optimality via pre-adaptation in simple neural systems. *Pages 409–429 of:* Langton, C. G., Taylor, C., Farmer, J. D., & Rasmussen, S. (eds), *Artificial Life III*. Addison-Wesley, Redwood City.

Valiant, L. G. (1984). A theory of the learnable. *Communications of the ACM*, **27**(11), 1134–1142.

Valiant, L.G. (1994). *Circuits of the Mind*. Oxford University Press, New York.

Van Essen, D., Anderson, C. H., & Felleman, D. J. (1992). Information processing in the primate visual system: an integrated systems perspective. *Science*, **255**, 419–423.

von der Heydt, R., Peterhans, E., & Baumgartner, G. (1984). Illusory contours and cortical neuron responses. *Science*, **224**, 1260–1262.

Zeki, S. (1993). *A Vision of the Brain*. Blackwell Scientific, Oxford.

Index

address, 136, 147, 148
algorithmic computation, in brains, 369
allele, 35, 36, 39
almost
  never, 159
  surely, 159
almost everywhere, 196, 225
arms race, 111
arrows, 54–58, 61, 69, 74
artificial evolution, 35
artificial intelligence, 20, 23
artificial life, 22, 92
Artificial life, 123
attractor, 188
  global, 188
attractor neural networks, 380
augmented cost function, 268
autecology, 342
autocatalysis, 73, 85
autonomous system, 177
axiom A, 211
axon, 383

Baas N., 101
Bak P., 113
balance of nature, 349
basin of attraction, 188
behavioural ecology, 358
Behavioural studies, 358
bifurcations, 174, 176, 204, 212, 214, 219, 232
  flip, 179, 223
  fold, 220
  Hopf, 222
  pitchfork, 221
  transcritical, 223
biodiversity, 33, 360
blind constructivism, 93
blood groups, 36
Boerlijst M., 118

boolean networks, 117
bounded set, 144
box-counting, 192
brain, size and connectivity, 368
Brownian motion, 134

Caltech Concurrent Computation
  Program, 290, 334
capacity dimension, 192
Cariani P., 100
carrying capacity, 353
cell body, 383
cell membrane, 383
cellular automata, 19, 114–118, 357
cellular automata simulations, 362
cellular automaton, 13, 28, 29, 44
chaos, 168
chaos, ecological, 353
chaos, edge of, 44, 45
chaotic edge, 46
chess, Deep Blue, 368
circle map, 179
classification, 347
classification of formal languages, 14
classification, neural, 375
closed loop system,
  asymptotically stable, 265
closed set, 144
code, 135, 136, 147
codimension, 220
codominant, 36
coevolution, 108
collage theorem, 147
community, 348
compact set, 144
compact support, 155
competition, 37, 357
complete metric space, 144
complex computer, 304, 306, 308, 309, 322

complex problem, 304, 306, 309
complex processor, 304
complex system, 303–305, 307–310, 313, 314, 319, 322, 323, 326
complexity,
  time, space, 372
  algorithmic, 372
  definitions and theory, 2
  interacting agents, 2
  iterative, 2
compound leaf, 17, 18
computation,
  nature as, 370
computation, quantum, 369
computation, neural, 371
computational graph, 309, 320, 322
computed torque method, 255
connected component, 56, 58, 64
connection point, 62
connectionism, 370
connectivity, 26–29, 33, 34, 37, 38, 44, 46, 53, 55, 60, 61, 63–70, 72, 73, 78, 82–84, 86, 87
consciousness, reducible, 370
conservation, 38, 348, 360
context-free, 16, 18
context-free language, 14, 19
context-sensitive language, 14, 18
continuation, 219, 223
contraction
  map, 136, 142, 143
  ratio, 136
contraction mapping principle, 143
contraction ratio, 142
control theory,
  categories, 250
  stochastic, 275
control theory, determinisitic optimal, 267
control theory, non-linear, 250
control,
  computation and complexity, 280
  open loop, 267
controllability, 258, 273
controller,
  stabilising feedback, 254
convergence, 142
Conway J.H., 97
Conway J. H., 94
correlation dimension, 194
critical phase changes, 33
critical point, 79, 80
critical state, 112–117
criticality, 25, 28, 29, 33, 34, 37, 38, 44, 45
Crutchfield J.P., 101
cycle, 55, 57, 61, 66

Darley V., 101
data domain, 306, 308, 316, 317
data locality, 299, 319, 323, 325, 327
data parallel, 291, 295–299, 301, 303, 309, 327
Dawkins R., 94, 102
decomposition,
  domain, 319
  irregular data, 316
  maximally efficient, 318
  optimal, 307, 321
  phase or data, 322
  scattered, 320, 321
  static domain, 319
decomposition, binary, 331
decomposition, data, 331
dendritic spines, 383
dendritic tree, 383
dependency graphs, 52, 54
deterministic sub optimal control, 278
diameter, 146
diffeomorphism, 178
differential geometry, 257
digraph evolution, 63, 67, 73, 76–79, 82, 83
digraphs, 56, 57, 61, 67, 69, 76, 82, 85
dimension, 140, 141
  capacity, 192
  correlation, 194
  fractal, 192
  Hausdorff, 141, 149, 150, 191
  information, 194
  Lyapunov, 195, 215
  scaling, 150
dimensions, 176
directed graph, 27, 28, 33, 45, 59, 73, 85
directives, 298, 325
disease spread, 358
dispersal, 357
dispersal processes, 32, 33, 38
distance, 142
distribution of node degrees, 61
diversity, 33
Diversity, 346
domain decomposition, 309
dominant, 36, 39
double jump, 62–68, 72, 77, 81, 84
drift-free system, 265
duration, 74
dynamic programing equation, generalised interpretation, 271
dynamic programming, 268, 269
dynamic programming, computational complexity, 283

ecology, 340
ecology, landscape, 32, 46

ecosystem management, 360
edge of chaos, 116–117
edges, 54, 57, 59, 60, 63
embedding, 195
emergence, 53, 60, 62, 64, 65, 67, 73, 76, 78, 83–87
emergent phenomena, 99
emergent properties, 12, 23, 26, 32, 35
encapsulate, 368
entropy, 176, 214, 215
environment, 20, 25, 32, 37
environmental gradient, 357
equilibrium, 36, 349, 350
equilibrium (fixed) point, 176, 178, 182, 203
Erdos, 53, 63
ergodicity, 378
ergodicity,
 broken, 380
evolution, 33, 35, 37, 38, 45, 46, 101, 341
evolution, programming, 42
exponential dichotomies, 210
extrinsic procedures, 298

Farmer J.D., 121
feedback, 250
feedback control,
 research areas, 280
feedback controller,
 stabilising robust, 278
 stabilising state, 273
feedback,
 control law, 255
 linearisable, 263
 linearisation, 255
 open, closed loop, 253
 time-varying periodic, 265
feedback, definition, 252
feedback, smooth, 265
feedbackfeedback,
 membrane permeability, 385
Fermi Pasta Ulam, 173
Fermi–Pasta–Ulam, 185
Fibonacci, 15, 18
finite difference scheme, 281
finite difference simulation, 310
finite horizon problem, 273
finite state automata, 43
fire, 33, 46, 342, 358
fire spread, 358
fixed point, 143
flows, 177
fluid equations, 173
Fontana W., 119
forest dynamics, 358
Fortran, 294, 295, 297–300, 319, 325, 326

fractal, 132
 brain, 153
 deterministic, 132
 dragon, 153
 mathematical, 128
 measure, 131, 157
 non-deterministic, 132
 physical, 128
 random, 132, 157, 159, 162
 random measure, 134
 random set, 133
 set, 129, 142, 146
 statistical, 132
fractal dimension, 192
fractal dimension , 314, 315
fractals, 191
frustration, 379
functional approximation, neural, 374

game of life, 94–101, 114
game theory, 267, 276
GARP, 43
Gaussian elimination, 300, 310
general systems theory, 87
generic, 196, 219
genetic algorithm, 39
genotype, 35, 36, 41, 102, 107
geographic information systems, 361
geometric dimension, 310, 313, 315
giant component, 60, 65, 77
Gibbs canonical ensemble, 378
Goldman equation, 387
grain size , 310, 312
graph, 27, 28, 33, 45
graphical evolution, 65, 83, 87
growth, 13, 19, 30, 35
growth of complexity, 105, 118

Haken, 26
Hamilton–Jacobi–Bellman equation, 270
Hamiltonian, 318, 320, 321
Hamiltonian circuit, 55, 72, 73
Hamiltonian coherence, 67
Hamiltonian cycle, 55, 61
Hardy–Weinberg law, 36
Hardy-Weinberg law, 36
Hausdorff dimension, 191
Hausdorff distance, metric, 144
Hausdorff metric, 189
headroom, 292
heat equation, 183
heterozygous, 35, 36
heuristic, 330
heuristics, 328
High Performance Distributed Computing, 301

High Performance Fortran, 297–300, 319, 325
Hodgkin–Huxley equations, 387
Hogeweg P., 118
Holland J.H., 121
homoclinic orbits, 176, 226
homotopy methods, 223
homozygous, 36
hyperbolicity, 204, 210, 219, 226, 227, 234, 235, 237
hypercube architecture, 313
hypercycles, 33, 34

identification procedure, 258
immersion, 198
in-degree, 54
infinite horizon problem, 273
infinite horizon value function, 273
information constraint, 278
information dimension, 194, 315
information superhighway, 361
information theory, 328
initial value problems, 210
integrable map, 181
integral, 154
interactive evolution, 102–108
interconnection topology, 309
intractability, 94, 101, 112, 118
invariant tori, 176
inverted pendulum, 273, 283
ion equilibrium, 385
ion pump, 387
irreversibility, 171
island biogeography, 350
island chain, 181
iterated function system, 128
iteration, 2

Jacobian matrix, 178

KAM theorem, 181
Kauffman, 45, 67, 73, 85
Kauffman S., 117, 119
Koch
  curve, 129
  measure, 131, 132
  random curve, 133
  random measure, 134

L-systems, 13, 15, 16, 18, 21
landscape ecology, 32, 356
landscapes, 356
Langton C.G., 101, 116
language paradigms, 295
languages, adapting existing, 294
Lem S., 119–120
limit cycle, 30–32, 182, 222, 239

linear systems methods, 263
linking hypothesis, oscillators, 393
Lipschitz constant, 136
load balance, 320
local connectivity, 70
locality, 291, 299, 319, 323, 325, 327
locus, 35
logistic growth, 353
logistic map, 179
loosely synchronous, 297, 322, 326, 327
lower semicontinuous, 189
Lyapunov dimension, 195, 215
Lyapunov exponents, 176, 214
Lyapunov stability, 213
Lyapunov–Schmidt reduction, 220

machine architecture, 325
machine epsilon, 187, 209, 218, 239
manifold, 171, 177, 197, 199, 203
  approximate inertial, 234
  centre, 174, 212
  centre stable and unstable, 212
  inertial, 170, 174, 214, 230
  invariant, 173, 212
  stable and unstable, 205, 207, 208, 226
Markov chain, 58, 282
Markov chains, 58
matrix, 298, 304, 308, 309, 320
matrix algorithm, 306
matrix Riccati equations, 280
maximum entropy, 328
McCulloch and Pitts neuron, 371
Mead, Carver, 292
measure, 131, 154
  Lebesgue, 158
  mass, 131, 154
  probability, 158
  support, 131, 132, 154
  unit mass, 154
measure of diversity, 346
message passing, 292, 295–297, 300, 301, 326
message passing interface, 296
metacomputing, 291
metric, 142
metric space, 142
  complete, 142
metropolis criterion, 329
MIMD, 292–297, 300, 309, 313
minimum time problem, 271
mobots, 78
molecular dynamics, 169, 172, 185
Monge–Kantorovitch distance, 155
Monitoring, 361
Monte Carlo, 309, 318, 329

moon landing problem, 269

nanotechnology, 121
natural selection, 37
neighbourhood, 28, 30
Nernst equation, 386
networks, 52–54, 57, 59, 66, 69, 75, 78, 79, 86
neural code,
 firing rate, 390
neural computation, 369
neural hierarchy, 382
neural morphology, 382
neural network, loading, 375
neural networks,
 for stabilising feedback, 266
neural output,
 graded, 384
 spikes, 384
neuro-anatomy, 383
neuroidal tabula rasa, 372
neuron,
 compartment model, 388
neutral evolution, 105–108
nodes, 54–57, 59–62, 64–66, 69–75, 82
non-equilibrium, 24
non-equilibrium processes, 25
non-linear dynamic optimisation, 281
non-smooth analysis, 271
non-wandering set, 211
normal distribution, 158
Northeast Parallel Architectures Center, 290, 334

observability, 258, 273
observability r, 262
observability,
 codistribution, 262
observation, 53, 62, 66, 70, 72, 73, 82, 84
open set condition, 141
Open Set Condition, 150
optimal control, 267
optimal control methods, 266
optimal control,
 equations for, 268
 solution and analysis, 268
optimal foraging, 359
optimisation problems, 306, 327
optimisation, land-use, 361
optimisation-based design, 267
order parameter, 26, 44
ordination, 347
organism, 12, 20, 23, 30, 45
origin of life, 33
oscillators,
 phase locking, 381
 synchronisation of populations, 381
 transmission delays, 381
out-degree, 54
output feedback,
 computation and complexity, 284
overhead, 312, 324

pantokreatika, 119
parallel language, 291, 294
parameter space, 152
partition function, 379
path, 55–57, 71
pendulum, inverted, 253
phase change, 26, 28, 29, 34, 45, 46
phase changes, 53, 62, 76, 78, 83, 86
phase space, 168, 177
phase transition, 322
phase transitions, 83–85, 291
phenotype, 35, 37, 102, 107
physical optimisation, 306, 327, 328
phytogeography, 341
policy space iteration, 282
Pontryagin minimum principle, 268, 280
portability, 294
predictability, 349, 361
prevalence, 197
prevalent, 220
Prigogine, 25
probability distribution, 133, 134, 158
probability theory, 157
problem, 291, 304, 306, 312, 315, 321, 324–327
problem architecture, 291, 294, 322, 325
problem locality, 291

QR factorisation, 216, 218, 227

random
 measure, 162
 set, 159
random digraphs, 28
random problem, 373
Rasmussen S., 119
Ray T., 94, 109–112, 119, 121
reachability, 259
reachability rank condition, 265
reachability,
 rank condition, 261
recessive, 36, 39
reductionist system, 100
regular language, 14
Rent's rule, 315
resilience, 352
robust control, 276
robust design, 257
rotation number, 179
rule transition network, 69

Santa Fe Institute, 378
satellite monitoring, 361
scale, 13, 23, 25, 38, 349, 353
scaling
  factor, 140
  law, 135–138, 146, 156, 159–163
  operator, 137, 138, 156, 160, 163
  property, 129, 131, 133, 134
scaling law, 128
scaling operator, 128, 146
schema theorem, 40
seed, 137
self-organisation, 12, 13, 24, 25, 45, 358
self-organised criticality, 84, 112, 117
self-replication, 33, 34
semiflows, 178
Seppen K., 113
sequence of nodes, 71
shadowing, 176
shadows, 235
SIMD, 292, 293, 297, 298, 300, 309, 313, 325
similitude, 140, 142
Simon H.A., 123
simplex algorithm, 280
Sims K., 102
simulated annealing, 307, 318, 320, 327, 329, 332
simulation, 19, 33, 38, 300, 302, 305, 306, 310–312, 315, 316, 319, 323, 358
simulation, cellular automata, 357
simulations, 357
simulations, cellular automata, 362
singular value decomposition (SVD), 218
singular values, 215
slaving, 170
smoothnes requirements,
  Lipschitz continuity, 265
software models, 325
solitons, 173
soma, 383
space–time domain , 309
spatial self-structuring, 34
speedup, 311
spin glass, 378
spin glasses, 318, 321
spreading of blockages, 80
stabilisation,
  feedback, 264
stabilised march, 218
stabilising state feedback controller, 273
stability, 345, 351
stability boundary, 204, 219
stable subspace, 205

standard domain decomposition, 312, 321
starfish outbreaks, 358
state feedback law, 265
state space, 168, 257
state transition graph, 68
state transition networks, 59
states,
  indistinguishable, 262
statistical mechanics
  equilibrium, 172, 185
  non-equilibrium, 172, 185
stochastic behaviour, 356
stochastic deadlock, 83
stochastic differential equation model, 275
structural stability, 204
Sylvester equation, 206
synapse, 383
synapse,
  temporal properties, 389
synaptic transmission, 384
synaptic vesicles, 388
synecology, 342
synergetics, 170
system connectivity, 314

taxonomy, 340
televirtuality, 302, 303
temperature, 291, 321, 322, 328–330, 332, 333
theorem,
  Frobenius, 259
  small gain, 258
Tierra, 43, 94, 108, 112, 113, 119, 121
TODO principle, 23
transition rules, 69
tree, 56–58, 60, 70
treppeniteration, 216
two point boundary value problems, 210, 226
two-dimensional mesh, 313

undecidability, 96
unfolding, 220
unit roundoff, 187, 218
universal computation, 44, 45
unpredictability, 96
unstable subspace, 205
upper semi continuous, 189

value space iteration, 282
van der Pol oscillator, 182, 239
Vapnik-Chervonenkis dimension, 375, 402
verification theorem, 270
vertices, 54

virtual reality simulations, 302
Von Neumann J., 118

wetware, 369
Wolfram S., 115

X-29 aircraft, 250